Cognitive Evolution

Cognitive Evolution provides an in-depth exploration of the history and development of cognition, from the beginning of life on Earth to present-day humans. Drawing together evolutionary and comparative research, this book presents a unique perspective on the evolution of human cognition. Adopting an information processing perspective – that is, from inputs to outputs – with all the mental processes in between, Boles provides a systematic overview of the evolutionary development of cognition and of its sensation, movement, and perception components.

The book is supported by long-established evolutionary theories and backed up by a wealth of recent research from the growing field of cognitive evolution and cognitive neuroscience to provide a comprehensive text on the subject.

Cognitive Evolution is an essential read for advanced undergraduates and postgraduate students of cognitive and evolutionary psychology.

David B. Boles is Emeritus Professor of Psychology, University of Alabama, Tuscaloosa.

"Connecting the great advances in cognitive science to their roots in human evolution must be one the greatest missed opportunities in the current psychology curriculum. *Cognitive Evolution* traces pattern recognition, memory, language and consciousness across species. For example, who knew that for pigeons the duration of immediate memory was two to six seconds, for monkeys ten to twenty, and for humans twenty to sixty seconds. This book is full of insights into what is already known and what is yet to be studied. I highly recommend it."

— **Michael Posner, Prof. Emeritus, University of Oregon, USA**

"This book combines a basic short course in biological evolution with an extended treatment of primate evolution emphasizing fossil and living hominids, their anatomy, habits and cognitive capabilities. Rich in detail, engagingly written and well illustrated, this volume is an excellent resource for upper level undergraduates and graduate students in the cognitive sciences. It serves also as a reminder of how even seemingly simple cognitive tasks depend on a brain architecture and circuitry of immense complexity with a deep evolutionary heritage."

— **Dr. Thurston Lacalli, Biology Department, University of Victoria, Canada**

Cognitive Evolution

David B. Boles

Routledge
Taylor & Francis Group

LONDON AND NEW YORK

First edition published 2019
by Routledge
2 Park Square, Milton Park, Abingdon, Oxon, OX14 4RN

and by Routledge
52 Vanderbilt Avenue, New York, NY 10017

Routledge is an imprint of the Taylor & Francis Group, an informa business

British Library Cataloguing-in-Publication Data
A catalogue record for this book is available from the British Library

Library of Congress Cataloging-in-Publication Data
Names: Boles, David B. (Writer on psychology) author.
Title: Cognitive evolution / David B. Boles.
Description: 1 Edition. | New York, NY : Routledge, 2019. | Includes bibliographical references and index.
Identifiers: LCCN 2018060196 (print) | LCCN 2018061333 (ebook) | ISBN 9780429028038 (eBook) | ISBN 9780367028541 (hardback) | ISBN 9780367028558 (pbk.) | ISBN 9780429028038 (ebk)
Subjects: LCSH: Cognition.
Classification: LCC BF311 (ebook) | LCC BF311 .B585 2019 (print) | DDC 155.7—dc23
LC record available at https://lccn.loc.gov/2018060196

ISBN: 978-0-367-02854-1 (hbk)
ISBN: 978-0-367-02855-8 (pbk)
ISBN: 978-0-429-02803-8 (ebk)

Typeset in Sabon
by Apex CoVantage, LLC

Visit the eResources: www.routledge.com/9780367028558

Cover source: File:La cueva de las manos Santa Cruz.jpg, CC BY-SA 4.0, CarlosA.Barrio.

For Joan, Walter, and Madeline.

by Madeline Boles, age 6, published with permission

Contents

Figures

Tables

Boxes

Introduction

Cognitive Evolution grew out of the needs of instructors and students for a textbook explaining how we as a species acquired our mental lives. On the one hand, cognitive psychology has had a rich history of developing descriptions of human mental processes and their organization, placing them in an information processing framework covering inputs, outputs, and all the processes in between. On the other hand, evolutionary scientists have developed a large literature using fossils, artifacts, and comparisons between species to explore the evolution of such processes. But for the most part, the two hands have not worked together to create a cognitive evolution textbook.

When I began teaching an evolutionary psychology course at the University of Alabama in 2001, I noticed the absence of such a textbook among others that took a social approach to evolutionary psychology. In the years since it has become clear to me that most evolutionary courses have covered primarily social material such as the evolutionary mechanisms relating to mating, kinship, competition, and cooperation. While there is certainly nothing wrong with that approach – in fact, it can be quite informative and engaging – it incorporates only a fraction of the existing knowledge base of evolutionary psychology. A whole world of cognitive evolution research has not been covered. Existing in parallel, I came to suspect, was another world of college and university professors coming from cognitive, perceptual, or neuroscience backgrounds, who might like to teach an evolutionary psychology course from a different perspective to a universe of students who could profit from it.

Thus *Cognitive Evolution* was born. Most chapters of the book debuted in my classroom in 2007 but have been continuously updated since. I think readers will find that *Cognitive Evolution* offers a consistent, coherent approach to the evolution of our mental lives that is unmatched by any other textbook.

Before describing how this book can best be used, I would like to add a bit about its reception. For several years I conducted anonymous surveys of my students on a chapter-wise basis, having a graduate student collect and keep them until grading was complete. Although most items were open-ended, one was a 1-to-7 rating as to how satisfied the student was with the chapter. Most recently, across the first 16 chapters of the book, the mean rating has been 5.7 and the median 6.0. It has been a well-received book.

How to use this book

The book is intended to be used by upper-level undergraduate and graduate students. It is organized into three major sections: Introduction to Evolution, Sensation and Movement, and Perception and Cognition. It is meant to be presented as a coherent whole, although it would be possible to incorporate some of its chapters into a broader evolutionary psychology course.

When I vetted the chapters with colleagues, one biology professor questioned whether it was really necessary to include the Introduction to Evolution. The answer is an unequivocal yes. Unlike biology courses, nearly every one of which is permeated by evolutionary background, most psychology courses have developed for decades with hardly a nod to evolution. Worse, the state of our public education system, not to mention our religious education system, is such that in many districts, students receive little or no evolutionary content. Without the Introduction to Evolution chapters, many students have insufficient intellectual background to contend with the misconceptions about evolution that abound in popular society.

The good news, however, is that the introductory chapters can be covered quickly in the classroom. I have typically spent only two or two and a half weeks to cover all five, and that even includes a handful of supplementary readings added for good measure. The content is sufficiently straightforward to allow that.

The second section of the book, on Sensation and Movement, consists of four chapters. In information processing terms, it covers inputs and outputs, including the mechanical, chemical, and visual senses, and postural, hand, and leg movements. When covering these along with supplementary materials, I have allowed two and a half to three and a half weeks.

The real meat of the book, however, is in the seven chapters on Perception and Cognition. By the time students tackle this material, they understand not only the principles of evolution, but how information enters the human processing system and is acted upon. They are fully ready to consider what happens in between. This is the section that takes best advantage of the explosion of cognitive-related evolution research over the past 20 years. For that reason, and because of the plethora of supplementary readings that can be brought to bear, I have typically taken seven to seven and a half weeks to cover this material.

I have recently added to the book a final, 17th chapter, summarizing its major findings in nine "firsts". This requires only a single session to cover, and it can be omitted if desired.

Pedagogical mileage will likely vary, particularly with the nature of the course. Having taught the material as a seminar, I often leave time at the end for student project reports, but that time could instead be accounted for by interspersing additional supplementary material. A lecture-based course, on the other hand, will have different needs and different timing. Although I have not personally done it, I can easily envision developing a more general lecture course on evolutionary psychology using this textbook supplemented with social cognition chapters from other sources.

Mileage will also vary with the level of the student. Although the book was written with upper-level undergraduates in mind, I have also used it with Ph.D.-level graduate students with no complaints or problems. Yet I have also tried to write it with sufficiently accessible language that any reasonably well-educated person should be able to use it profitably. I would be pleased to learn that all who use the book will to some degree expand their understanding of how evolution has guided our thinking and perceiving.

Finally, I should note that the book is unabashedly anthropocentric. While it is not shy about taking on comparisons between species, it doesn't aspire to coverage of cognitive evolution across the animal kingdom. To state the case more positively, this is a psychology book for those interested in human psychology. Most of the questions it asks fit under the umbrella of "how did you and I come to be?" I think that is a reasonable question, and it is assuredly the question that psychology students with only limited exposure to the topic want answered.

Acknowledgments

I would like to thank my wife, Joan Barth, and children Walter and Madeline for their unbounded patience and support while I have repeatedly written and revised this text. My daughter Madeline deserves special recognition for contributing her art.

I would also like to express my appreciation of my late mentor Steve Keele, and of Mike Posner, Mary Rothbart, and Russell Fernald, for supporting the unorthodox collection of major preliminary exam topics that set me on the path toward writing this book. I also thank the now-forgotten fellow student, a year or two ahead of me in graduate school at the University of Oregon (it may have been Tram Neill or Terry Hines), who convinced me to follow my own star in that endeavor, instead of taking a more orthodox, "mainstream" cognitive psychology approach. Without that little push, all of this may never have happened.

I would like to thank a number of Alabama colleagues, both from the university and from the Geological Survey, who read preliminary drafts of some of the chapters and provided feedback. They include, in alphabetical order, Fred Andrus, Jim Bindon, Joe Chandler, Fran Conners, Michael Dillard, David Kopaska-Merkel, Tony Loewald, Ed Merrill, Brad Okdie, Kelly Pivik, Leslie Rissler, Jason Scofield, Ed Stevenson, Carl Stock, and Emily Wakeman. I also appreciate the more recent chapter reviews written by Terry Hines and by numerous anonymous members of my undergraduate and graduate seminars.

Also deserving of thanks are those who helped develop the book for publication. These include Elena Bellaart, Christine Cottone, Sophie Crowe, Chris Johnson, Ceri McLardy, Beverly Peavler, and three anonymous experts who contributed full-book reviews through Routledge.

Finally, I would like to express gratitude to the personnel connected with two resource repositories. First are the library staff at the University of Alabama. This book was made possible only by their provision of extensive electronic journal resources covering virtually every need for a PDF of an article, and by their prompt interlibrary loan deliveries of articles that weren't available electronically. Second are the staff of Creative Commons, the online repository of public domain and freely licensable graphics. Just a few years ago, before their efforts produced explosive growth in useable files, this would have been a much more drab and less informative book. It goes without saying, of course, that my thanks extend to the multitude of persons who have made their graphic works available through the Creative Commons website.

Section I

Introduction to evolution

1 Life begins

No one knows how it started. No one is sure exactly what started. But science is led by best guesses, and a good guess is that at some point about four billion years ago, perhaps a half billion years after Earth's formation, a complex molecule became self-replicating. From that humble, nearly invisible beginning, increased complexity was inevitable due to descent with modification, acted upon by natural selection. We'll return to that idea later to consider it more carefully. First, though, let's look at the question of how life on Earth originated.

How did life begin?

In one view, life began with self-replicating ribonucleic acid, or RNA – a position called the "RNA world" hypothesis (Vázques-Salazar & Lazcano, 2018). RNA is composed of a chain of *nucleotides*, each composed of a nucleobase (cytosine, guanine, adenine, or uracil, as shown in Figure 1.1), a sugar, and a phosphate. RNA is similar in structure to deoxyribonucleic acid, or DNA, the complex molecule that carries our genetic code. RNA is a simpler molecule, though, both because it is a much shorter chain and because it exists as a single, rather than a double, strand.

RNA is also less stable, liable to alteration by radiation or even, over time, by water. That observation leads some to wonder whether a more stable alternative was the first self-replicator – perhaps an RNA precursor or a chemical near-relation (Lunine, 2005; Zhang, Peritz, & Meggers, 2005). Yet even proponents of other molecules usually concede that life funneled through RNA world at some point. Among other arguments, the nucleobases most frequently used by RNA, namely cytosine and guanine, are the ones less susceptible to errors during copying. That suggests that at some point RNA was subject to natural selection that reduced the frequency of error-prone nucleobases (Mallary, 2004).

It has been pointed out that sooner or later, RNA would have required a wrapping to protect it and to provide a suitable microenvironment for replication. Initially, sheets of the mineral mica might have served the purpose, with the sheets serving as an incomplete external barrier that could have nurtured the evolution of a more complete biological one (Hansma, 2010). Others, however, envision a fatty shell as the initial wrapping (Mallary, 2004). Fancifully, this could have been a bubble generated by a foaming sea that eventually became a membrane.

But there is also the problem of *metabolism* (the cellular process by which energy is made available). How did a self-replicating molecule harness energy to replicate if it was encased in a bubble or was alone in the so-called *primordial soup* (the thin solution of chemicals comprising the early ocean)? Here, the mica hypothesis again has something to offer, because the mechanical energy involved in one sheet moving against another may have both broken and formed chemical bonds (Hansma, 2010, 2014).

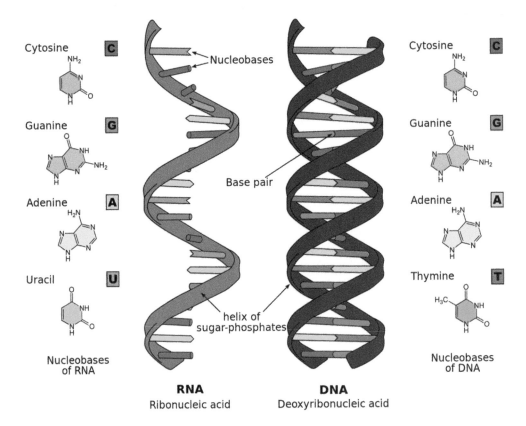

Figure 1.1 The structure of RNA and DNA.

Source: Difference DNA RNA-EN.svg, CC BY-SA 3.0, Sponk.

The beginning of life was a sufficiently complex process that scientific best guesses as to its place of origin are incomplete at best. Only recently, after millions of years of human cognitive evolution, have we had the intellectual tools to identify a range of possibilities. First, what if life originated . . .

. . . In a warm little pond?

Writing in 1871, long before the discovery of RNA and DNA, Charles Darwin suggested that life originated in a "warm little pond" containing what he thought were the necessary components of ammonia and phosphorus salts, acted upon by electricity as well as solar heat and light (Follmann & Brownson, 2009). Modern experiments have varied the ingredients to more closely resemble Earth's primitive chemical composition, and have been rewarded by the creation of nucleotides, the building blocks necessary to synthesize RNA. An important condition is that the "pond", actually a solution in laboratory glassware, must exist in a reducing atmosphere, meaning an atmosphere with free hydrogen and little or no oxygen (Parker et al., 2011). Oxygen interferes with the formation of nucleotides because it readily reacts with the simpler carbon-based molecules from which they are built. Thus today's atmosphere, with almost no free hydrogen and abundant oxygen, would have been quite unsuitable for the creation of life.

Geological evidence tells us, however, that the original primordial atmosphere was in fact hydrogen-rich and oxygen-free. Earth presumably formed from rocks attracted by gravity, ultimately derived from a parent nebula but in part coming together during a "heavy bombardment" stage in Earth's early history (Grimm & Marchi, 2018). A study of the gases that would have been released from such rocks indicates the creation of an atmosphere rich in hydrogen, methane, ammonia, nitrogen, and water vapor, but having no free oxygen (Zahnle, Schaefer, & Fegley, 2010).

Eventually, as a light gas, hydrogen was stripped off by solar heating and radiation. But that process took time, diminishing only about 2.5 billion years ago (Zahnle, Catling, & Claire, 2013). Before that, there was a long window of opportunity for an oxygen-free, reducing atmosphere to generate RNA and cellular life.

Early RNA would not necessarily have had the ability to self-replicate. Today's RNA generally requires protein to do so, leading to the "chicken-and-egg" problem of which came first: RNA or protein, each requiring the other for its own reproduction (Bernhardt, 2012).

Recently, though, RNA molecules have been discovered that can split into two components, each component replicating the other without the aid of proteins. In part because the components are small, replication occurs relatively efficiently and is self-sustaining. The copies subsequently reassemble into complete RNA molecules. Importantly, occasional "mutations" occur during replication, resulting in different variations of the components. Over time, those variations that are most efficient at replication are found to outcompete the others, resulting in a kind of evolution where mutations beneficial to self-replication are retained and others are lost. By analogy, natural selection may have operated in a similar way on early RNA components, producing ever-more-efficient self-replication (Lincoln & Joyce, 2009).

This, or whatever alternative process allowed RNA replication, is thought to have occurred by 3.85 billion years ago, because a form of carbon characteristic of life is found in Greenland rock at least that old (Lunine, 2005; Westall, 2004). It may have occurred even earlier, 4.1 billion years ago, in that similar carbon has been found in an Australian zircon of that age (Bell, Boehnke, Harrison, & Mao, 2015).

Following the stripping off of light elements, the atmosphere would have been only mildly reducing, although more so in the immediate vicinity of volcanic vents due to their release of hydrogen among other gasses (Parker et al., 2011). Eventually, microorganisms evolved that were capable of photosynthesis, consuming water and giving off oxygen. That is thought to have started in bulk no later than 2.6 billion years ago, producing a substantial increase in atmospheric oxygen by 2.2–2.3 billion years ago (Zahnle et al., 2013). Today's high 21% oxygen level was reached about half a billion years ago.

However self-replication originated, an accompanying problem concerns the concentration of chemical components necessary for reactions to occur. For life to transpire, conditions had to create RNA in greater abundance than was possible in the chemically thin gruel of a lake or ocean. Thus, the "warm little pond" of Darwin's notion, a place where chemicals could be concentrated through evaporation. Tidal pools are another possibility.

But perhaps life did not begin in Darwin's "warm little pond". Could it instead have begun . . .

. . . On clay surfaces?

Another proposal is that chemicals were concentrated on wet clay, which was abundantly produced by early volcanic activity. Clay's advantage over ponds and pools is that it not only concentrates but scaffolds: that is, it allows molecules to organize on its surface. Experiments find that volcano-derived clay aids the formation of short chemical chains, and that RNA binds efficiently to it (Ferris, 2005). Short RNA chains can also grow much longer on clay surfaces (Ferris, Joshi, Wang, Miyakawa, & Huang, 2004). In addition, clay provides protection against reactive chemicals, and it aids in RNA self-replication (Franchi & Gallori, 2005).

Or did life begin . . .

. . . Around black smokers?

In 1977, scientists were stunned to discover whole ecosystems organized around "black smokers" – volcanic vents at the bottom of the sea. Life was thriving there under unimaginable pressures and temperatures, the simplest life form being *thermophilic* (heat-loving) *archaea*, a type of single-cell microorganism. Although many species of archaea live in less extreme environments, thermophilic types usually require sulfur for growth. They are often *anaerobic* (they live without oxygen), and they can exist in isolated locations around local black smokers. All these facts suggest that archaea could have evolved in the extreme conditions of Earth's early history.

It remains to be seen, however, whether black smoker chemistry can plausibly have led to RNA. High temperatures impede the folding of RNA into its required three-dimensional structure, and decompose both it and its building blocks (Islas, Velasco, Becerra, Delaye, & Lazcano, 2003; Moulton et al., 2000). The problem may be especially great in the alkaline water around black smokers (Bernhardt, 2012). Although salt provides some protection against these effects (Marguet & Forterre, 1998), it is not clear whether salt concentrations were sufficient in the ancient sea, or whether an early membrane could have enclosed enough salt to allow high-temperature RNA synthesis.

There is also some evidence against a thermophilic origin of life. The analysis of protein sequences across life's domains, including archaea and bacteria, suggests that the earliest-evolving proteins were temperature sensitive and most likely formed at about 20°C, or 70°F (Boussau, Blanquart, Necsulea, Lartillot, & Gouy, 2008). That is more compatible with Darwin's warm little pond than with the 50–80°C (120–180°F) temperatures favored by thermophilic organisms. These results suggest that life originated at relatively low temperatures, and that today's heat-loving archaea had ancestors that later adapted to high-temperature environments.

Finally, as a last exercise, did life originate . . .

. . . In space?

Reports of organic compounds in meteorites first surfaced in the 19th century but were discredited because of contamination due to handling and storage. By the second half of the 20th century, however, sophisticated analyses showed that some uncontaminated meteorites contain *amino acids*, the building blocks of proteins, as well as long-chain hydrocarbons. These molecules are assumed to have been created by nonbiological processes operating on the parent bodies from which meteorites derive.

Although amino acids are not RNA, the formation of complex organic molecules in extraterrestrial matter allows us to speculate that RNA or its precursors could have fallen to Earth in meteors, dust grains, or comets. Nucleobases have in fact been found in meteorites, including uracil, an important component of RNA but not DNA (Pearce & Pudritz, 2015). Thus space-originated nucleobases could have helped create RNA world.

An experiment exposing common space materials to radiation like that found in nebulae produced a rich array of organic compounds, including amino acids and the *full set* of RNA and DNA nucleobases (adenine, cytosine, guanine, thymine, and uracil). The space materials were powdered meteorites from which organic compounds had been removed, and formamide, an amide of formic acid. The implication is that important compounds needed to originate life may have been synthesized in space and then fallen to Earth (Saladino et al., 2015).

Even more intriguingly, in 1996 NASA scientists announced fossil evidence of microbial life in a meteorite discovered 12 years previously on Antarctic ice. The meteorite had presumably been blasted from the planet Mars by an asteroid impact, for it chemically resembled other meteorites in which the levels of the inert gases argon, krypton, neon, and xenon matched concentrations in Mars's atmosphere (Gibson, McKay, Thomas-Keptra, & Romanek, 2000; Norton, 1998). The apparent fossils were segmented bodies (Figure 1.2) resembling fossil bacteria on Earth, but smaller (Gibson, McKay, Thomas-Keptra, & Romanek, 2000).

This extraordinary claim was immediately subjected to intense scientific scrutiny. Some findings, such as organic material in the meteorite and bacteria-like bodies near its surface, were found to be attributable to terrestrial contamination. Specifically, the bodies resembled a bacterium, called Actinomycetales, commonly found in Antarctica (Steele et al., 2000). Certainly, terrestrial contamination seemed quite possible, since the meteorite had lain in Antarctica an estimated 13,000 years before its discovery (Gibson et al., 2000).

In a 2008 article, Schulze-Makuch and colleagues reviewed the evidence. They noted that two other Martian meteorites contain evidence of borehole-like structures, on a microbial scale, similar to those found in basaltic rocks on Earth, and that the Earth

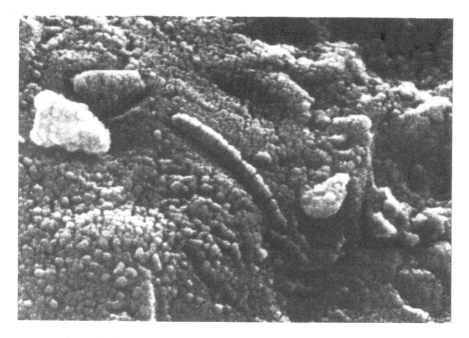

Figure 1.2 Evidence for life on Mars? Some have claimed that this 380-nanometer object is a fossilized bacterium.

Source: ALH84001 structures.jpg, public domain, NASA.

structures test positive for cellular material. However, terrestrial contamination could still not be ruled out (Schulze-Makuch, Fairen, & Davila, 2008).

These observations counsel caution in interpreting Martian meteorites, but another observation is equally important. As others have noted, even if life came from space in cellular form, that simply pushes the question of life's origins to an older time and another place. It seems at least as plausible that life emerged here by building on amino acids and other complex molecules delivered from space.

So at the end of the day, how did life emerge? It's clear from the diverse views just presented that no one really knows. Certainly, there is no scientific consensus. A diversity of views does not mean that anything is possible, however. Warm little ponds, clay surfaces, black smokers, and space are among a very few possibilities that sensibly could have sparked life. Possibly, some combination generated it – for example, with space providing some of the hydrocarbons that evaporation concentrated in ponds, and with clay or mica scaffolding the first RNA.

Membranes and metabolism

We've considered some possibilities for how life began, but how did it continue to develop? Let's approach this question first by thinking about two requirements for life mentioned earlier: A membrane to protect the organism and to make self-replication possible, and metabolism to provide the self-replicator with energy.

One of the problems in replicating RNA, as already noted, is that it is easily degraded by solar radiation and water. While clay or mica surfaces could have played an early protective role, eventually RNA needed a membrane to protect it from its environment. The membrane could have started as a simple structure based on *lipids* (fatty or waxy molecules). Some lipids spontaneously form *bilayers* (films) and *vesicles* (sacs) that can enclose other compounds. Lipid structures also permit a difference in electrical charge between the inside and outside, a possible precursor of transport mechanisms that exchange chemicals through living membranes (Trevors, 2003). *Fatty acids*, a class of lipids, have been found to form vesicles capable of admitting and retaining sugars and charged nucleotides. Under certain limited conditions, nucleotides self-replicate inside the vesicles (Mansy et al., 2008).

To survive and reproduce, cells need energy to drive the chemical reactions that build and maintain structures. Because acidic solutions have abundant hydrogen ions – also called protons – and those are an energy source, some suggest that early oceans were acidic, a tongue-in-cheek "primordial vinaigrette" hypothesis (Bernhardt, 2012). Energy may also have been made available by chemical reactions *catalyzed* (accelerated) by RNA itself (Botta, 2004). However, RNA is an inefficient catalyst, so if it served that purpose, it would soon have been replaced by a more efficient alternative. *Peptides* (short chains of amino acids) admitted through the cell membrane from the primordial soup may have filled the catalytic role. Alternatively, peptides may have been manufactured by RNA within the cell, although that is considered less probable. That is because of the recursion involved – RNA producing peptides, providing the energy allowing RNA to produce peptides – making it unlikely to have emerged in a single evolutionary step.

DNA world

Although they had a membrane and a means of metabolism, the first protocells were still RNA-based. At that point, RNA's role was one of information storage, its sequences of nucleobases providing the instructions for manufacturing proteins. But soon, RNA world gave way to DNA world, with all life thereafter based on the strands of the famous double helix.

During the transition to DNA world, RNA gave up much of its information storage role to a molecule that was chemically more stable, less subject to copying errors, and, due to its length, able to carry instructions for the creation of more complex proteins. The DNA-protein partnership, combining superior information encoding and superior energy production, outcompeted RNA-based life (Lunine, 2005). RNA was thenceforth relegated to a helper role, mediating protein creation based on DNA instructions.

While there is no evidence indicating how RNA gave rise to DNA, an evolutionary succession is logical because of the similarity of the two molecules. Both have backbones of phosphate and sugar, and while the sugars differ, they are chemically closely related. To the backbones of each are attached only four types of nucleobases, three of which are identical between RNA and DNA. It was not a major evolutionary trick to convert RNA or its chemical precursors into DNA.

Whatever scenario transpired, a period of rapid evolution must have followed in order to specify the huge number of proteins used subsequently in cell metabolism (Trevors, 2003). At some point, in some branches of life, light-absorbing pigments were introduced into the cell as a means of capturing solar energy to drive chemical reactions (Botta, 2004).

One-celled life

Today's one-celled organisms are either *prokaryotes* (cells having no nuclei) or *eukaryotes* (literally "true kernels" – cells having nuclei). Based upon similarities and differences among genes appearing in the three domains of bacteria, archaea, and eukarya, there is a consensus that prokaryotes evolved first (Boeckmann et al., 2015). However, there is no consensus about whether the first cells were bacteria or archaea, the two prokaryote domains. Some studies favor bacteria (Baldauf et al., 2004; Williams et al., 2015) and others archaea (Caetano-Anollés et al., 2014; Raymann, Brochier-Armanet, & Gribaldo, 2015). It may be that there was so much *horizontal gene transfer* (movement of genetic material from one organism to another without a parent-child relationship) between the early domains that a resolution will never be possible (Thiergart, Landan, & Martin, 2014).

In round figures, prokaryotes are thought to have emerged a little less than four billion years ago, and eukaryotes about two billion years ago (Doglioni, Pignatti, & Coleman, 2016; Hedges, Marin, Suleski, Paymer, & Kumar, 2015). The eukaryotes are of particular interest to us because they were ancestral to animals and then humans.

The eukaryotes, by definition, have nuclei, composed of an inner double membrane enclosing DNA and proteins. Nuclei thus emerged around two billion years ago. It is thought that another early addition to eukaryotes were *mitochondria*, small structures responsible for energy production that exist in multitudes inside today's cells. Mitochondria bear a strong resemblance to miniature bacteria and are believed to have initially been incorporated into eukaryotes in a *symbiotic* (mutually beneficial) relationship: Mitochondria provided energy, and eukaryotes provided nutrition and protection (Lane & Martin, 2010). The relationship stuck, and mitochondria became heritable along with the rest of the cell's components.

What did the earliest cells look like? It is difficult to be sure, but filament-like shapes are apparent within the first large-scale evidence of life on Earth. These are *stromatolites* – mats of sediment formed by primitive bacteria, dating to 3.5 billion years ago and found in chert. The mats formed in very shallow water, and similar living structures can be seen today in shallow water in Western Australia (Figure 1.3), the same place fossil ones are found (Cowen, 2005; Westall, 2004).

Figure 1.3 Stromatolites in Western Australia.

Source: Stromatolites in Sharkbay.jpg, CC BY-SA 3.0, Paul Harrison.

The earliest stromatolites show no evidence of exposure to oxygen and some evidence of reduction, suggesting that the microorganisms that created them evolved in a hydrogen-enriched, reducing atmosphere (Westall, 2004). A likely scenario is that the earliest cells forming the mats were anaerobic, but at a later date such cells emitted oxygen as a byproduct, similar to today's cyanobacteria. Initially, the oxygen they emitted was scrubbed out of the atmosphere by dissolved iron in the oceans, which oxidized (rusted) and precipitated out to form ocean floor deposits. Only when oxygen production began to outpace the weathering-out of iron from rock did oxygen significantly increase in the atmosphere (Cowen, 2005; Kasting & Catling, 2003; Lunine, 2005).

Remarkably, single-cell organisms were the *only* life forms on Earth for over three quarters of its biological history. The first true *metazoa*, animals in which multiple cells have specialized roles such as forming body walls, stiffening internal structure, and digestion, appeared only 665 million years ago. The long hegemony of the single cell testifies to the difficulties posed by multicellular organization and to the extensive changes evolution had to create to make it possible. We turn next to a discussion of how evolution works.

The workings of evolution

Even knowledgeable writers sometimes use terminology suggesting that evolution "knows" where it is going. Such terminology is a shortcut, an easy way to communicate a direction of change without engaging in a constant struggle to make it clear that evolution is not a conscious thing and therefore accomplishes nothing by premeditation. Even when suggestive wording is avoided, it is easy for a reader to make the inference anyway. An example might be the statement that evolution had to create extensive changes to make multicellular life possible. Although the sentence does not say outright that evolution had a particular direction or intent, it is easy to assume that it did – that it strived toward a "higher", more complex organism.

In fact, that is not at all how scientists view the workings of evolution. If evolution seems to strive toward complexity, it is an illusion arising from the fact that it started with the simplest possible organism and, by creating change, could only move toward increasing complexity. There is no reason other than the improbability of the starting point that evolution could not begin with a highly complex organism and create simplicity. In fact, it has repeatedly been observed that once evolution created complex organisms, it became possible for simplification to occur. Thus, vertebrates lost at least nine gene families that existed in ancestral invertebrates (Danchin, Gouret, & Pontarotti, 2006). In more familiar examples, ancestral snakes and cavefish respectively lost their legs and eyes (Porter & Crandall, 2003).

Descent with modification and natural selection

The engine of evolution is nothing more or less than *descent with modification*, operated upon by *natural selection*. Descent with modification means that the genetic instructions of offspring differ from those of the parents in some way. In sexual organisms, this is accomplished in part through a reassortment of genetic material, half coming from each parent.

More fundamentally, however, it comes about by imperfect copying. For example, *genetic recombination* is an error during the reproduction of chromosomes resulting in the movement of genetic instructions from one chromosome to another. Or within a chromosome, there may be a copying error that results in a slightly altered set of instructions. Such changes, known as *mutations*, are often fatal to development; in that case, they are not carried down to successive generations. Even if a mutation is not fatal, it may involve an inactive portion of the instructions and have no effect. But in some cases, a mutation creates a new *allele* (alternative form) of an active gene. If the animal carrying the new allele survives, it and the allele become subject to natural selection.

Natural selection affects the descent of alleles by favoring some over others. Again, there is no premeditation to this process. Rather, certain capabilities are favored or disfavored by the environment, and the resulting difference in survival affects transmission to successive generations. Examples abound, but in keeping with our single-cell theme, consider the extreme heat around black smokers. This environment favors microbes whose alleles give them greater heat tolerance over those whose alleles give them less tolerance. The ability to resist heat is therefore likely to be passed to successive generations. Indeed, after some sufficient number of generations, the descendants become unable to live in a cold environment.

In black-and-white terms, favored individuals reproduce; unfavored ones die without reproducing. "Survival of the fittest", the label used by Herbert Spencer when referring to Darwin's (1859) description of natural selection, and later adopted by Darwin himself, ensures that those organisms best adapted to their environments are more likely to leave progeny. However, it need not be a black-and-white, live-or-die distinction. For example, there are and have been only tiny fitness differences between Y chromosomes in the human population, so that virtually any man has been able to have children. Nevertheless, all living men descend from the same ancestral Y chromosome that existed on the order of 100,000 years ago (Gibbons, 1997; Yuehai et al., 2001). Thus, very small fitness differences eliminated all other Y chromosomes over that span of time.

Speciation

Why, then, isn't there only one species of animal, conqueror of all competitors and supremely fit to rule its environment? The primary reason is that there is no single environment. At a macro level, water, land, and air make wildly differing demands, and within each, there are a multitude of niches, each of which favors some animals over others. A koala chewing Australian gum tree leaves, a gazelle browsing on African grasses, a tiger consuming an Asian deer, and a grizzly bear feeding on North American salmon have all adapted to particular environments affording different types of food. Within each of these geographical areas, countless other species grasp different threads of the web of life.

Speciation, the creation of new species, often occurs due to physical separation between two parts of an animal population. During the voyage of the ship *Beagle*, Charles Darwin collected numerous variations of finches inhabiting different islands of the Galapagos chain west of South America. The islands provided similar environments, but separation by water had prevented interbreeding between their inhabitants. Over time, natural selection had created different variations on different islands. Today, a number of them are considered separate species.

This kind of speciation is called *allopatric*, from Greek roots literally meaning "other native village". It results in animals well adapted to specific environments and therefore often "unconquerable" by other species. Allopatric speciation is believed to account for most of the speciation that has occurred. Besides the water separating islands, geographic barriers that can divide populations and lead to speciation include mountains, rivers, and energy-impoverished zones of ocean water such as those separating black smokers from the surface.

At the other extreme, *sympatric* (literally "together native village") speciation occurs in populations not physically separated. Sympatric speciation can happen if different individuals in the population feed on differing food sources and mating is linked to those sources. For example, the apple maggot fly has evolved from a North American fly infesting hawthorn fruit. When apple trees were introduced to the continent in the early 19th century, some flies that normally infested hawthorns instead laid their eggs in apples, probably because of a preexisting geographically related genetic variation that was not yet a difference between species (Jiggins & Bridle, 2004). Subsequently, "hawthorn" and "apple" females showed a preference for mating with males feeding on their preferred fruit. This mating preference largely isolated the two groups from one another. Today, the groups show only low levels of genetic exchange, strongly suggesting the beginnings of speciation (Feder et al., 1994). Furthermore, a wasp that is parasitic of the flies shows corresponding change, with apple wasps genetically diverging from hawthorn wasps (Forbes, Powell, Stelinski, Smith, & Feder, 2009).

A female preference for mating with some males over others is an important aspect of *sexual selection*, an adjunct to natural selection. Selection of desirable qualities in mates makes them more prevalent over successive generations. A classic example is the preference of many female birds, such as pea hens, for extremes in feather coloration and form, like those in peacock tails. This is not just a frivolous form of selection, because more substantial traits often accompany decorative ones. More colorful and robust males tend to be healthier and better fed, and therefore they tend to pass down health-related and possibly foraging-related traits as well (Boogert, Fawcett, & Lefebvre, 2011).

Is evolution random?

To what extent is evolution a random process? The answer to this depends on which component of evolution is considered, descent with modification or selection. The modification part of descent with modification is largely random in that mutations occur at random places in the *genome* (the set of genetic instructions embodied in DNA). Although some of the nucleobases in DNA are more susceptible to change than others, there is no biochemical equivalent of "pick me!" signaling an intention to change in a particular direction. Genetic recombinations are also random.

Selection, however, is anything but random. To argue otherwise is to argue, for example, that fish were equally likely to evolve on land or sea. The truth is that the environment places strong directional (i.e., nonrandom) demands on organisms. Over time, natural selection, aided by sexual selection, creates animals that fit their environment, whether it involves the creation of fin-like structures in both fish and whales, or the creation of wing-like structures in both birds and bats. Although our ability to predict the specific products of selection is very limited, general predictions are quite possible. Thus, while no one predicted that hawthorn maggot flies would like apples, it was inevitable that some North American insect would come to occupy the new high-energy environmental niche provided by the apple's introduction.

Although life began as single cells within about 0.5 billion years of Earth's formation, it took another 3.3 billion years for descent with modification, acted upon by natural selection, to produce metazoan life. Presumably, speciation played a large role by creating a multitude of "experiments" that ultimately led to more complex organisms. From that point in history, we leave the realm of weak evidence and speculation behind and can rely on a rich fossil record to trace life's course. Naturally, we need to accurately date the fossils to reconstruct the path, and to that topic we soon turn.

Conclusion

Science can only guess at the ultimate origins of life, but an understanding of the natural processes permitting it is beginning to take shape. A first requirement was replication, a process implying chemical synthesis involving not only raw materials, but also energy (metabolism) and a protective envelope (a membrane). The first self-replicating molecule may have been a short RNA strand formed in a "warm little pond", on a clay surface, around a "black smoker", or in space, all representing locations where energy was available if not abundant. However it originated, the molecule soon needed a membrane to protect it from the environment, perhaps a bubble based on lipids that could enclose other compounds.

Although RNA catalyzes a number of chemical reactions, it is not an efficient energy producer. At some point, RNA was relegated to a support role, with the chemically similar DNA molecule taking over information coding and, through its coding of proteins, the production of energy.

The earliest true organisms were prokaryotes, emerging about four billion years ago, although it is unclear whether bacteria or archaea came first. Stromatolites dating to 3.5 billion years ago, formed in shallow water, record the past presence of primitive bacteria. Eukaryotes, cells with nuclei, emerged around two billion years ago and were the ancestors of animals including humans. Mitochondria were incorporated into eukaryotes in the role of energy producer, probably at first in a symbiotic relationship but later in a heritable one.

The engine of evolution is descent with modification, acted upon by natural selection. Modification is largely random, involving reassortment and recombination of genetic material inherited from parents, or mutations occurring to chromosomes prior to or during copying. Natural selection, however, is driven by environmental characteristics and is decidedly nonrandom. Even small fitness differences can produce large effects over time, as in the elimination of all but one ancestral human Y chromosome line within the last 100,000 years.

Speciation usually occurs allopatrically, with geographic barriers such as water expanses or mountains separating two groups, which then evolve independently. However, it may also occur sympatrically, in the same location, if there are other barriers to interbreeding, such as differing food preferences. The evolution of complex organisms is the product of countless genetic "experiments" producing species that could better exploit their environments.

Science stands on far firmer ground with evolution than it does with life's origins. As we shall see, evolving life left abundant evidence of descent with modification, acted upon by natural selection. Some of it would be in the rock around us, and some in every living cell.

2   Life gets complicated

The single-cell aspect of Earth's earliest life imposed severe limits on the complexity that could be attained through evolution. Organisms would have to become multicellular for life to become complicated.

Sponges were the first true multicellular organisms, dating to about 665 million years ago. As immobile filter feeders, they lacked tissues such as muscles or nervous systems (Peterson & Butterfield, 2005). However, they possessed specialized cells, including tubular cells with large pores that allowed water to pass into the animal, and cells with *flagella* (whip-like structures) that pumped it out. Other cells provided structural support through embedded stiff rods called spicules. Recently, silica-based spicules from early sponges were found in Iran, dating to about 635 million years ago (Antcliffe, Callow, & Brasier, 2014).

A major leap forward occurred with the emergence of *cnidarians* (pronounced "nigh-dare'-ee-ans"), comprised of the jellyfish, hydras, sea anemones, and related species, about 605 million years ago (Peterson & Butterfield, 2005). They had true tissue serving as skin on the outside, and a digestive cavity on the inside, the forerunner of the gut (Cowen, 2005; Steinmetz, Arman, Kraus, & Technau, 2017).

The earliest body fossils of possible metazoans that are *not* sponges date from about 600 million years ago. One in the genus *Vernanimalcula* ("small spring animal") is found in rock known as the Doushantuo Formation in southwest China. Averaging only 150 micrometers across, it is bilaterally symmetrical (having mirror-imaged sides) and reportedly has an identifiable mouth and digestive tract (Petryshyn, Bottjer, Chen, & Gao, 2013). However, these interpretations are controversial, and *Vernanimalcula*'s status as a metazoan is not universally accepted (Cunningham, Vargas, Yin, Bengtson, & Donohue, 2017).

Vernanimalcula is almost exactly 600 million years old. The confident assignment of age is a key issue in evolutionary science. In this chapter we consider how fossils form, their dating using geological and molecular clocks, and their role in helping to construct the tree of life. We also examine the proliferation of species during the Cambrian explosion, including the emergence of chordates, which were the ancestors of all vertebrates including humans.

Fossil formation

A fossil most commonly forms when an organism is covered by sediment, shutting out oxygen and protecting some parts – most often bones and teeth, but sometimes soft tissues as well – from rapid decay. Initial burial is followed by a long period in which the parts are gradually replaced by minerals precipitating from water permeating the sediment. It is not uncommon, however, for original bones and teeth to survive as fossils. Very rarely, soft tissues themselves survive without mineralization, as in cases of preserved tissue from the dinosaurs *Tyrannosaurus rex* and *Brachylophosaurus canadensis* (San Antonio et al., 2011; Schweitzer et al., 2009). Fossils are usually discovered in sedimentary rocks and volcanic ashfalls, and only uncommonly in volcanic lava, which typically consumes organic material. However and wherever they form, however, fossils are rare, and their record should be assumed incomplete (Box 2.1).

Geological clocks

Dating a fossil found in rock starts with considering its position on site. A basic consideration is that rock is generally found in layers, with the oldest on the bottom and the newest on top. Of course, layers can become inverted from folding of the earth's crust. Newer rock can also be at the same level as older rock, as when a river cuts through and deposits sediment, or when molten rock is injected into sedimentary layers. Careful geological study can identify such cases, however, and ascertain whether the oldest-on-bottom, newest-on-top relationship holds. But how do we date the layers? That is the fundamental problem solved by geological clocks.

Assuming undisturbed geology, dating a fossil is straightforward if it is found in rock sandwiched between layers of volcanic rock or ash. *Potassium-argon dating* can be used on the two volcanic layers, and the fossil's age must then lie between those dates. This method is possible because minerals forming in a volcanic layer must once have been liquid and therefore heated enough to drive off any argon gas. However potassium remained behind as a solid, nongaseous element, with a tiny percentage (about one in

---

Box 2.1   The fossil record

In a world ideal for paleontologists, all animals would fossilize, and assembling their record would involve only digging down to the right level in the right place to see what is there. The discoveries from such a dig, arranged in time, could be likened to a movie, recording a continuous series of changes in species across time. Unfortunately, the world is not ideal, and the fossil record more resembles a series of snap-shots than it does a movie. The snap-shots are widely spaced in time, and the series is almost always grossly incomplete.

There are many causes of the incomplete fossil record. Most importantly, animals usually decay without leaving a recognizable trace. It's not even the case that some constant percentage of individuals fossilize, because many animals live in environments disfavoring rapid burial and mineral replacement of body parts. Chimpanzees provide a good example. Ancestral chimpanzees left almost no fossils, and none of great age, because most lived in hot, damp forest environments with acidic soils, factors promoting rapid decay (McBrearty & Jablonski, 2005; Orwant, 2005).

Nevertheless, an incomplete record can still yield good science. For instance, finding jawless fishes in older deposits, and jawed fishes in younger ones, suggests that jawed fishes evolved from jawless ones and not vice versa. Hypotheses formed in this way can then be tested through new fossil discoveries. If the hypothesis is correct, new fossils will likely be discovered over time to fill gaps in the record. Yet the approach is flexible: Should newly discovered fossils prove inconsistent with a hypothesis, it can be revised or replaced, and tested with future discoveries.

10,000 atoms) existing naturally in the form of potassium-40, a radioactive isotope. The isotope began decaying over time into the gas argon-40, becoming trapped in the cooled, solidified material. Thus, by measuring the existing ratio of potassium-40 to argon-40 and using its known conversion rate, the age of the volcanic layer can be determined. Because the radioactive half-life of potasssium-40 is 1.25 billion years, the method is appropriate even for the oldest volcanic materials on earth (Brown, 1992). However a number of alternative techniques exist, based on other radioactive isotopes (for brief, partial reviews, see Brown, 1992; Levin, 2010).

Dating the Doushantuo Formation in which *Vernanimalcula* was found proved more problematic because it is sedimentary, with no closely spaced volcanic layers allowing precise "sandwich" dating. Initial attempts to date it used volcanic layers well above and below the formation, providing a wide range of possible ages, somewhere between 539 million and 748 million years (Barfod et al., 2002).

The problem was eventually solved by determining the age of the sediment itself. When water deposits certain minerals such as phosphorite, minute quantities of uranium are included. The radioactive decay of uranium follows a set sequence producing predictable levels of lead isotopes. Thus dating proceeds by measuring the levels of the isotopes. When the method was applied to Doushantuo phosphorite, the precise age of 599.3 (± 4.2) million years was obtained (Barfod et al., 2002).

Geological clocks are essential tools for dating newly discovered fossil remains. Even when a rock layer is unsuitable for isotope methods of dating, it is often possible to indirectly derive a date by identifying ordinary fossil plants and animals within it. If those species have been dated elsewhere using isotope methods, the date can often be applied to the layer of interest.

Of course, the value of geological clocks increases if more than one method is applied and the results converge on similar values. But they are not the only dating methods.

Molecular clocks

One of the most exciting developments in late 20th century paleontology was the invention of *molecular clocks*. Of great benefit when the fossil record is poor, these estimate when the ancestors of two species diverged from one another based on quantified differences in DNA. The forerunners of molecular clocks were clocks based on blood protein differences. These scored an impressive early victory by showing that primate divergences were much more recent than most paleontologists assumed (Pilbeam, 1972).

The fundamental assumption of molecular clocks is that genetic differences between animals are directly related to when they branched off a common ancestor. Closely related descendants from a common ancestor should show relatively few differences in their genes; a thousand generations later, more differences should be apparent; and a million generations later, differences should be pronounced. This expectation is a direct outcome of the natural processes that over time damage and miscopy DNA. The fundamental assumption is therefore so well grounded as to be beyond serious dispute. Instead, the controversial aspect of molecular clocks is how to *calibrate* them. In other words, how can genetic differences be translated into a time scale that can be used to estimate how long ago species diverged?

The linear molecular clock

The simplest approach to estimating age from molecular differences is to assume a linear relationship between genetic differences and time. Of course, to arrive at an actual estimate of the time at which two species diverged, it is necessary to calibrate the clock using fossil record estimates of one or more divergence times.

Figure 2.1 shows an example from primate ancestry, drawn from early work by Wildman, Uddin, Liu, Grossman, and Goodman (2003). It uses a 25-million-year-old divergence between humans and Old World monkeys, and a 14-million-year-old divergence between humans and orangutans, as inferred from the fossil record.

As these ages predict, and as shown in the figure, Old World monkeys genetically differ from humans by a greater value (5.9%) than do orangutans (3.7%). A line between those points calibrates the clock, and it is then used to estimate that gorillas, differing from humans by 1.8%, diverged from our ancestral line five million years ago; and chimps, with a 1.6% difference, four million years ago.

Of course these estimates are critically dependent on the calibration points. The ages of the two used by Willman and colleagues are now believed to be underestimates, resulting (as we will see) in misestimated ages for the gorilla and chimpanzee divergences.

Calibration aside, comparison to established portions of the primate tree show that a linear clock provides a fairly good fit to primate data (Drummond, Ho, Phillips, & Rambaut, 2006). That is probably because primate species diverged over a relatively short period of time, within a single biological order.

More problematic are attempts to extend linear clocks across biological taxa broader than a class, such as a phylum or kingdom (Table 2.1). Extended across a kingdom, linear clock estimates in excess of one billion years have been obtained for the common ancestor of all animals (Wang, Kumar, & Hedges, 1999). Such "deep" estimates are rejected by most scientists because, as we have seen, the oldest metazoan fossil is only about 635 million years old.

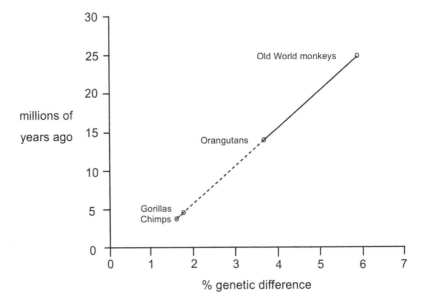

Figure 2.1 Example of a linear clock using two calibration points (Old World monkeys and orangutans) to estimate the divergence times of gorillas and chimpanzees from the human ancestral line, based on genetic differences.

Source: Original figure.

Table 2.1 The major biological taxa, which can be remembered using the mnemonic "Kings play cards on fat green stools", as applied to modern humans. Note that lower levels are members of higher levels, so for example, all primates are also chordates (phylum Chordata).

Kingdom: Animalia (animals)
Phylum: Chordata (animals with a notochord or spinal column)
Class: Mammalia (mammals)
Order: Primates
Family: Hominidae (apes and humans)
Genus: *Homo* (ancient and modern humans)
Species: *sapiens* (modern humans)

The nonlinear molecular clock

Such deep misestimation of dates by the linear clock means that substantial nonlinearity must exist when comparing distantly related animals. In other words, the clock must run fast for some animals and slow for others. Rutschmann (2006) provided a useful listing of causes of nonlinearity in the molecular clock. One is *differences in generation time*, the fact that some species breed at much shorter intervals than others. Over time, the chance that DNA will be miscopied must be higher for a mouse than for a human, simply because mice breed far faster and therefore run through many more generations. A second factor is *differences in metabolic rate*. Animals with high-energy expenditure show faster rates of DNA synthesis and therefore have more opportunities for miscopying errors. Finally, a third is *differences in DNA repair*. To some extent DNA can repair itself if damaged, and the ability to do this varies over species (Rutschmann, 2006). As it happens, mice are good examples of all three factors. Not only do they breed at rapid intervals, they have a high metabolic rate and they are not as efficient in repairing DNA. No wonder, then, that mice accrue genetic errors much more quickly than do humans (Welch & Bromham, 2005), a straightforward indication that their molecular clock runs fast relative to ours.

The alternative to a linear clock is what has picturesquely been called a "relaxed" or "sloppy" clock (Rutschmann, 2006). A number of different versions exist, but all share the assumption that the clock runs at different rates in different branches of the ancestral tree (Welch & Bromham, 2005). Estimation of multiple rates requires considerably more calibration points, so the clock might more properly be called a "constrained clock" because it stresses the importance of fossil evidence. Nevertheless, the approach arguably results in a substantially more accurate clock. In particular, it was a relaxed clock that indicated metazoans first evolved about 665 million rather than one billion years ago. The same approach provided the 605-million-year-old emergence date for cnidarians (Peterson & Butterfield, 2005). Closer to home, a recent relaxed clock approach using multiple fossil-based anchor points indicated that the ancestors of chimpanzees and humans diverged 7.5 million years ago (Wilkinson et al., 2011), older than linear clock estimates.

But could it be that the fossil record is so incomplete that metazoans really did evolve over one billion years ago, as the linear clock states? That has been a hotly debated aspect of the clock controversy. Rocks of such age certainly exist, even sedimentary ones, yet extensive searches have failed to yield unambiguous evidence of metazoan fossils within them. Initially promising "fossils" have been subsequently discounted as caused by geological and not biological processes (Budd & Jensen, 2004). But possibly early metazoans were so soft-bodied that they simply did not fossilize.

Peterson and Butterfield (2005) ingeniously argued that there exists an indirect indication in the fossil record of the first appearance of metazoans. They pointed out that the onset of multicellular animals should have impacted preexisting organisms, because those organisms would have been food sources for metazoa. As it happens, the record of *acritarchs* (tiny acid-resistant fossils showing evidence of a cell wall) is extensive during the period in question. Many acritarchs are believed to have been similar to today's algae as found in plankton, certainly a food source for small multicellular animals. Significantly, the form of acritarchs remained virtually unchanged in the fossil record for hundreds of millions of years – until about 635 million years ago. They then began a rapid evolution of form, with some developing outward spines, probably as a defense against consumption. The implication seems clear that metazoan consumers had recently evolved, creating a selective pressure for creation of defensive structures. The dating of these events at 635 million years ago corresponds to reasonable expectations for the first large populations of *motile* (independently moving) metazoans, as opposed to the earlier *sessile* (attached) sponges. Motility presumably enabled the consumption of acritarchs (Peterson & Butterfield, 2005).

A 30-million-year gap before the first fossils of metazoans and their first effect on acritarchs, from 665 to 635 million years ago, is much easier to explain than a gap proposed by the linear clock to be over 10 times bigger. In the beginning, metazoan populations were small, rendering improbable the preservation and future discovery of their earliest remains (Budd & Jensen, 2004). Likewise, they did not yet visibly affect other organisms. Later, as metazoans became abundant, they left a more extensive and thus more discoverable fossil record, and they had a pervasive effect on acritarchs.

Thus, fossils are generally younger than an animal's first appearance. This consideration, combined with the metazoan and acritarch fossil records, and the results of the relaxed clock, make a metazoan origin 665 million years ago seem quite reasonable.

Early metazoan dating, however, has not been laid to rest. More recent relaxed clock approaches, while not suggesting a metazoan origin as far back as one billion years, still peg it at a distant 740–780 million years (dos Reis et al., 2015; Erwin et al., 2011). If ultimately accepted, this relatively deep estimate would implicate a missing-fossil interval of 105–145 million years rather than 35 million years.

Molecular clocks are valuable, but as the metazoan case shows, their estimates are only approximate. Geological clocks are usually much more precise, as exemplified by the dating of the Doushantuo Formation within 4.2 million years. Nevertheless, both have their uses. Geological clocks date fossils, while molecular clocks date the first appearance of new categories of life. Neither substitutes for the other.

The tree of life

Both geological and molecular clocks have proved invaluable in reconstructing the tree of life, a concept entertained by Charles Darwin. His insight was that all life could be envisioned as a vast tree, rooted in the distant past and with many branches and twigs extending to the present day.

Before the invention of geological and molecular clocks, the tree of life was reconstructed by examining anatomical similarities and differences among species, with similar ones assumed to have branched off from a recent common ancestor. This approach is systematized by *cladistics*, the grouping of species by novel characteristics absent in less related species. A *clade* is a branch on the tree of life defined by one or more characteristics. For example, the mammalian clade is characterized by fur, which all mammals but no other animals have. Of course, a clade can be a subset of a larger clade. *Artiodactyls* are a clade within mammals, consisting of hooved mammals with an even number of toes (Cowen, 2005).

Characteristics of species are frequently termed either *primitive characters* or *derived characters*. A primitive character is one possessed by the earliest-evolving member of a clade. Fur, for example, is a primitive character of mammals. A derived character is one possessed by one subgroup but not by all subgroups within a clade. Thus, being hooved with an even number of toes is a derived character within mammals. Notice that the terms are relative to the clade under consideration; being hooved with an even number of toes is a primitive character within artiodactyls.

However, there has sometimes been vigorous debate over the definition of clades. Even something as intuitively obvious as a mammalian clade can be surprisingly difficult to define, especially when applied to the fossil record (Cowen, 2005). For present purposes such debates are unimportant, for they are mostly over the labels applied to branches on the tree of life. The branches themselves are much less controversial.

Today, reconstructing the tree of life is increasingly accomplished by combining cladistics with analyses of genetic differences between species. Divergences are then dated using sloppy clocks calibrated with the fossil record.

Figure 2.2 expresses findings using the combined approach, stitched together from a number of independent research projects. Although it omits date estimates, we can use its structure to continue exploring the evolution of complex animal life. It certainly captures much of what has been described thus far. Eukaryotes (eukarya) were among the first groups to emerge (top left of the figure; note that the bacteria-archaea ordering is controversial), followed after several branchings by metazoa (middle), which further branched into sponges and then cnidarians (labels toward the right). Humans emerged many branchings later, among placental mammals at the very bottom of the figure.

Details of the tree of life are subject to constant revision as additional research accrues. For example, if archaea are eventually determined to have preceded bacteria, an alteration would be required to the top part of Figure 2.2. Also, it has recently been suggested that ctenophores, not sponges, should be the first metazoans represented (Whelan, Kocot, Moroz, & Halanych, 2015). However, the largest and most carefully controlled study conducted to date supports the sponges-first depiction (Simion et al., 2017).

The Cambrian explosion

An extraordinary feature of the fossil record is that within a short period of 10–20 million years, most of the living metazoan phyla suddenly appeared for the first time. This apparent rapid evolution of life forms, called the *Cambrian explosion*, covered all the major branchings of the tree of life in Figure 2.2 from the *bilateria* (bilaterally symmetric animals) through the *Chordata* (chordates). Occurring in the narrow range of 520 or 530 to 540 million years ago (in round figures), it is named for the geological age called the Cambrian Period (in round figures, 490–540 million years ago).

The Cambrian explosion has caused wonder, doubt, and endless theorizing. The wonder, of course, is that so many animal forms appeared so quickly. As Figure 2.2 indicates, the branching of the metazoa down through the chordates includes the ancestral lines of animals as diverse as insects, clams, starfish, and all vertebrates collectively. That the ancestral lines of such dissimilar life forms seemingly emerged within 10–20 million years is remarkable to say the least.

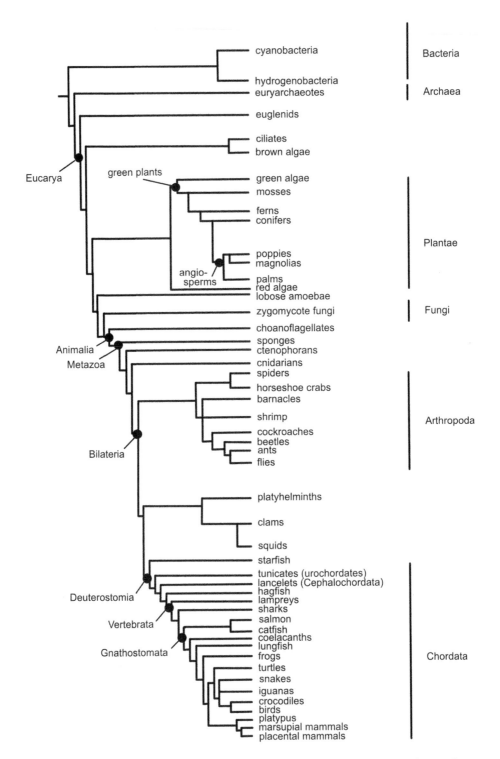

Figure 2.2 The eukaryote tree of life, showing some commonly known plant and animal groups. Origin at top left, progressing to existing phyla on the right. Note that the bracket widths are not meant to directly reflect time from a common origin.

Is it a little too remarkable? The Cambrian explosion has had its share of doubters. Early debate over the explosion concerned the adequacy of the fossil record. Perhaps there had been previous forms, but they were too soft-bodied to fossilize. If that was the case, then attempting to date the origins of the phyla from fossils would be like trying to reconstruct an historical event from a photograph taken afterward.

There is no conclusive way to show that the fossil record is adequate, but it is now clear that sedimentary rocks of the Cambrian and late Precambrian periods exist and do preserve soft-body evidence. What is notable is the lack of hard-body evidence until the three-to-six million years immediately preceding the Cambrian explosion (Valentine, Jablonski, & Erwin, 1999). After that, the proliferation of relatively large animals with hard-body parts does seem to qualify as an explosion (Butterfield, 2003). If the emerging phyla became ancestral to most living animals, it is because hard-body parts proved exquisitely adaptable. A skeleton, whether external as in the shells of ancestral insects and crustaceans, or internal as in sponges or ancestral vertebrates, is invaluable in providing protection from predators, firm attachment points for muscles, and a framework for repair after damage (Cowen, 2005).

Early hypotheses on the explosion's cause cited rising oxygen levels as enabling the evolution of large, complex bodies. Because oxygen penetrates poorly into external shells, large-body evolution could not occur until it reached a sufficient concentration that it could be brought into the body through small openings such as gills (Cowen, 2005). This supposition is supported by a correspondence between increasing metazoan consumption of oxygen, and increasing oxygen levels. Thus, mollusks appeared before arthropods, which appeared before fishes. Oxygen consumption increased across the three groups, reflecting atmospheric increases in oxygen (Zhang & Cui, 2016).

It is also possible that evolving hard parts simply took time and, once it began, was so successful that hard bodies radiated worldwide. By this account, the Cambrian explosion is self-explanatory. Unfortunately for this explanation, the explosion was not limited to "hard" animals but also included a proliferation of soft-body forms (Valentine et al., 1999).

An ecological view was expressed by Logan, Hayes, Hieshima, and Summons (1995). They suggested that the earlier proliferation of small metazoans feeding on plankton-like life forms created a seabed environment that could be exploited by new phyla. Fecal pellets dropping to the seabed would have created a carpet of high-energy foodstuff while at the same time improving the oxygenation of the water due to the removal of carbon (which was therefore not incorporated into carbon dioxide). The strength of this hypothesis is that it provides one reason why most of the new life forms of the Cambrian explosion had *benthic* (bottom-dwelling) lifestyles (Valentine et al., 1999).

Chordates

The Cambrian explosion phylum leading to vertebrates was *Chordata*, the chordates. These relatively complex animals have a hollow neural tube behind a *notochord* (supporting rod), which stiffens the body and provides muscle attachment points. The neural tube is segmented into three compartments, namely a forebrain, a hindbrain, and probably a midbrain, the beginnings of a modular nervous system (Holland & Chen, 2001). A living example thought to closely model the chordate ancestor is amphioxus (genus *Branchiostoma*), also known as the lancelet, a sea-bottom burrowing filter feeder. It is a member of the subphylum cephalochordates. Its importance is its descent from a common ancestor shared by another chordate subphylum, the vertebrates, but not by other chordate subphyla (Hickman, Roberts, & Larson, 1997). Thus, its anatomical and behavioral characteristics are important in reconstructing early vertebrates.

The cephalochordate-vertebrate common ancestor is generally accepted as having evolved from earlier chordates by *neoteny*, the retention of juvenile characteristics in adulthood (Lacalli, 1994). Most invertebrates pass through a larval stage before metamorphosing into adults. Very often, the larvae swim, but the adults are sessile. Thus, assuming it results in a viable organism, a mutation producing sexual maturity prior to metamorphosis would result in a swimming, breeding adult, accounting for swimming in the common ancestor.

Amphioxus resembles a translucent fish, a few centimeters in length, with no jaw or fins, and with a single eye (Holland & Chen, 2001). Marked behavioral and structural asymmetries are present despite membership of the chordates in the bilateria. As seen from behind, newly hatched amphioxus move by spiraling counter-clockwise. Among other anatomical asymmetries, the mouth opens on the left and the mid-gut diverticulum runs along the body's right side. These asymmetries, and similar ones in other chordates, indicate that lateral differences, so important to primates as we shall see, first emerged in chordates' immediate ancestors and may be considered a chordate character (Li et al., 2017).

By the Cambrian explosion's end, the chordate phylum was well established. Over about 150 million years, the animal ancestors of humans had evolved from microscopic collections of cells to small fish-like creatures. Their neural tubes and anatomical asymmetries would have important implications for the evolution of sensation, perception, and cognition. After another 150 million years, some of their descendants would start leaving the water.

Conclusion

The earliest metazoa, in the form of sponges, originated about 665 million years ago. Defense-related changes in the shape of tiny fossil acritarchs 30 million years later suggest that they were responding to predation – presumably by metazoa. After another 30 million years, true tissues first appeared in the cnidarian ancestors of modern jellyfish.

Accurate dating is a key issue in evolutionary science. Geological clocks typically exploit relative levels of radioactive isotopes to date rock. Molecular clocks use genetic differences between species to date ancestral divergences. In primates, a linear molecular clock is reasonably accurate, but it breaks down when extended across broad biological taxa, with the clock found to run at different rates for different animals. Thus a "relaxed" or "sloppy" clock is often used, employing fossil evidence to calibrate different branches of the ancestral tree. Even so, controversies emerge, as in the dating of the earliest metazoa.

The tree of life is reconstructed using both cladistics (the grouping of species by novel characteristics), and genetic similarity to map its branches. Its details are constantly revised as research accrues.

Most living metazoan phyla suddenly appeared during the Cambrian explosion (roughly 520 or 530 to 540 million years ago). Proposed explanations include an inadequate fossil record preceding the explosion; rising oxygen levels enabling the existence of large, complex bodies; the evolution of hard parts simply taking time but proving highly successful; and the carpeting of the seabed with fecal pellets that provided high-energy foodstuff for new benthic (bottom-dwelling) species.

An early branch of animals arising during the explosion and leading to vertebrates was the chordates, in which a hollow, segmented neural tube accompanies an internal supporting rod. Amphioxus, a modern species likely similar to the common ancestor of early cephalochordates and vertebrates, resembles a small fish, has a single eye, and shows behavioral and anatomical asymmetries.

3   Vertebrates to early mammals

The evolution of chordates (phylum Chordata) during the Cambrian explosion paved the way for *vertebrates* (subphylum Vertebrata), or animals with a brain and a vertebral column composed of cartilage or bone. In this chapter we first consider early vertebrates and their characteristics, before describing the emergence of tetrapods. Their descendants included the amniotes as well as the diapsid and synapsid lines of the dinosaurs and mammals.

Early vertebrates

Early vertebrates were fish-like. The Chenjiang rock layers of southern China, dating to about 520 million years ago, provide one of the most complete fossils of a probable one called *Myllokunmingia fengjiaoa*, illustrated in Figure 3.1 (Benton, 2005; Shu et al., 1999). It clearly shows the presence of upper and lower fins.

Identification of *Myllokunmingia* as a vertebrate is largely based on evidence of gill pouches as well as a distinct head with what may have been a cartilaginous skull (Holland & Chen, 2001). However, some doubt remains because there is no evidence that the notochord extended to the head, or that there was a distinct brain. These absences leave open the possibility that the animal was a prevertebrate *craniate* (an animal with a skull, in early forms made of cartilage). Yet either way, it must be very close in form to an ancestral vertebrate.

Metaspriggina walcotti, from the Burgess Shale in Canada, likewise had a head, with cranial cartilage, paired eyes, gill bars, and a notochord. It is similarly identified as an early vertebrate (Morris & Caron, 2014). That it came from the Walcott Quarry section of the Burgess Shale tightly dates it to 505 million years ago, give or take a couple of million years (Caron & Rudkin, 2009).

In whatever specific animal a brain first formed, it was a highly significant event, for the vertebrate brain is what made high-level cognition possible.

Biomineralization

"Harder" evidence of vertebrate status might be taken as *biomineralization*, the incorporation of minerals into structures such as bones or teeth. If so, uncontroversial vertebrates existed a little later. About 493 million years ago, *Anatolepis* had primitive teeth, formed from skin that incorporated minerals to form *dentine* (bonelike tissue, but without cells). That clearly identifies it as a vertebrate, in which both teeth and skin develop from epidermal cells. Thus, within about 40 million years, natural selection worked from the chordate body plan to produce identifiable hard parts in early vertebrates (Benton, 2005).

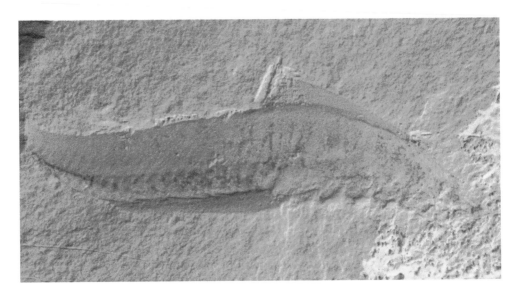

Figure 3.1 Fossil of *Myllokunmingia fengjiaoa*, a probable early vertebrate.

Source: Adaptation of Myllokunmingia big.jpg, CC BY-SA 3.0, Degan Shu, Northwest University, Xi'an, China.

All of these complex animals were sea-dwelling, as evolution had not solved the problems raised by land habitation. The earliest fossils of land-dwelling, air-breathing animals, millipedes that fed on decaying vegetation, date back only to about 420 million years ago (Rodriguez et al., 2018).

Nevertheless by the time of *Anatolepis*, evolution produced several important adaptations that vertebrates would build upon. First, the significance of hard parts cannot be overestimated. Teeth allowed more effective predation, and the later development of bone increased body integrity while protecting the nervous system and providing a ready supply of calcium (Maisey, 1996).

The notochord, the chordate innovation that supported the neural tube, was incorporated by vertebrates into the spinal column where both became surrounded by bone. In later-evolving vertebrates such as reptiles, birds, and mammals, the notochord is most apparent during embryonic development, transforming to a pulpy mass within the spinal column in adulthood.

Housed within the spinal column of even the earliest vertebrates was the spinal cord – an important nervous system innovation, as it served as the conduit for information passing from peripheral sensors to the brain, and from the brain to the muscles. These would become key adjuncts to a knowing and acting cognitive brain.

Hox genes

Another major vertebrate adaptation was a multiplication of genes controlling the development of body segments, or as has been written, "distinguishing anterior from posterior, where segments will form in the embryo, where limbs or fins will form, and which way is up" (Maisey, 1996, p. 31). These *Hox* genes, an abbreviation of *homeobox* (i.e., "like a box", so named because the DNA sequence is short and can be enclosed by a box in written form) are believed to have originated no later than the emergence of cnidarians (Ferrier,

2016; Reddy, Unni, Gungi, Agarwal, & Galande, 2015). In early jellyfish they may have helped specify the "bell" at one end, and the mouth and tentacles at the other. Later, during vertebrate evolution, Hox coding was responsible for several innovations including the transition between fins and limbs, the formation of ribs, and variations in the number of vertebrae (Pascual-Anaya, D'Aniello, Kuratani, & Garcia-Fernàndez, 2013).

Nonvertebrate chordates have a single cluster of Hox genes, strung like beads on a string along a single chromosome (or more accurately, because chromosomes are paired, along each of a single pair of chromosomes). Vertebrates have four clusters, each strung along a different chromosome. In humans, they are chromosomes 2, 7, 12, and 17 (Hughes, da Silva, & Friedman, 2001). Within each cluster, a Hox gene's position determines its influence on a corresponding body segment. Thus, in each of the four chromosomes involved in vertebrates, the Hox gene affecting head development is at one end, while the other end has genes affecting the body's hind parts (Hickman et al., 1997).

Having four vertebrate clusters instead of one allowed for increased structural complexity because it permitted more elaborate developmental instructions for different body segments. One way to quadruple the clusters would have been for the entire genome to double not just once but twice, a "two-round" theory of genome duplication. Thus a single cluster on one chromosome would have become four clusters on different chromosomes. Genetic analysis supports the two-round theory, with the first duplication dating to the origin of vertebrates and the second to a point early in their evolution (Cañestro, Albalat, Irimia, & Garcia-Fernàndez, 2013). Once Hox genes multiplied, mutation followed by natural selection would have caused divergence among the duplicates, allowing a corresponding increase in the complexity of body structure.

A related vertebrate innovation was the appearance of a *neural crest* during development, regulated by Hox genes. In the embryo, the crest consists of a cluster of cells arranged in ridge-like formations to the sides of the neural tube. Cells from it migrate to various parts of the body during development to form the brain and many of the nerves and *ganglia* (functional groupings of nerve cells) – as well as skull bones, teeth, connective tissue, smooth muscles, and the heart among other organs and tissues (Trainor, Melton, & Manzanares, 2003). The result is a more complex design, including a more complex nervous system to serve a more complex body plan.

Visual system changes

Particularly important among vertebrate adaptations were changes to the visual system. Unlike earlier-evolving animals, vertebrates have muscles outside the eye allowing its repositioning. The eye also has a small spot of acute central vision called a *macula*. Eyes occur in pairs, although this adaptation appeared independently in many invertebrates as well (Rowe, 2004). These characteristics improve vision in multiple ways, by permitting more rapid image acquisition (i.e., by eye movements instead of whole body movements), by increasing image resolution (i.e., by way of a macula), and by increasing the size of the visual field (i.e., with the two eyes pointing in different directions in most vertebrates). Improved vision in turn called for an increase in the size and complexity of visual processing areas in the brain, eventually leading to advances in visual cognition including improved object recognition.

Gill arches as predecessors to jaws

Another early vertebrate found in Chenjiang rock was *Haikouichthys*. Unlike *Myllokunmingia*, it possessed clear *gill arches*, thin support structures for the gills made of cartilage (or bone in later fishes). Gill arches were a significant development in vertebrates, because jaws are believed to have evolved through their modification. Specifically, the most *anterior* (forward) gill arch developed a joint permitting the mouth to be opened wider. That allowed more water to be pumped through the gills, thus providing greater oxygenation. The joint in turn became the basis for evolving a jaw. Thus jawless fishes first evolved, followed by jawed fishes somewhat over 400 million years ago (Cowen, 2005; Schilling, 2003).

Tetrapods

In 2004, a crew member on a paleontological expedition to the Canadian Arctic found a snout sticking out of a cliff. It was *Tiktaalik*, an animal with thick, fleshy, almost hand-like fins (Figure 3.2). Although not classified among *tetrapods* – vertebrates with four feet or legs – it seemed a likely near ancestor of them. Remarkably, the expedition had been mounted to discover just such a creature. The search, begun five years earlier, ended with the excavation of three nearly complete fossils, the largest almost nine feet in length (Morelle, 2006).

Scientists had long known that such an animal must have existed. Previously discovered fossils of *Panderichthys* (Figure 3.3), dating to 385 million years ago, revealed a creature with bony gill covers that facilitated the flow of water, indicating a predominantly aquatic lifestyle. In effect it was a predatory fish. It was lobe-finned, and from such fleshy structures "hands" – including our own – are believed to have evolved (Ahlberg & Clack, 2006). In comparison, *Acanthostega* (365 million years ago) lacked bony gill covers though it retained internal gills, indicating less water flow and implying that the animal breathed partially out of the water. This conclusion matches the limb structure, in which the limbs were broad and flat, suitable for propping up the head in shallow water. Eight digits per limb are clearly visible (Laurin, Girondot, & de Ricqlès, 2000; Shubin, Daeschler, & Coates, 2004). With the earlier animal finned and the later one not, there

Figure 3.2 A reconstruction of *Tiktaalik*.

Source: Tiktaalik roseae life restor.jpg, public domain, Zina Deretsky, National Science Foundation.

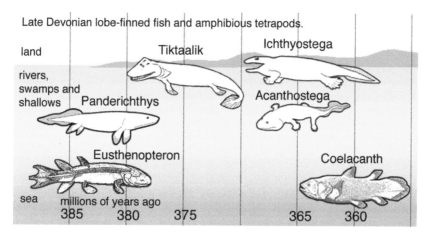

Figure 3.3 The transition from fish (bottom left) to tetrapods (upper right), with Tiktaalik in a transitional role. A lobe-finned coelacanth fish is also shown for comparison.

Source: Fishapods.png, CC BY-SA 3.0, dave souza.

must have been a transition between the two forms: A "predicted *Tiktaalik*" that was searched for, found, and subsequently described (Daeschler, Shubin, & Jenkins, 2006).

At 375 million years old, *Tiktaalik* is a transitional form between fish and tetrapods. Traditionally, tetrapod evolution was attributed to a shift from water to land, but modern research points to an intermediate phase in which animals propped themselves up with fleshy fins in shallow water and mud flats. That gave access to the small prey available there, while providing protection from large, deeper-water predators. The intermediate animals, the fossils tell us, were still fish even though they were partially air breathing.

Besides prop fins, transitional forms are believed to have had a prominent tail fin used when free swimming. Unfortunately, the tail of *Tiktaalik* was not preserved in the fossils, despite the fanciful ones shown in Figures 3.2 and 3.3. Nevertheless a tail fin was later present on the true tetrapod *Acanthostega* (360 million years ago). That animal also had separated digits, although it was still not a permanent land dweller.

The lineage represented by these species was a freshwater one, evolving in and beside lakes and streams (Ahlberg & Clack, 2006; Shubin, 2008, pp. 40–41). There is little support for the alternative, that the water-to-land transition was first made by sea-dwelling animals. Sea-related rock depressions initially interpreted as 395-million-year-old tetrapod trackways in Poland (Niedzwiedzki, Szrek, Narkiewicz, Narkiewicz, & Ahlberg, 2010), instead appear to have been fish nests or traces of feeding. Indeed, all three confirmed tetrapod trackways from the Devonian period (covering 419–359 million years ago) are from freshwater environments (Lucas, 2015).

The early tetrapods were amphibians, living a larval phase in water and an adult phase on land like tadpoles and frogs do today. The problems involved in the transition to land were formidable. Besides the obvious one of breathing air, there were also those of supporting the body without the buoyancy of immersion, moving across land, feeding, and retaining water. Lobe-finned fishes had a "leg up" in making the transition because their fleshy fins could serve as both body supports and thrusters.

Other adaptations were useful as well. Early tetrapods were heavily scaled, aiding water retention (Cowen, 2005), and they may already have been capable of some air breathing. Among lobe-finned fishes, lungfish as opposed to coelacanths are the closest relatives of tetrapods, suggesting that tetrapods evolved from something very similar – and lungfish can breathe some air (Irisarri & Meyer, 2016; Takezaki & Nishihara, 2017). Finally, lobe-finned fishes could raise their upper jaw and not just lower their lower jaw, a *pre-adaptation* (an evolutionary change later built upon by other changes) for feeding in the shallows up to water's edge without scooping up mud (Cowen, 2005).

During the transition, prey were initially captured through water suction, the feeding method favored by most aquatic vertebrates. Obviously, however, this method is useless for hunting out of water, so biting eventually evolved. We know roughly when this occurred, because it is visible in the fossil record: The adjoining edges of skull bones show different strain patterns in sucking versus biting species. Thus *Eusthenopteron* (the earliest species in Figure 3.3) was a sucker, and *Acanthostega* was a biter (Markey & Marshall, 2007). *Tiktaalik* was also a biter, using a sideways snapping motion to capture prey much like modern crocodiles (Hohn-Schulte, Preuschoft, Witzel, & Distler-Hoffmann, 2013).

One peculiarity of early tetrapods is that the modern plan of five digits per limb did not predominate. More common were six, seven, or eight digits per limb (Bradshaw, 1997; Coates, Jeffery, & Ruta, 2002; Saxena, Towers, & Cooper, 2017), a condition known as *polydactyly* (having more than five digits per limb). The five-a-limb plan did not become fixed in most tetrapods until it became specified by Hox genes (Kherdjemil et al., 2016). As Bradshaw (1997) pointed out, if the number of digits had not been reduced in the human ancestral line, we might now be counting in base 12 – or 14, or 16! But regardless of their number, the evolution of digits would eventually have profound implications for the manipulation of objects, as well as their creation using memory representations, spatial transformations, and other cognitive tools.

Why did tetrapod ancestors begin to venture onto land? The reason is similar to that famously given by trailblazing mountaineer George Mallory when asked why he wanted to scale Mount Everest: "Because it's there". Land was an environment lacking predators, unexploited by animals larger than millipedes, and covered with potentially nutritious vegetation. It therefore represented an environmental niche – indeed, many niches – waiting to be filled. Venturing onto land had huge value to the first pioneers.

Amniotes

As advanced as tetrapod amphibians otherwise were, they resembled their fish ancestors by laying eggs in water. The major innovation of their descendants the *amniotes* was the development of embryos within a membrane. In egg-laying species the membrane and associated shell allow gases to be exchanged with the environment, but the embryo to remain hydrated. That permits the eggs to be hatched on land rather than in water (Cowen, 2005). In fact, it appears that the early amniotes were fully terrestrial, having relatively complex, multichambered lungs for breathing air (Lambertz, Grommes, Kohlsdorf, & Perry, 2015).

The first known amniotes date from about 325 million years ago. This estimate is based on characteristics other than fossilized eggs, which are missing from the record for tens of millions of years after amniotes had to have evolved. It is thought that early eggs fossilized very poorly because they had little or no mineral content (Piñeiro, Ferigolo, Meneghel, & Laurin, 2012). Instead, dating amniote origins depends on finding evidence of other characteristics of known amniotes. Those include an *anapsid* skull (a skull lacking openings behind the eyes; see Figure 3.4) with a narrow snout; vertebrae allowing head swiveling; and upright rather than sprawling limbs. Based on these and other characteristics, one of the earliest amniotes was *Westlothiana lizziae*, found in West Lothian, Scotland (Prothero, 2007, pp. 234–237; Smithson, Carroll, Panchen, & Andrews, 1994). It inhabited a freshwater lake environment but was terrestrial and dated to about 325 million years ago (van Tuinen & Hadly, 2004). These were much smaller animals than the early transitional tetrapods. They resembled small lizards and were an

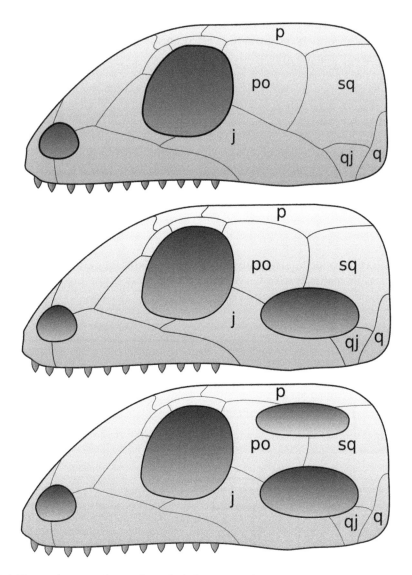

Figure 3.4 Top to bottom: Anapsid (no-holed), Synapsid (one-holed), and Diapsid (two-holed) skulls.

Source: Adaptation of Skull anapsida 1.svg, Skull synapsida 1.png, and Skull diapsida 1.png, CC BY-SA 3.0, Preto(m).

evolutionary step away from amphibians, although they were not quite reptiles. They were likely cold-blooded and ate small invertebrates including insects (Oftedal, 2002), using a weak bite to disable their prey (Kemp, 2005).

Synapsids and diapsids

Within short order (on an evolutionary time scale), the anapsid amniotes gave rise to *synapsids* (tetrapods with one opening in the skull behind the eye) about 310 million years ago, and a brief time later to *diapsids* (tetrapods with two openings; see Figure 3.4). These adaptations lightened the skull and provided attachment points for heavy jaw musculature, allowing a more powerful bite (Kemp, 2005; Lee, Reeder, Slowinski, & Lawson, 2004). The distinction between the two taxa is not just a technical one because each gave rise to an important group of vertebrates. Specifically, diapsids gave rise to dinosaurs and synapsids to mammals.

The evolutionary pathways, however, were a bit more complicated than that. First, early synapsids gave rise to therapsids, which in turn became the immediate ancestors of mammals (Figure 3.5). The synapsid single opening in the skull was then lost as the cheek bones became jaw muscle attachment points. As a result, living mammals lack the opening but nevertheless are considered synapsids by descent (Bard, 2017).

A second complication is that after the synapsids split off, the remaining amniotes can be regarded as primitive reptiles of which the diapsids were only one group – though a spectacularly successful one. Thus there were other early reptiles (e.g., parareptiles) with anapsid skulls (Lee et al., 2004). Figure 3.5 illustrates these relationships.

Dinosaurs

Our main narrative concerns synapsids, but brief consideration of the diapsids, and more particularly the dinosaurs, is in order. As Figure 3.5 indicates, dinosaurs were a relatively

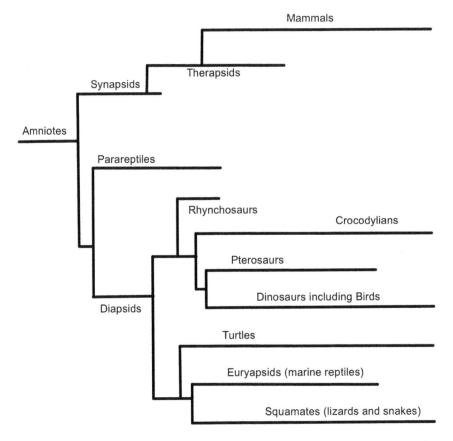

Figure 3.5 Amniote, synapsid, and diapsid relationships. Lines reaching fully to the right are living taxa.

Source: Original figure.

late-emerging group, first appearing about 225 million years ago (Cowen, 2005). In the popular mind modern crocodiles and alligators are often imagined to be the remnants of the dinosaurs but are actually part of a group that split from the dinosaurs' ancestors a few million years earlier. It is nevertheless correct to refer to them as the dinosaurs' nearest living relatives. That claim pales, though, in comparison to the recent discovery of actual living, breathing dinosaurs. They are called birds.

Discovery of the feathered *Archaeopteryx* in 1860 gave firm evidence of a transitional form between earlier dinosaurs and birds. However, it was Thomas Huxley, Darwin's early supporter, who made the first systematic argument that birds and dinosaurs are related (Ibrahim & Kutschera, 2013). Examining the pelvic bones of a dinosaur in 1868, he quickly concluded that they were birdlike in character. Other similarities were found in the vertebrae, leg bones, and toes. Huxley (1870) stated:

> there can be no doubt that the hind quarters of the Dinosauria wonderfully approached those of birds in their general structure, and, therefore, that these extinct reptiles were more closely allied to birds than any which now live.

It wasn't until the late-20th-century discovery of numerous feathered dinosaurs in China that a consensus emerged. Birds are not just related to dinosaurs, but descendants of them, and indeed, cladistically one and the same as them (Cracraft et al., 2004). This affinity received striking confirmation when peptide sequences were isolated from well-preserved fossils of a 68-million-year-old *Tyrannosaurus rex*. They were found to be similar to those of the common chicken, and unlike those of mammals. Opposite results were found for sequences derived from a mastodon (160,000–600,000 years old), which more closely resembled other mammals and not chickens (Asara, Schweitzer, Freimark, Phillips, & Cantley, 2007). Subsequent analysis comparing the peptides to 21 living species largely reproduced the expected "tree of life" while confirming the relationships between mastodons and mammals, and between dinosaurs and birds (Organ et al., 2008). Recently, these results were replicated using a leg bone from a second dinosaur species, an 80-million-year-old duck-billed hadrosaur called *Brachylophosaurus canadensis*. Molecularly it was similar to the *Tyrannosaurus*. Both were found closely related to the ostrich, chicken, and other birds, less so to alligators, and more remotely to other taxa including mammals (Schroeter et al., 2017; Schweitzer et al., 2009).

The large meat- and plant-eating dinosaurs beloved of children and familiar to museum-goers everywhere evolved and radiated over a period of 160 million years. They descended from small reptiles whose weight was distributed vertically over the rear legs, producing a *bipedal* (two-legged) stance (Lee et al., 2004). While many dinosaurs re-evolved a *quadrupedal* (four-legged) stance, bipedalism was favored by many of the later *theropods* (large carnivorous dinosaurs with short forelimbs and large jaws). They included such well-known forms as *Allosaurus* and *Tyrannosaurus*. Many millions of years later, human ancestors would habitually mimic the dinosaur's bipedal form of locomotion in a striking example of *convergent evolution* (the evolution of similar form or behavior in relatively unrelated species, due to similar environmental pressures).

The dinosaur age ended in cataclysm 66 million years ago when a six-mile diameter asteroid impacted Earth, leaving a vastly larger crater in the Yucatán peninsula of Mexico and neighboring seabed. Unmistakable evidence for the event includes melted rock inside the crater, of an age corresponding with the dinosaur extinction, as well as a similarly aged thin global layer of debris containing high levels of the element iridium. Iridium is rare on Earth but common in meteorites. Associated with the layer are small glass spheres as well as "shocked" quartz, the results of an extreme pressure pulse. It is clear the asteroid blasted out an enormous quantity of partially melted rock, the lightest material dispersing in the atmosphere to fall worldwide as the iridium-rich layer (Cowen, 2005; Schulte et al., 2010).

The immediate consequences were a formidable tsunami, evidenced as disordered sediments up the east coast of the United States, and a years-long blockage of sunlight from dust lingering in the upper atmosphere. In addition, the blasted rock was high in sulfur content, creating acid rain (Cowen, 2005; Schulte et al., 2010). A simulation suggests that atmospheric sulfates caused global mean temperature to drop a staggering 27°C (81°F) within three years of the impact. As this implies, the *mean* air temperature on Earth's surface was below freezing for several years (Brugger, Feulner, & Petri, 2017).

Even this might not have caused extinction, but in the prototypical model of bad luck, at the same time gigantic volcanic eruptions were occurring in India. Within a 750,000-year period straddling the asteroid event, lava and ash covered an area at least the size of Texas, and perhaps as large as Texas and Alaska combined, to a depth of over 1.5 miles. The eruptions contributed significantly to atmospheric dust and to increases in poisonous trace elements (Kemp, 2005; Schoene et al., 2015).

Together the two events led to the lowering of global temperatures and a massive die-off of plants. Ocean plankton greatly declined, and about 80% of land plant species disappeared, at least in North America (Kemp, 2005). The extinction of all large land animals followed.

Much of the remaining debate is over how quickly extinction occurred – within a few years according to the majority view emphasizing the asteroid impact, but as many as 250,000 years according to the minority view stressing volcanic eruptions and the poisoning of embryos within their eggshells. Dinosaurs may also have been in decline for 10 million years or more preceding these events (Cowen, 2005; Sakamoto, Benton, & Venditti, 2016).

Many aquatic animals, crocodilians, birds, insects, and other small land-dwelling creatures survived. With respect to large land dwellers, the age of the dinosaurs was over. The future belonged to mammals.

Early mammals

Mammal-like synapsids evolved at the same time as the dinosaurs, about 225 million years ago (Oftedal, 2002). Unfortunately, identifying the first true mammal is difficult despite numerous fossils covering the transition. That is because mammalian characteristics evolved gradually at varying times, producing a smooth transition with no clearly marked border between mammal and non-mammal (Sidor & Hopson, 1998).

For example, *endothermy* ("warm-bloodedness", the internal generation of heat to maintain body temperature) probably evolved in mammalian ancestors as early as 250 million years ago. Certain fossils of the time reveal ribs shaped to accommodate a diaphragm, nasal bones associated with warming air before it entered the lungs, and the skeletal build of an active lifestyle, all signs of endothermy. Less clear is the origin of mammary glands and the nursing of offspring. However, judging from similarities in lactation across monotreme, marsupial, and placental mammals, the secretion of milk probably derived from their common ancestor (Oftedal, 2002). If so, it may date from about 220 million years ago (Foley, Springer, & Teeling, 2016).

Other mammalian characteristics, such as the pairing of a large brain relative to body size – with its implied consequences for cognition – as well as middle ear bones separated from the jaw, did not appear until 195 million years ago in the tiny two-gram creature *Hadrocodium* (Kemp, 2005). For its part, fur is first known with certainty from a "halo" around the fossil skeleton of *Megaconus*, a mammal-like animal from 165 million years ago (Zhou, Wu, Martin, & Luo, 2013).

The multiple-character nature of mammals, accompanied by uncertainty over the onset of specific characters, makes identification of the first mammal difficult if not impossible. In any case, our ancestral line depended specifically on the emergence of placental mammals, the *eutheria* ("true beasts"). These have fetuses that develop with the aid of a fully formed placenta and are born live, unlike marsupials that develop without a fully formed one and in most species without a placenta of any kind. Eutherian ancestors branched off from marsupial mammals perhaps 170 million years ago (Foley et al., 2016).

The earliest known eutherian was *Juramaia*, a small climbing mammal weighing a little over half an ounce, from 160 million years ago (Luo, Yuan, Meng, & Ji, 2011). Closely related was *Eomaia* from 125 million years ago. Its fossil is extraordinary in preserving a pronounced fur halo (Figure 3.6) that makes its mammalian nature instantly clear.

Eomaia was probably partly *terrestrial* (living on the ground) and partly *arboreal* (living in trees) and was adapted for grasping and climbing. Of course, as an internal soft tissue the placenta did not fossilize, but the animal can be classified based on similarities to living eutheria in *dentition* (the form and patterning of teeth), wrist bones, and ankle bones (Ji et al. 2002).

Like *Juramaia*, *Eomaia* was very small, weighing just under an ounce (Ji et al., 2002). It is no accident that it and other early mammals were tiny. *Archosaurs* (the dinosaurs and their kin) ruled the large-animal habitats, while smaller lizard-like reptiles dominated the small-animal niches. But small warm-blooded mammals could and did thrive in the cool of night. Indeed, environmental pressure from a nocturnal existence encouraged the evolution of sharp senses and intelligence (Cowen, 2005). When large land animals disappeared after the extinction event 66 million years ago, new niches opened for the survivors.

Figure 3.6 Eomaia, a 125-million-year-old eutherian from China.

Conclusion

Vertebrates originated about 500 million years ago as sea-dwelling animals with brains and mineralized body parts. Teeth allowed more effective predation, and bones increased body integrity and protected the nervous system. The emergence of Hox genes, aided by a full-genome duplication, allowed body segmentation and a new avenue for evolutionary change, in that one segment could undergo modification without affecting others. Another factor increasing vertebrate complexity was the neural crest, from which cells migrate during development to form nerves, bones, smooth muscles, and other localized structures. Other vertebrate adaptations included external eye muscles, a macula, and eye pairing, all substantial improvements of vision.

Tetrapods in the form of four-legged amphibians emerged from fresh water over a period exceeding 20 million years. They lived a primarily aquatic existence 385 million years ago but breathed at least partially out of water 365 million years ago. "Hands" evolved from fleshy lobe fins, and jaws evolved to open upward, allowing feeding at water's edge without ingesting mud. *Tiktaalik*, a transitional form, captured prey with a sideways snapping motion much like modern crocodiles.

Amniotes appeared by about 325 million years ago. Their major innovation was to lay eggs incorporating a membrane, which allowed gas exchange while keeping the embryo hydrated. This permitted eggs to hatch on land so that from this point, many vertebrates could be terrestrial in lifestyle. Amniote skulls were anapsid, having no holes behind the eye.

Synapsids and diapsids, creatures with one versus two holes behind the eye, emerged shortly afterward and were the ancestors of mammals and dinosaurs, respectively. Although dinosaurs were once thought to be extinct, we now know that birds are their living representatives. The large dinosaurs disappeared in a cataclysmic extinction 66 million years ago when a six-mile diameter asteroid collided with Earth. Its dust combined with that from coinciding volcanic eruptions to create a years-long blockage of sunlight and a calamitous decline in global temperatures. The ensuing die-off of plants doomed large land animals.

Many smaller animals survived, however, including mammals. Identifying the first mammals is difficult because of their multiple-character nature, but warm-bloodedness and lactation probably emerged by 250 million and 220 million years ago, respectively. The pairing of a relatively large brain with middle ear bones that had separated from the jaw likely evolved by 195 million years ago. Fur is known from 165 million years ago.

Eutheria, or "true beasts", evolved by 160 million years ago. Eutheria have fully formed placentas and undergo live birth. Like other early mammals, the initial ones were tiny, an ounce or less in weight, but from such slight beginnings would evolve the large mammals of post-dinosaur Earth.

4 Later mammals through primates

The extinction of large land animals 66 million years ago left a vacuum to be filled by evolution. Smaller animals survived, including lizards, salamanders, birds, insects, and mammals, among others. The increasing size of mammals would prove to be a major theme of the post-extinction period. Yet the event did not cause an explosive diversification. Relaxed clock models indicate that all of today's eutherian orders had emerged long before the large dinosaurs' demise. Although some diversification occurred in the immediate period afterward, much of it was within orders now extinct (Springer et al., 2017).

Thus before the extinction, mammals were already a varied class. Their teeth confirm a diversity of diets across species, with foods varying between insects, meat, fruit, and seeds. But they were very small, averaging only 150 g, or a little over five ounces, before the extinction. That increased within a million years to one kg (2.2 pounds), about seven times heavier than before (Novacek, 1999).

In this chapter we examine the slow emergence of primates from this mammalian background, and their subsequent diversification into today's taxa. Those include strepsirrhines; tarsiers; and the large anthropoid group including New World monkeys, Old World monkeys, and both the lesser apes and great apes. It should be kept in mind as these species are described that all have undergone evolution since their ancestors diverged from the primate ancestral tree. As a result, none is a true representation of its earlier ancestors, any more than a kangaroo resembles early mammals. However, just as it is possible to deduce characteristics possessed by early mammals from common characteristics among living mammals (e.g., fur and endothermy), it is often possible to deduce from common characteristics among a group of related primates, characteristics possessed by their common ancestor.

Early primates

Only in cartoon worlds did cavemen and dinosaurs coexist. The reality is that they were separated by tens of millions of years. *Primates*, however, did overlap the end of the large dinosaurs. According to relaxed clock models, they originated about 83 million years ago, give or take nine million years, with nearly all estimates dating well before the dinosaur extinction (e.g., Herrera & Dávalos, 2016; Perelman et al., 2011; Pozzi et al., 2014; Springer et al., 2017).

Despite originating in the geological epoch known as the Late Cretaceous, primates have fossils dating back only to the Paleocene (Table 4.1), implying a slow emergence. The earliest known was *Purgatorius*, from 65 million years ago. The recent discovery of its tarsal (foot) bones shows that it had mobile, grasping feet, as expected of a primitive primate but unlike primates' closest relatives, the flying lemurs and tree shrews. To an untrained eye, *Purgatorius* would have been unremarkable, perhaps squirrel-like. Recognizing that, Chester and colleagues wryly noted that "the divergence of primates

Table 4.1 Geological epochs immediately before and following the dinosaur extinction.

Epoch	Dates
Late Cretaceous	101–66 million years ago (mya)
Paleocene	66–56 mya
Eocene	56–34 mya
Oligocene	34–23 mya
Miocene	23–5.3 mya
Pliocene	5.3–2.6 mya
Pleistocene	2.6–0.01 mya
Holocene	0.01 mya–present

from other mammals was not a dramatic event" (Chester, Bloch, Boyer, & Clemens, 2015, p. 1491).

Early primates are generally accepted to have been *nocturnal* (active at night), a lifestyle that helped them survive alongside dinosaurs even though there were probably nocturnal dinosaurs (Santini, Rojas, & Donati, 2015; Schmitz & Motani, 2011). Characteristics of today's primates include grasping hands and feet with *opposable thumbs* (ones that can rotate along their long axis, allowing contact with the other fingers of the same hand), the modification of claws to relatively flat nails; eye convergence toward the front of the face with resulting stereoscopic vision; and large brains for a given body size (Rasmussen, 2002b). As we will see, opposable thumbs, stereoscopic vision, and an enlarged brain improved object manipulation and then, many millions of years later, object creation itself.

Opposability and nails are beautifully represented in a nearly complete skeleton of *Carpolestes simpsoni* (Sussman, Rasmussen, & Raven, 2013). This species originated about 58 million years ago. Following the aforementioned list of characters, it too was a primate (Chester et al., 2015). Figure 4.1 shows an artist's reconstruction.

Figure 4.1 Carpolestes simpsoni.

Source: CarpolestesCL.png, CC BY-SA 3.0, Sisyphos23.

This animal's big toe was curved, indicating opposability, and it had a flattened end segment accommodating a nail. These were adaptations for grasping branches, probably because *Carpolestes* fed at their slender ends as an *omnivore* (both animal and plant eater) though one emphasizing fruit (Bloch & Wilcox, 2006). Claw loss probably resulted from their declining effectiveness as clinging devices as body sizes increased. Flattened digits, substituting nails for claws, were increasingly favored because of their increased gripping surface (Soligo & Martin, 2006).

The strepsirrhine-haplorhine divergence

An early branching of the primate tree of life led to important consequences today, namely a split between the suborders of *Strepsirrhini* (or strepsirrhines, i.e., wet-nosed primates) and *Haplorhini* (or haplorhines, i.e., dry-nosed primates). Following the split, the haplorhines can be viewed as a "remainder clade" that includes human ancestors. As Figure 4.2 indicates, lemur-like animals were among the Strepsirrhini, and indeed, these were the ancestors of living strepsirrhines, among them lemurs (Figure 4.3), sikafas, pottos, and galagos (Figure 4.4).

Strepsirrhines are found in continental Africa, Madagascar and nearby islands, and southeast Asia. With few exceptions, all species have a *dental comb* (outward-projecting lower canines and incisors, used as a comb while grooming), a divided lower jaw with a joint at the chin (as in most mammals), elongated noses, and a divided upper lip closely attached to the gums. Many are nocturnal and have a breeding season as opposed to individual cycles. Across species, their food sources include leaves and other vegetation, insects, and fruit (Cartmill, 1992).

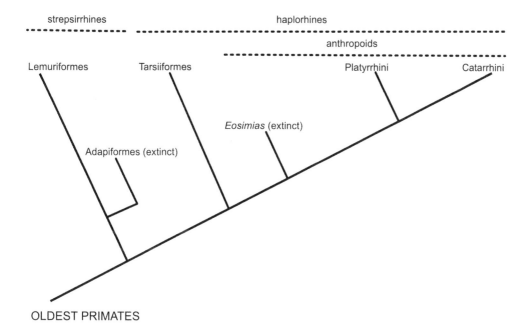

Figure 4.2 Cladogram illustrating early branching of the primates. Earlier splits are to the lower left; later ones to the upper right.

Source: Original figure.

Figure 4.3 Ring-tailed lemur.

Source: Lemur catta 001.jpg, CC BY-SA 3.0, Alex Dunkel (Maky).

Figure 4.4 Northern greater galago.

Source: Garnett's Galago (Greater Bushbaby).jpg, CC BY 2.0, Mark Dumont.

The split of the Strepsirrhini from other primates is thought to have occurred about 73 million years ago (Table 4.2), although their earliest fossils date from several million years later (Shoshani, Groves, Simons, & Gunnell, 1996). The Eocene yields evidence of about 40 genera of strepsirrhines. By that time there were both nocturnal and *diurnal* (active in daytime) species. They moved by leaping and jumping (Simons, 1992).

The tarsier-anthropoid divergence

The splitting off of the line leading to tarsiers (Figures 4.2 and 4.5) occurred about 72 million years ago (Table 4.2), leaving the remainder clade *Anthropoidea* (or anthropoids, i.e., monkeys, apes, and man). Together with the Strepsirrhini, tarsiers have traditionally been called *prosimians* ("before monkeys and apes"), a grouping losing favor as it has become clear that tarsiers are both dry-nosed and genetically less similar to the Strepsirrhini than they are to anthropoids (Pozzi et al., 2014; Shoshani et al., 1996). Nocturnal, clinging *insectivores* (insect eaters), tarsiers live on islands in southeast Asia.

Table 4.2 Approximate dates of primate divergences from the human ancestral line (which follows the remainder clades). (Derived in part from Chatterjee, Ho, Barnes, & Groves, 2009; Herrera & Dávalos, 2016; Perelman et al., 2011; Pozzi et al., 2014; Schrago & Voloch, 2013).

Divergence	Living examples	Remainder clade	Date
strepsirrhines	lemurs, galagos	haplorhines	73 mya
tarsiformes	tarsiers	anthropoids	72 mya
platyrrhines	capuchins, howlers	catarrhines	45 mya
Old World monkeys	macaques, baboons	apes	32 mya
lesser apes	gibbons, siamangs	great apes	21 mya
orangutans	Bornean, Sumatran	African apes	17 mya
gorillas	eastern, western	chimps/humans	10 mya
chimpanzees	common chimp, bonobo	humans	7.5 mya

Figure 4.5 Tarsier.

Source: Tarsier Sanctuary, Corella, Bohol (2052878890).jpg, CC BY 2.0, yeowatzup.

The earliest known anthropoid is *Anthrasimias*, a 75 g (3 ounce) animal known only from teeth discovered in India. Dating to about 55 million years ago, it is inferred to have eaten fruit and insects (Bajpai et al., 2008). The best known of the early *Anthropoidea*, however, are two species of the genus *Eosimias*. These appeared by the mid-Eocene, about 45 million years ago. Most of their fossils are in the form of jaws, teeth, and foot bones. From these it is known they were better adapted than strepsirrhines for horizontal body postures. *Eosimias* was still well under half a pound in weight (Gebo, Dagosto, Beard, Qi, & Wang, 2000).

As seen in living examples, anthropoid adaptations include more convergent eyes, presumably for improved depth perception; a shift toward a diurnal lifestyle (Ross, 2000); and a fully fused chin joint.

The platyrrhine-catarrhine divergence

A particularly important split, involving descendants of the early anthropoids, was of the *Platyrrhini* ("broad-nosed" monkeys) from the *Catarrhini* ("narrow-nosed" monkeys). The platyrrhines are known as *New World monkeys* (i.e., living in the Americas). Conversely, the catarrhines are the *Old World monkeys* (i.e., living in Africa, Asia, and Europe) as well as the apes including humans. The platyrrhine-catarrhine divergence was about 45 million years ago (Table 4.2).

New world monkeys

Although our focus is mostly on the Old World lineage as ancestral to humans, it is worth spending a bit of time on New World monkeys. How they wound up in South America is one of the great mysteries of primate evolution. The order of primates evolved in the Old World, and no fossil primates are known from South America before the divergence or even for several million years afterward. Accordingly, it is assumed they arrived there from elsewhere (Dagosto, 2002).

But from where? Any answer must be consistent with two observations stemming from the movement of Earth's tectonic plates. First, 45 million years ago, there were multiple interruptions in the Central American land bridge between North and South America. Second, while closer than it is at present, Africa was nevertheless 1300 km (800 miles) away from South America (de Oliveira, Molina, & Marroig, 2009; Houle, 1999).

These considerations pose problems for all three of the hypotheses that might account for the existence of South American monkeys. These are that they arrived by migration from (a) North America, (b) Asia, or (c) Africa. The least geographically extreme possibility is that they came from North America, but the multiple gaps in the land bridge were a significant barrier. Worse yet, platyrrhines were anthropoids, and no anthropoids are known from North America at that time.

The second possibility, that they were from Asia, has exactly the same problems: They would have had to be present in North America, where no anthropoid fossils of the time are known, and then traverse the broken land bridge. But in addition, they would first have had to cross the Bering Strait between Asia and America (Dagosto, 2002).

By a process of elimination, that leaves an African origin. But is it really possible that they came across the Atlantic from Africa? At present this is actually the favored hypothesis (Dagosto, 2002; de Oliveira et al., 2009). It is supported in part by anatomical similarities between platyrrhines and African anthropoids (Schrago & Russo, 2003).

However, before we imagine a troop of African monkeys planning a westward adventure and busily building rafts from woven vines and branches, it's important to understand just what is proposed. The migration would have been entirely accidental, with a very small number of monkeys surviving the journey on a mat of vegetation – a "floating island" – swept offshore by a storm and carried across by wind and current (Schrago & Russo, 2003). One such island, observed on a Canadian river in 1881, measured 60 m by 23 m (200 ft by 75 ft) and was only a piece of what had been a larger floating island. The piece reportedly had upright trees reaching 15 m (50 ft) in height (Houle, 1999).

While our first reaction to the possibility of a land animal migration across large expanses of ocean may be to consider it "improbable, unobservable and consequently untenable" (Censky, Hodge, & Dudley, 1998, p. 556), one has actually been observed. Following a highly active hurricane season in 1995, iguanas appeared on Antigua, a Caribbean island, for the first time. Local fishermen named the exact date of arrival, having seen lizards on a log mat in the bay (Censky et al., 1998). The possibility of even longer migrations is posed by the discovery of North American Douglas firs on Hawaiian beaches, and a South American log on the beaches of Tasmania (Coyne, 2009, p. 105).

A migration of 800 miles from Africa to South America need have happened only once across many millions of years. All that was required was the survival of one or more mating pairs of a single species, from which all of the New World monkeys descended. Calculations accounting for prevailing winds and currents indicate that 45 million years ago, a floating island could have crossed the narrowest gap between Africa and South America in about nine days (Houle, 1999).

Could a small group of monkeys have survived a journey of that length? Probably so, for modern mammals of the requisite size can survive about 13 days without food and water, and vegetation on the floating island would likely prolong the survival period (Houle, 1999). A transatlantic journey from Africa to South America is therefore the most likely scenario for the origin of platyrrhines, the New World monkeys.

As their name indicates, living platyrrhines have flat noses with widely separated nostrils. They are generally small and arboreal, and most are diurnal. They typically have long tails that are often *prehensile* (able to grasp objects). A wide range of locomotion occurs across species, including clinging and leaping, quadrupedal branch-walking, and suspension and *brachiation* (swinging from branch to branch). Food sources also vary widely across species and include leaves, fruit, tree gum, and insects (Rosenberger & Hartwig, 2002).

The classification of the New World species has been controversial in recent years and has given rise to competing sets of family names. However, molecular-clock approaches make it clear that the group containing marmosets and tamarins, and the group containing capuchins and squirrel monkeys, are more closely related to each other than either are to a third group containing spider monkeys, muriquis, and howler monkeys (Chatterjee, Ho, Barnes, & Groves, 2009; Pozzi et al., 2014). Several species are illustrated in Figures 4.6–4.11.

Early catarrhines

Ancestral Catarrhini, meanwhile, were continuing their *radiation* (expansion in the number of species) in the Old World regions of Africa, Asia, and Europe. We are fortunate

Figure 4.6 First related group of New World monkeys: Common marmoset.

Figure 4.7 First related group of New World monkeys: Emperor tamarin.

Figure 4.8 Second related group of New World monkeys: Capuchin monkey.
Source: Capuchin Costa Rica.jpg, CC BY-SA 3.0, David M. Jensen (Storkk).

Figure 4.9 Second related group of New World monkeys: Common squirrel monkey.
Source: Saimiri sciureus-1 Luc Viatour.jpg, CC BY-SA 3.0, Luc Viatour / www.Lucnix.be.

Figure 4.10 Third group of New World monkeys: Ornate spider monkey.

Figure 4.11 Third group of New World monkeys: Black howler monkey.

to have substantial fossils of an early catarrhine, *Aegyptopithecus*, dating from 29 to 33 million years ago (Simons, Seiffert, Ryan, & Attia, 2007; Rasmussen, 2002a). A nearly complete male skull was found in Egypt in 1966 (Figure 4.12). It was from an individual estimated to weigh 3–6 kg (7–14 pounds), possessing the heavy limbs of a slowly moving arboreal quadruped. Its upper teeth were apelike as were its feet, and it was probably *frugivorous* (eating fruit; Cowen, 2005; Rasmussen, 2002a).

Early catarrhines existed for millions of years before their descendants split into two groups, one ancestral to the living Old World monkeys, and one ancestral to the apes including humans. *Aegyptopithecus* is a potential candidate for a common ancestor of both groups. Of course, our primary story line continues with the apes. However, the Old World monkeys are important because they will figure prominently in later cognitive comparisons.

Old World monkeys

Like early catarrhines, today's Old World monkeys are native to Africa, Asia, and Europe. A number of species were subjected to genetic analysis by Xing et al. (2005), with results generally supporting previous cladistic analyses: Old World monkeys exclusive of apes comprise the family *Cercopithecidae*, and there are two subfamilies, the *Cercopithecinae* and the *Colobinae*.

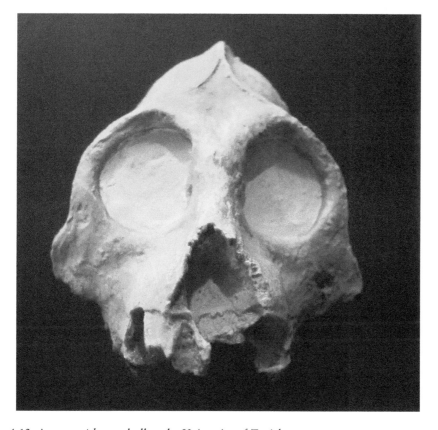

Figure 4.12 Aegyptopithecus skull at the University of Zurich.

Source: Aegyptopithecus face (University of Zurich)-1.JPG, CC BY-SA 3.0, Guérin Nicolas.

The Cercopithecinae consist of about 70 species, most living in Africa south of the Sahara, although some live in Asia. One (the Barbary macaque), lives on the island of Gibraltar, technically in Europe. Generally speaking, they are diurnal and omnivorous, but many are primarily frugivorous, and they have cheek pouches for storing food. They live in highly varied climates and terrains, so that some are arboreal while others are terrestrial. They have short to medium-length tails and typically have *ischial callosities*, callused skin patches on their buttocks that serve as padding while sitting. Figures 4.13 and 4.14 illustrate two well-known species.

The Colobinae comprise nearly 60 species living in Africa and Asia. They are primarily *folivores* (leaf eaters) and have a specially adapted stomach with multiple chambers to ferment food. Unlike cercopithecines, who have well-developed thumbs, colobines

Figure 4.13 Rhesus macaque.

Source: Macaca mulatta in Guiyang.jpg, CC BY 2.0, Einar Fredriksen.

Figure 4.14 Guinea baboon.

Source: Male Guinea Baboon in Nuremberg Zoo.jpg, CC BY 2.0, Jakub Friedl.

have small or vestigial thumbs. Nearly all are arboreal, and while they inhabit diverse climates, they do not live in dry or desert areas as do some cercopithecines. They usually have very long tails. Figure 4.15 shows one such species.

The ape divergences

Finally, we come to the apes, comprising the superfamily of *Hominoidea* (hominoids). Apes differ from monkeys in not having tails or well-developed ischial callosities, although lesser apes have reduced ones. They also have stiffer backs, with broader and shallower chests and pelvises (Walker & Shipman, 2005).

Early apes

Apes diverged from monkeys about 32 million years ago, but their first substantial fossils date to about 20 million years ago. The skeleton of *Proconsul* (Figure 4.16), from Kenya, reveals a much larger body than earlier primates, ranging from 9 to 90 kg (20–200 pounds) depending on the species and gender (Harrison, 2002). It had an unspecialized skeleton typical of arboreal, frugivorous, quadrupedal primates, but its skull was apelike (Cowen, 2005), and of course it was tailless (McNulty, Begun, Kelley, Mnthi, & Mbua, 2015; Simons, 1992).

Figure 4.15 Grey langur.
Source: MNP Grey Languer.JPG, CC BY-SA 3.0, Marcus334.

Although *Proconsul* is a strong candidate for an ancestor of the living apes including man, its unspecialized skeleton has led some to propose an alternative. *Morotopithecus*, from Uganda, was a large-bodied (40 kg, 90 pounds) ape also dating to a little over 20 million years ago. Its shoulder was highly mobile, consistent with brachiation or arm-hanging but contrasting with the quadrupedalism of *Proconsul* (Gebo et al., 1997; MacLatchy, 2004). Certainly, brachiation or arm-hanging is a plausible character for an ancestral ape. Phylogenetic analysis, largely based on teeth, does suggest that *Morotopithecus* was a closer relative than *Proconsul* of living apes including humans (Stevens et al., 2013). However, nothing like a complete skull is available to make the case more convincing.

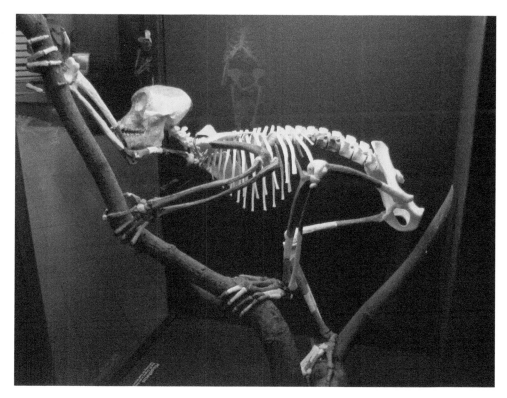

Figure 4.16 Proconsul reconstruction at the University of Zurich.

Source: Proconsul skeleton reconstitution (University of Zurich).JPG, CC BY-SA 3.0, Guérin Nicolas.

It is worth asking why apes arose at all, when Old World monkeys already existed in similar arboreal environments. Wouldn't the smaller bodies of monkeys allow them to outcompete any emerging apes? Kevin Hunt (2016) points to a key ape adaptation, namely the ability to assume eccentric grasping postures. That allows them to suspend from multiple slim supports to forage at the extreme ends of fruit tree branches (Figure 4.17). When this adaptation began to appear, ape and monkey body sizes were likely similar, but apes thrived because of it. According to Hunt, the ensuing 20-million-year competition resulted in a splitting of environmental resources. Monkeys began consuming most of the fruit near the core of trees and developed the ability to consume unripe fruit. Apes, on the other hand, consumed ripe fruit at the tree periphery. A difference in body size evolved, with monkeys remaining small so they could feed while balanced on branches in sitting postures (e.g., Figure 4.15). Apes became large not only to physically displace monkeys, but to reach slender branch supports at greater distances.

Lesser apes

Among living species, the small-bodied gibbons and siamangs (5–11 kg, 12–25 pounds) are considered "lesser" apes as opposed to the great apes (orangutans, gorillas, chimpanzees, and humans). Their divergence from the human ancestral line was about 21 million years ago (see Table 4.2). The four genuses are composed of agile brachiators. Gibbons

Figure 4.17 Orangutan in multi-branch suspension.

Source: Sumatran orangutan (Pongo abelii), CC BY-SA 2.0, Lip Kee.

(Figure 4.18) live in portions of an area roughly defined by northeast India, to southern China, to the island of Java. Siamangs live in part of the same range, specifically in Malaysia and Sumatra. These rainforest apes lack cheek pouches for food storage, but males among the siamangs and some species of gibbons have throat sacs used in making vocal calls. They are omnivores, typically emphasizing fruit.

Great apes

Our understanding of relationships between the great apes, including humans, has undergone a revolution since the advent of molecular testing. Previously, a single branch was assumed to have diverged from the human ancestral line, representing a common ancestor of orangutans, gorillas, and chimpanzees. According to that view, none would

Figure 4.18 White-handed gibbons.

Source: Hylobates lar pair of white and black 01.jpg, CC BY-SA 3.0, MatthiasKabel.

be more closely related to us than another. However in the late 1960s, testing of blood proteins showed chimpanzees to be more closely related to us than are orangutans and gorillas (Pilbeam, 1972). DNA testing further clarified the relationships, demonstrating a genus-by-genus divergence from the human line.

The first great ape divergence was of the genus *Pongo*, the orangutans (or orangutangs), about 17 million years ago (Table 4.2 and Figure 4.17). There are two similar species, the Bornean and Sumatran orangutans, their names betraying where they live. They are the only Asian great apes other than humans. Forest dwellers, they are almost completely arboreal, engaging in hanging and branch-swinging behaviors but often supporting part of their weight (up to 90 kg, or 200 pounds) on their feet (Figure 4.17). On the ground they walk quadrupedally, putting weight on their fists instead of knuckles as in other great apes. They are generally solitary but also strongly territorial, and are primarily frugivorous although they do eat other plant material, insects, and fungi.

The genus *Gorilla* was the next to branch off the human ancestral line, about 10 million years ago (Table 4.2). There are two living species, the eastern gorilla and the western gorilla (Figure 4.19).

Gorillas live exclusively in the "mid-belt" region of Africa, east and west, usually in forested areas. They are largely terrestrial, walking by putting weight on their knuckles, although female and young animals do climb trees to sleep. They are chiefly folivorous but also eat fruit, insects, and seeds. Males can naturally weigh up to 200 kg (440 pounds), making them the largest living primates.

The last primate genus to diverge from the human ancestral line was *Pan*, the chimpanzees, about 7.5 million years ago (Table 4.2). There are two living chimpanzee

Figure 4.19 Western gorilla.

Source: Western Lowland Gorilla at Bronx Zoo 2 cropped.jpg, CC BY-SA 3.0, Fred Hsu.

species (Figures 4.20–4.21), the common chimp *Pan troglodytes*, and the pygmy chimp *Pan paniscus*, also known as the bonobo (pronounced buh-NO-boh). Common chimps weigh 40–65 kg (90–145 pounds) in the wild, while bonobos are somewhat smaller, weighing 30–45 kg (65–100 pounds).

A question that sometimes arises is whether common chimpanzees or bonobos better model our last common ancestor. Since it was their joint ancestor that broke away from our line, they are in fact equally related to us, a supposition confirmed by genetic comparisons (Prüfer et al., 2012). Thus they, and us, are co-equal representatives of our common ancestor.

The genetic comparisons also indicate that the two species diverged from each other about two million years ago (Raaum, Sterner, Noviello, Stewart, & Disotell, 2005; Stone et al., 2010). That split may coincide with a lowering of the Congo River due to dry conditions, allowing bonobo ancestors to cross to the south of the river while common

Figure 4.20 Common chimpanzee.

Source: Schimpanse zoo-leipig.jpg, CC BY-SA 3.0, Thomas Lersch.

Figure 4.21 Bonobos or pygmy chimpanzees.

Source: Bonobo-04.jpg, CC BY-SA 3.0, Photo by Greg Hume.

chimpanzee ancestors remained to its north. If so, geographic isolation occurred when full river flow resumed, permitting speciation (Takemoto, Kawamoto, & Furuichi, 2015). Nevertheless limited genetic exchange, less than 1% of the genome, likely occurred between the two species down to around 100,000 years ago (de Manuel et al., 2016).

Chimpanzees live in central and west Africa, are partly arboreal and partly terrestrial, sleep in trees, and knuckle walk when on the ground. They are omnivorous, and thus while eating significant quantities of fruit, they also consume insects, fungi, and even small mammals. Both species live in swampy rainforest, but the common chimp is more widely distributed and also inhabits woodland areas and *savanna* (grassland with sparse stands of trees).

The final divergence of apes from the human ancestral line, in the form of chimpanzees, holds great significance for us. From that point forward, the *hominins* (bipedal apes) were on their own.

Conclusion

Although the dinosaur die-off 66 million years ago produced some mammalian diversification, it was mostly within orders now extinct. The largest effect on the ancestors of living mammals was to allow an increase in size, about sevenfold within a million years.

Most closely related to flying lemurs and tree shrews, primates originated some 17 million years prior to the extinction. Early ones were nocturnal, probably aiding their survival alongside dinosaurs. *Purgatorius*, dating to 65 million years ago, had a mobile, grasping foot. Other primate features emerging early on included opposable thumbs, stereoscopic vision, and relatively large brains. *Carpolestes*, from 58 million years ago, had a nail in place of a claw on its big toe. It probably fed at the slender ends of branches, on a mixed diet emphasizing fruit.

Initial divergences from the human ancestral line included the Strepsirrhini (e.g., lemurs and galagos) about 73 million years ago, and tarsiers about 72 million years ago. Today strepsirrhines live in continental Africa, Madagascar and vicinity, and southeast Asia. They have a dental comb and a varied diet, and many are nocturnal. Tarsiers live on islands in southeast Asia, are nocturnal, and eat insects and fruit.

The remainder clade, the anthropoids, had adaptations that included more convergent eyes and a largely diurnal lifestyle. They gave rise to the Old World and New World monkeys.

The platyrrhini or New World monkeys (e.g., marmosets, capuchins, spider monkeys) diverged from the human ancestral line about 45 million years ago. As their name implies, they have flat noses. They usually have long tails that are often prehensile, and across species show a wide range of locomotion and feeding styles. How they got to the Americas is an enduring problem, with current thinking favoring a crossing from Africa on natural rafts of vegetation.

The remaining catarrhines or narrow-nosed monkeys resided in the Old World (Africa, Asia, and Europe). *Aegyptopithecus*, from 29 to 33 million years ago, weighed 7–14 pounds, had heavy limbs, and probably moved slowly in trees. Around that time Old World monkeys diverged from the human line, and they now comprise the Cercopithecinae (e.g., macaques and baboons) and the Colobinae (e.g., langurs and proboscis monkeys). Both are very diverse subgroups, defying simple characterization. However, Cercopithecinae tend to have short to medium-length tails, relatively large thumbs, and ischial callosities, while Colobinae usually have long tails, small or vestigial thumbs, and lack ischial callosities.

Lesser apes (gibbons and siamangs) split off about 21 million years ago. They live in Asia and are agile brachiators.

In contrast to an older view that our fellow great apes split off our ancestral line as a group, molecular data are conclusive that multiple divergences occurred. Thus orangutans diverged about 17 million years ago, gorillas about 10 million years ago, and chimpanzees 7.5 million years ago. Orangutans live in Borneo and Sumatra in southeast Asia. They are almost completely arboreal and have a varied diet emphasizing fruit. Gorillas live in the "mid-belt" of Africa, and are largely terrestrial. They also have a varied diet but emphasize leaves. The chimpanzees (the common chimp and the bonobo) live in central and west Africa. They are partly arboreal and partly terrestrial, and are omnivorous. They are also equally related to us.

5   Humans

Through much of the 20th century, the term *hominid* was reserved for humans and our immediate ancestors and relations. That made for a very uncertain clade, because the oldest indisputably hominid fossils dated from three million years ago (Pilbeam, 1972, p. 104), while the living apes were thought to be descended from a common ancestor that broke from the human line 14–17 million years ago (Pilbeam, 1972, pp. 88, 98). When during that period did we transition from a more apelike to a more human form?

Recent advances have provided both a more satisfying and a more troubling way of defining the human clade. Recognition that the ape genera broke off one-by-one from our ancestral line, with chimpanzee ancestors last doing so 7.5 million years ago, has allowed tight definition of the clade as consisting of humans, our ancestors, and our nearest relations back to the chimpanzee split. With a date that recent – and as we shall see, with fossils nearly that old – it seems likely that human characteristics emerged relatively soon thereafter.

However, using the definition "since the chimpanzee split", while reducing uncertainty, is also troubling. The complete genomes of humans and chimpanzees have been decoded, and it is clear that we are genetically very similar. Thus we share over 99% of our coding for amino acid sequences. Responding to this similarity, some proposed that we rename the two chimpanzee species as *Homo troglodytes* and *Homo paniscus*, in effect admitting them to humanity (Wildman et al., 2003).

In contrast, others downplayed any special similarity between humans and chimpanzees and instead advocated creating a new hierarchy of taxa that took into account the separate divergences of the Asian and African apes. Under this scheme the family *Hominidae* would refer to all great apes including humans (and orangutans), the subfamily *Homininae* to all African apes including humans (excluding orangutans), and the tribe *Hominini* to all bipedal apes.

Whatever its merits, that proposal has now been widely accepted and the terminology settled. Accordingly, for our purposes, the term *hominin* (derived from *Hominini*) refers to humans and their bipedal ancestors and relatives back to the chimpanzee-human split. Nevertheless it should be kept in mind that historically, the term *hominid* was commonly used, even recently, to refer to the identical classification.

In this chapter we will consider the characteristics of hominins from the Miocene through the Pliocene, Pleistocene, and today, with special attention to key species. Recent discoveries of late-living, small-brained hominins will be described. Trends over time will be identified, including a generalized, disproportionate increase in brain size relative to body size. Finally, possible ancestral connections will be explored.

The first (Miocene) hominins

Within the past quarter century, four spectacular finds have taken the human lineage back almost to the chimpanzee-human divergence, placing it firmly within the Miocene geological epoch. The earliest species, named *Sahelanthropus tchadensis*, is remarkable for the preservation of a crushed but nearly complete skull nicknamed "Toumai" (Figure 5.1). The nickname means "Hope of life" in the Goran language used in the area of Chad, north central Africa, where the fossil was found.

Whether *Sahelanthropus* should be considered an early hominin or a non-hominin ape has been controversial (Wolpoff, Hawks, Senut, Pickford, & Ahern, 2006). That is not surprising considering the 7-million-year age of the specimen. Close to the chimpanzee-human divide, species on either side would be nearly identical. Arguments for hominin status include the presence of a relatively vertical, short face, a prominent brow ridge similar to that found in most later hominins, reduced canine teeth, thickened tooth enamel, and a downward-opening, oval-shaped *foramen magnum*, the opening on the bottom of the skull allowing entry of the spinal column (Zollikofer et al., 2005). The tooth roots likewise suggest that Toumai was a hominin, for they share some characters with later hominins that are imperfectly shared with African apes, and not shared at all with orangutans or with *Proconsul* (Emonet, Andossa, Mackaye, & Brunet, 2014).

That constellation of traits has produced general, though not complete, consensus that *Sahelanthropus* was a hominin. The downward-opening foramen seems particularly significant, as a possible indication of bipedalism. In quadrupedal apes, and even in brachiating and clambering apes like orangutans, it opens angled a bit toward the rear, allowing the head to more easily tilt upward relative to the spine. Toumai's thickened tooth enamel also seems consistent with bipedalism, because it implies reduced fruit consumption and thus less time spent in trees. If *Sahelanthropus was* bipedal, the freeing of the hands from contact with the ground was a watershed event that allowed new possibilities for carrying and eventually manufacturing objects – which in turn would have profound cognitive implications.

Figure 5.1 Skull of *Sahelanthropus tchadensis*, or "Toumai".

Source: Sahelenthropus tchadensis-MGL 95214-Download P4150633-white.jpg, CC BY-SA 3.0 FR, Rama.

However, Toumai's brain was small, only 365 cc (about ¾ of a pint) and thus about the size of a bonobo's. Also, the foramen magnum was toward the rear of the skull's base, not the center as is typical of later hominins.

An alternative approach to classification is to mathematically model species divergences using as many quantified characters as possible. Strait and Grine (2004) modeled divergences using 198 characteristics of the *cranium* (bony braincase) and teeth of a number of living ape species, fossil hominins, and living humans. The resulting cladogram (Figure 5.2) was largely replicated in a later study using 380 characteristics (Dembo, Matzke, Mooers, & Collard, 2015). It shows the apes diverging from the human line in the expected order – gibbons, orangutans, gorillas, then chimpanzees – followed immediately by *Sahelanthropus tchadensis*. This result lends support to the interpretation that Toumai was a hominin.

Ardipithecus

The second major Miocene find was of *Ardipithecus*, shown by the cladogram as branching off the human ancestral line soon after *Sahelanthropus*. The oldest examples date to about 6.3 million years ago and are from Ethiopia. Initially, only one species was described, *Ardipithecus ramidus* (White, 2002). Subsequently, a case was made that the earliest examples should be designated a separate species called *Ardipithecus kadabba*, with *ramidus* reserved for later ones (Haile-Selassie, Suwa, & White, 2004). Others, however, consider these subspecies, not separate species (Harcourt-Smith & Aiello, 2004).

The Miocene fossils of *Ardipithecus* consist of a jaw bone, teeth, parts of an upper limb, and a foot bone. Later ones, about 4.3 million years old, include a partial skeleton including a pelvis, femurs, and portions of a cranium. Many believe bipedalism is indicated (Lovejoy, Suwa, Spurlock, Asfaw, & White, 2009), but others are skeptical

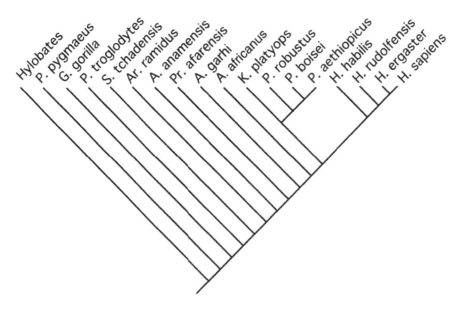

Figure 5.2 Cladogram of living apes, fossil hominins, and humans, based on cranial and dental characteristics.

Source: Licensed from Elsevier under STM Permissions Guidelines.

(Gibbons, 2009). A key point of contention is that a 4.4-million-year-old specimen had a long, splayed, grasping big toe (Figure 5.3; Lovejoy, Latimer, Suwa, Asfaw, & White, 2009). That may be more consistent with a tree-dweller than a habitual biped (Hawks, 2009). As we shall see, both earlier and later hominins left footprints showing very little splay, suggesting that *Ardipithecus* may have been a side branch of hominin evolution not on our direct ancestral line (Begun, 2010).

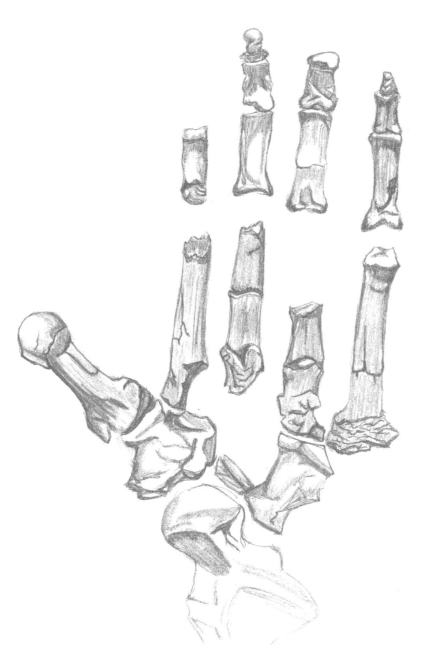

Figure 5.3 The foot of *Ardipithecus*, showing its splayed big toe.

Source: Ardipithecus Fuß.jpg, CC BY-SA 3.0, Tobias Fluegel.

Orrorin

The third spectacular find from the Miocene was of a species named *Orrorin tugenensis*. It is missing from the analysis of Strait and Grine because a cranium has not been found. Instead, its fossils consist primarily of limb bones and teeth from several individuals. Discovered near Tugen, Kenya, in a *tuff* (a layer of rock formed from compacted volcanic ash), they are dated at between 5.7 million and 6.0 million years.

Senut (2006) summarized several bipedal features of *Orrorin*'s *femur*, the thigh bone that forms a ball joint with the pelvis. One such feature concerns the femoral neck, the short span of bone between the shaft and the femoral head (Figure 5.4). *Orrorin*'s neck is noticeably longer than that of chimpanzees. Its length is characteristic of early hominin bipedalism because it allowed the bottom of the femur to angle back toward the body's midline, centering body weight in a more columnlike fashion over the legs. Interestingly, following modification of the pelvis during human evolution, the femoral neck no longer needs to be so long to achieve the same effect and so it has again shortened in modern *Homo sapiens* (Holliday, Hutchinson, Morrow, & Livesay, 2010; Richmond & Jungers, 2008).

Some unknown hominin species – with only chronology leading to the speculation it might be *Orrorin* – was responsible for the fourth spectacular Miocene discovery, a trackway of footprints recently found near Trachilos, Crete. Dating to 5.7 million years, the prints are critically important in showing only modest splay of the big toe, much less than that of *Ardipithecus*. The find is notable not only for its antiquity but its location, the island of Crete not having previously been implicated in hominin evolution. However, at the time, the island was likely connected to the Greek mainland and could have been reached by walking (Gierlinski et al., 2017).

The discoveries of the Miocene hominins have been nothing short of revolutionary, and they settle an old controversy (Box 5.1).

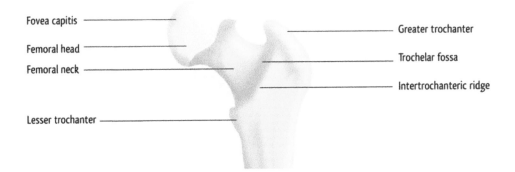

Fovea capitis

Femoral head

Femoral neck

Lesser trochanter

Greater trochanter

Trochelar fossa

Intertrochanteric ridge

Figure 5.4 Top portion of a modern human femur, showing the short femoral neck.

Source: Adaptation of Fumur Posterior annoted.png, CC BY-SA 3.0, Frank Gaillard.

Box 5.1   The missing link

The term *missing link* was coined in the 19th century, soon after the publication of Charles Darwin's *Origin of Species* (Lewin & Foley, 2004). It was recognized that if the nonhuman apes split off from the human ancestral line millions of years ago, there should surely be fossil evidence of it. The lack of such evidence led to the conclusion that there was a missing transitional form that sooner or later would be discovered by scientists working in the field. However, for some members of the public who doubted evolutionary accounts of human origins, the missing link became a rallying point. As long as it remained unknown, the gap between apes and humans was something science had not fully explained.

For scientists, on the other hand, the missing link proved a means of capturing the public imagination, and a succession of fossil finds were popularly billed as the missing link (Reader, 1981). Thus the discovery of "Java man" (now called *Homo erectus*) in 1892 provided what then appeared to be the missing link. Yet, over the next century, further discoveries were made; for example, *Australopithecus africanus* in 1924, *Australopithecus afarensis* in the 1970s, and *Ardipithecus ramidus* in 1994 (Lewin & Foley, 2004). Early in this process, William Jennings Bryan, the famous anti-evolution politician and orator, is said to have asked, "If the missing link has been found, why are they still looking for it?" (BBN International, 2008). In truth, each of these was *a* missing link that pushed human origins ever earlier. But *the* missing link, the earliest possible transitional form with human characteristics, remained elusive.

For some researchers during the early 1970s, these discoveries seemed only to nibble at a much deeper problem. The australopithecines, the earliest hominins known at that time, dated back only three million years. The view then current, that the ancestor of all nonhuman apes diverged from our ancestral line prior to 14 million years ago, set off debate over which fossils of that age were closest to representing *the* missing link. Pilbeam (1972) argued that the closest was *Ramapithecus* (now considered *Sivapithecus*; Simons, 1992), an orangutan-like creature dating from 14 million years ago. However, the remains were fragmentary, and it was assumed that future fossil discoveries would more firmly establish the human-ape split as having occurred around that time.

Today, the reason those fossils have never been found is clear: They do not exist. Because the ancestor of chimpanzees split from our ancestral line about 7.5 million years ago, the first fossils with unmistakably human characteristics must be younger than that split. The Miocene hominins appear to neatly fill the role. The missing link is no longer missing.

Hominins of the Pliocene

In the Pliocene geological period (2.6–5.3 million years ago) first appeared four species in the cladogram in Figure 5.2, namely the australopithecine species *anamensis*, *afarensis*, and *africanus*, as well as *Kenyapithecus platyops*. To these could possibly be added two other australopithecines, *aethiopicus* and *garhi*. *Aethiopicus* originated in the Pliocene 2.7 million years ago, but was mostly of the early Pleistocene. *Garhi* is technically an early Pleistocene species due to its dated fossils of 2.5 million years of age, but the cladogram implies it diverged from our ancestral line at an earlier time, between *afarensis* and *africanus*.

Note that as Figure 5.2 indicates, certain australopithecine species are sometimes assigned to the genuses *Praeanthropus* and *Paranthropus* rather than *Australopithecus* (Harcourt-Smith & Aiello, 2004; Strait & Grine, 2004; Wood & Boyle, 2016). Also, two other

Pliocene australopithecines, *bahrelghazali* and *deyiremeda*, have been proposed but are not yet widely accepted as distinct from each other and from *afarensis* (Wood & Boyle, 2016).

Cladograms indicate at a glance what forms were (or are) intermediate to others, based on an integration of many different physical characters. They provide a useful overall summarization and signpost the order of emergence of lineages. However, it is important to recognize that cladograms do not themselves identify ancestral forms, any more than Figure 5.2 argues that chimpanzees descend from gorillas. Thus the branch points are only that, branch points leading to species.

The truth is that tracing species-to-species descents through fossils is a very tricky business, one prone to the emphases and biases of individual scientists. To an outsider, it may on occasion seem to fall to the level of "My fossil is more important than your fossil", or even simply "I'm right and you're wrong". Cladograms can certainly limit the possibilities. For example, Figure 5.2 would make it difficult to argue that *Australopithecus garhi* descended from *Kenyanthropus platyops* even though the fossils of *Kenyanthropus* are older. But cladograms by themselves can't prove descent.

Nevertheless, it is true that some species-to-species connections are more plausible than others, generating, if not a complete consensus, then a preponderance of opinion. Such cases will be highlighted as we progress through the Pliocene species.

Anamensis

Anamensis is known from jaws, teeth, and a partial femur found in Kenya and Ethiopia, dated to 3.8–4.2 million years ago. From associated fossils it is believed to have lived in a woodland environment. Given its location and characteristics, an *anamensis* to *afarensis* descent is considered plausible, although other possibilities exist (Wood & Boyle, 2016).

Afarensis

Afarensis has the next oldest set of fossils, dating to 3.0–3.7 million years ago. They come from Tanzania, Kenya, and Ethiopia (White, 2002; Wood & Boyle, 2016). The most famous example is the partial skeleton of "Lucy", including a femur with an early biped's distinctively long neck and a shaft that angled in to a horizontal base (Figure 5.5). The shape of the pelvis is also consistent with bipedalism. Any remaining doubt is all but eliminated by Mary Leakey's discovery of footprints at Laetoli, Tanzania (Figure 5.6). Imprinted in ash, allowing firm dating at 3.66 million years ago, the footprints comprise a trackway of two bipedal individuals apparently walking side by side, with a smaller individual stepping in their footprints (Potts, 1992; Masao et al., 2016). Like the 5.7-million-year-old Trachilos prints, but in contrast to *ramidus* anatomy, the big toe showed relatively little splay. The Laetoli prints are attributed to *afarensis* because that species is associated with the geological setting in which the prints were found, and because its foot was moderately arched like the ones that made the prints.

Afarensis is one of the best known hominins due to extensive fossil remains. It is estimated to have stood 1–1.5 m (3–5 feet) in height depending on gender (with males larger), and to have averaged 35–40 kg (75–90 pounds) in weight. Its brain is estimated at somewhat over 400 cc (almost a pint) in volume, although cranial remains are incomplete. It had noticeable brow ridges (Figures 5.7–5.8). Relatively small, unspecialized teeth suggest an omnivorous diet, and fossil pollen collected at Hadar, Ethiopia, one of the main sites, indicate it was able to live in a varied environment that included grassland, temperate woodland, tropical woodland, swamp, and evergreen forest, among others. Its hand was remarkably modern-looking, with a thumb extending beyond the knuckles (unlike chimpanzees) but with narrow fingertips (like chimpanzees). Its long

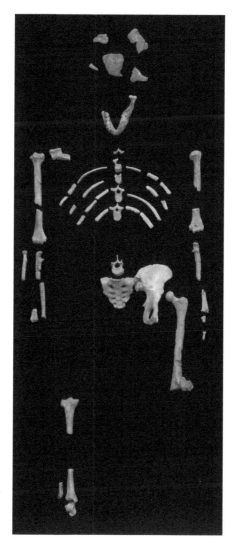

Figure 5.5 "Lucy", *Australopithecus afarensis.*
Source: Lucy blackbg.jpg, CC BY-SA 3.0, 120.

arms suggest it may have been partially arboreal, perhaps sleeping in trees at night like chimpanzees (Fleagle, 1992; Trinkaus, 1992; Wood, 1992). This supposition is supported by the fossil remains of a juvenile, showing evidence of bipedalism in the foot and leg, but of arboreal existence in the shoulder blade and hand (Alemseged et al., 2006). This species, or possibly *Kenyanthropus platyops*, is currently believed to be the earliest known hominin stone toolmaker (Lewis & Harmand, 2016).

Garhi

Garhi, the next australopithecine to split from the hominin line according to the cladogram (Figure 5.2), is known from a relatively late fossil cranium and palate found in Ethiopia, about 2.5 million years old. It differs from other species of the period in having

Figure 5.6 Replica of a portion of the footprints at Laetoli.

Figure 5.7 Cast of skull of *Australopithecus afarensis*.

Figure 5.8 Reconstruction of *Australopithecus afarensis*.

Source: NHM – Australopithecus afarensis Modell 1 a.jpg, CC BY-SA 4.0, Wolfgang Sauber.

large teeth, especially in the back of the jaw. Its brain size is estimated at 450 cc, about a pint in size (Asfaw et al., 1999). *Garhi* may have manufactured stone tools, found in the same area of Ethiopia and dated to about 2.6 million years ago, but the connection is not certain (Semaw, 2000).

Africanus

Africanus is the first-discovered australopithecine, described by Raymond Dart in 1925 from South African fossils. Next to branch off our ancestral line according to the cladogram, it dates to 2.4–3.0 million years ago (White, 2002; Wood & Boyle, 2016). Together, *Africanus* and *afarensis* comprise what traditionally have been called the *gracile* (slender) forms of *Australopithecus*, as opposed to the *robust* (heavily built) forms of *robustus* and *boisei*. However, as we will shortly see, this division is a bit of a misconception.

Africanus stood 1.1–1.5 m (3.5–5 ft) in height, averaged 35–40 kg (75–90 pounds), and had a brain 420–515 cc in volume. Its face was shorter than that of *afarensis*. It was certainly bipedal. In all likelihood it ate fruit and leaves, but carbon isotope analysis suggests that it also consumed the roots of grass or grasslike plants (or perhaps animals feeding on them). In turn, that suggests that *africanus* lived in a grassland or woodland environment (Wood, 1992; Wood & Aiello, 1998).

Hominins of the early Pleistocene

The early Pleistocene was marked by further australopithecine diversification. *Robustus*, *boisei*, and *aethiopicus* are usually considered to be later-living members of that genus, although some classify them as genus *Paranthropus*. The *robustus* and *boisei* fossils date to about 1.0–2.3 million years ago, with *aethiopicus* older at 2.3–2.7 million years (Wood & Boyle, 2016). *Robustus* has been found in South Africa, and the others primarily in Kenya and Ethiopia.

These species had massive jaws and very large back teeth. Most individuals had a *sagittal crest* (a ridge of bone running front to back on the top of the skull), serving as an additional attachment point for jaw muscles. These adaptations allowed consumption of a diet heavy in fibrous plants. The species were of course bipedal, and had brains between 410 and 530 cc in volume. Traditionally these were considered "robust" forms with heavily built bodies for their height of 1.1–1.4 m (3.5 to 4.5 ft). However, that was somewhat misconceived, because early weight estimates based on their outsized jaws were inflated compared to later, more accurate estimates based on skeletal remains. For *robustus* the skeletal estimate is 35–40 kg or 75–90 pounds, the same as for the "gracile" forms. *Boisei*, however, does appear to have been a bit heavier, at about 42 kg or 95 pounds. There are no skeletal remains of *aethiopicus* (Harcourt-Smith & Aiello, 2004; McHenry & Coffing, 2000; White, 2002).

In the cladogram, *robustus*, *boisei*, and *aethiopicus* form a separate clade along with *Kenyanthropus platyops* (Strait & Grine, 2004), reflecting a general consensus that they represent a separate, dead end branch of the hominin tree. *Platyops* actually dates from the Pliocene at 3.4–3.5 million years old and is known only from a highly distorted fossil skull (Wood & Boyle, 2016).

A recently discovered addition to the genus *Australopithecus* is *sediba*, from South Africa. Two partial skeletons and a partial cranium have been reported, dating to 2.0 million years ago (Wood & Boyle, 2016). A phylogenetic analysis of a limited number of dental characters suggests that *sediba* most resembles *africanus*, and that these two South African species may form their own clade (Irish, Guatelli-Steinberg, Legge, de Ruiter, & Berger, 2013).

The emergence of Homo

For our purposes, the most important event in the early Pleistocene was the emergence of the genus *Homo*, commencing with the species *Homo habilis* (Figure 5.9) and *Homo rudolfensis*. These were roughly contemporaneous, with fossils of *habilis* dated to 1.7 to 2.4 million years ago (Spoor et al., 2007), and those of *rudolfensis* to about 2.0 million years ago (Wood & Boyle, 2016). They differ in size, with *habilis* the smaller form (Dunsworth & Walker, 2002). Compared to the australopithecines, such as those in Figures 5.7 and 5.8, the face of *Habilis* was flatter and the molars were smaller. Comparatively, *rudolfensis* had broader rear teeth with thicker enamel.

There is much controversy surrounding the definition of *Homo*, but a large brain relative to body size figures prominently in most definitions. Over the previous five million years, hominin brain size was nearly static. Toumai, the earliest known Miocene

Figure 5.9 Replica skull of *Homo habilis*.

Source: Homo habilis-KNM ER 1813.jpg, public domain, José-Manuel Benito Álvarez.

hominin, had a brain about 365 cc in volume, and the australopithecines did not top 530 cc and averaged closer to 450 cc.

In contrast, the average brain size of habilis was 612 cc (about 1¼ pints). Although that appears larger, caution must be observed because brain and body size are strongly correlated: Bigger species have bigger brains, in part because a bigger body requires a bigger brain for sensory and motor purposes. Thus Martin (1981) found that brain mass in grams (g) was linearly related to body mass in kilograms (kg) raised to the 0.76 power, a relationship holding across a range of placental mammals weighing as little as 5 g (⅙ ounce) and as large as 50,000 kg (55 tons). This poses a problem when directly comparing brain sizes.

The encephalization quotient

Fortunately, there is a way to compare brain sizes across species independent of body size. An *encephalization quotient* (*EQ*) is calculated, involving an equation taking the brain-body size relationship into account. We could, for example, use Martin's finding and define EQ as brain mass divided by the 0.76 power of body mass.

However, Williams (2002) noted that a 0.76 exponent is problematic when applied to primates. Specifically, it results in (a) a disordered ranking of species, with apes showing less encephalization than many Old World and New World monkeys, and (b) substantial sex differences, with females having notably larger EQ values due to their generally smaller body size. A comparative analysis of three published exponent values was undertaken, using as validation criteria a rough ordering of species in learning ability (humans > apes > rhesus monkeys > squirrel monkeys) and the reduction of sex differences. Williams found that an exponent of 0.28, the lowest of the three values tested, best satisfied both criteria. Subsequently, others similarly found that an exponent of 0.3 fit better than exponents exceeding 0.6, based on a more sophisticated measure of cognitive ability integrating nine behavioral categories (e.g., detour, tool use, and object discrimination; Deaner, Isler, Burkart, & van Schaik, 2007).

Williams (2002) recommended using the following relationship, which is adopted here:

$$EQ = \text{brain mass in g/body mass in g}^{0.28}$$

In practice, brain size in cubic centimeters (cc) can be substituted for brain mass in grams, as the brain is about the density of water (i.e., one gram per cc). Of course, when calculating EQ, it must be kept in mind that sample sizes are generally small and within-species variability large. Further imprecision comes from estimating brain size using skulls reconstructed from crushed or incomplete states, and in estimating body mass from measurements of leg bones or even of correlated structures like the orbit of the eye. The resulting EQ values must therefore be considered imprecise. Nevertheless, general evolutionary trends are obvious (Table 5.1).

Table 5.1 Eocene primates, *Aegyptopithecus*, *Proconsul*, and the common chimpanzee compared to hominin species, arranged by genus and geological period. EQ = encephalization quotient (see text), kya= thousands of years ago, mya = millions of years ago.

	Age	Brain size	Body mass	EQ	Notes
Comparison primates					
Eocene primates	43 mya	8.3 cc	2.5 kg	1.0	3 species
Aegyptopithecus zeuxis	30 mya	30 cc	6 kg	2.6	
Proconsul nyanzae	18 mya	167 cc	15 kg	11.3	
Pan troglodytes	present	366 cc	45 kg	18.2	
Miocene hominins (5.3–23 mya)					
Sahelanthropus tchadensis	6.8–7.2 mya	365 cc	45 kg	18.2	
Ardipithecus ramidus	4.3–6.3 mya	325 cc	50 kg	15.7	also *kadabba*
Orrorin tugenensis	5.7–6.0 mya	n/a	35 kg	n/a	
Pliocene hominins (2.6–5.3 mya)					
Australopithecus anamensis	3.9–4.2 mya	n/a	42 kg	n/a	
Australopithecus afarensis	3.0–3.7 mya	438 cc	37 kg	23.0	
Kenyanthropus platyops	3.4–3.5 mya	n/a	n/a	n/a	
Australopithecus africanus	2.4–3.0 mya	452 cc	36 kg	24.0	
Early Pleistocene hominins (0.8–2.6 mya)					
Australopithecus aethiopicus	2.3–2.7 mya	410 cc	38 kg	21.4	
Australopithecus garhi	2.5 mya	450 cc	n/a	n/a	
Homo habilis	1.7–2.4 mya	612 cc	35 kg	32.7	
Australopithecus boisei	1.3–2.3 mya	521 cc	42 kg	26.4	

	Age	Brain size	Body mass	EQ	Notes
Homo ergaster	1.4–1.7 mya	871 cc	61 kg	39.8	African
Australopithecus robustus	1.0–2.0 mya	530 cc	36 kg	28.1	
Australopithecus sediba	2.0 mya	420 cc	33 kg	22.8	
Homo rudolfensis	2.0 mya	752 cc	56 kg	35.2	
Homo erectus	0.3–1.8 mya	987 cc	64 kg	44.5	mainly Asian
Middle and Late Pleistocene hominins (12 kya-0.8 mya)					
Homo heidelbergensis	0.1–1.0 mya	1158 cc	54 kg	54.8	aka archaic *sapiens*
Homo floresiensis	50–700 kya	417 cc	26 kg	24.2	Indonesian island
Homo naledi	253 kya	513 cc	46 kg	25.4	
Homo neanderthalensis	40–130 kya	1431 cc	76 kg	61.5	
Homo sapiens	0–195 kya	1350 cc	54 kg	63.9	

Table notes: Ages largely from Wood and Boyle (2016); for Eocene primates from http://eol.org; for *floresiensis* from van den Bergh et al. (2016) and Sutikna et al., 2016; for *naledi* from Dirks et al. (2017). Size and mass estimates for Eocene primates from Harrington, Silcox, Yapunich, Boyer, and Bloch (2016); for *Aegyptopithecus* and *Proconsul* from Begun and Kordos (2004); for *aethiopicus* from Kappelman (1996); for *sediba*, near-adult brain size estimate from Berger et al. (2010), and adult body size retrieved from www.profleeberger.com/AusedibaQandA.html; for *erectus and heidelbergensis* from Rightmire (2004); for *floresiensis* from Falk et al. (2006). Brain size estimate for *neanderthalensis* from Wood and Constantino (2004). Body mass estimate for *Pan* from Williams (2002); for *tchadensis* from Brunet et al. (2004); for *ramidus* from Suwa, Asfaw, Kono, Kubo, Lovejoy, and White (2009), and White et al. (2009); for *naledi* from Berger et al. (2015); for *neanderthalensis* from Cameron and Groves (2004). Most other values from McHenry and Coffing (2000).

When EQs are calculated in this way, it is clear that *habilis* did have a large brain relative to the australopithecines. Thus while australopithecine EQs are in the range of 21.4–28.1, the value for *habilis* was 32.7. The species packed a big brain for its size. The single known specimen of *rudolfensis* did as well: It had a substantially larger body but also a larger brain, resulting in an even greater EQ of 35.2.

In any case, the general consensus is that *Homo* commenced with *habilis*, meaning "able, handy, mentally skillful", a notable example of assignment of a species name because of cultural capabilities (Dunsworth & Walker, 2002). Of course, mental skills are synonymous with cognitive capabilities and are reflected in the abundant stone tools associated with *Habilis*.

Ergaster *and* erectus

The next Pleistocene members of *Homo* were two similar species, *Homo ergaster* (Figure 5.10) and *Homo erectus*. Not as old as the oldest *habilis*, they were much larger but also more encephalized (Table 5.1). *Ergaster* mostly comes from African sites in Kenya, and *erectus* from Asian ones in China and Indonesia, although some of the earliest *erectus* are found in Tanzania and the Republic of Georgia as well (Antón, 2003). It is thought that the non-African examples of *erectus* descended from African forms that left the continent at about the time of the species' origin 1.8 million years ago, then disappeared from it altogether about 800,000 years ago (Antón, 2003). *Erectus* survived in Java until about 310,000 years ago (Indriati et al., 2011).

Figure 5.10 Skull of *Homo ergaster*.

Source: Homo ergaster skull replica, World Museum Liverpool.jpg, CC BY-SA 3.0, Reptonix free Creative Commons licensed photos.

The differences between the species are subtle, with *ergaster* showing more primitive features of the cranium, jaw, and teeth. It was the first known hominin without arboreal adaptations, indicating full-time bipedalism. Some researchers downplay the differences and define only one species, *Homo erectus* (Wood & Constantino, 2004), with *ergaster* considered a subspecies (Lordkipanidze et al., 2013). A similar consideration applies to another claimed early Pleistocene species, *Homo georgicus*, likewise not clearly differentiated from *erectus* (Wood & Boyle, 2016).

For both species, meat likely played a larger dietary role than previously, providing the energy required by a large brain (about 850–1000 cc, or 2 pints). There is some evidence they used fire, but at a much later period than their time of origin (Antón, 2003).

Hominins of the middle and late Pleistocene

Dating somewhat later to the middle and late Pleistocene was *Homo heidelbergensis* (0.1–1.0 million years ago). It is also known as archaic *Homo sapiens*. As that name implies, it had a skull close to modern form (Figure 5.11). It still had a small brow ridge, but over each eye rather than continuously across the face as in *erectus*. Its brain was nearly as large as that of modern humans (about 1150 cc, nearly 2½ pints). Fossils of this species have been found at multiple sites in Europe, Africa, and Asia. Some assign the oldest specimens to a separate species, *Homo antecessor* (Wood & Constantino, 2004). Some younger hominin specimens, including *Homo rhodensiensis* (0.3–0.6 million years ago) and *Homo helmei* (90,000 years ago) may instead be *heidelbergensis* (Wood & Boyle, 2016).

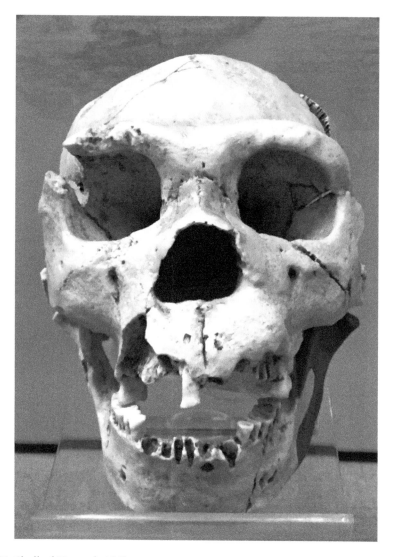

Figure 5.11 Skull of *Homo heidelbergensis*.

Source: Homo heidelbergensis Atapuerca 5 IMG 5649 BMNH.jpg, CC BY-SA 4.0, Photo: Erlend Bjørtvedt.

Naledi

Homo naledi is a very recently described form, discovered by spelunkers exploring a South African cave system. Reached by a torturous underground path, the fossil chamber was nearly inaccessible due to a narrow entrance chute. Its excavation waited on the results of a worldwide call for petite scientists with the requisite skills, and six women ultimately accomplished the job (Berger et al., 2015; Shreve, 2015).

An assessment of the fossils' relatively primitive characteristics led some to think that they dated back as far as two million years ago. Researchers were stunned, then, when a variety of dating techniques showed that they are only 253,000 years old (Dirks et al., 2017). Even more stunningly, *naledi* had a very small brain, averaging 513 cc across eight individuals, less than 40% of the modern human average (Berger et al., 2015). Even after taking into account *naledi*'s relatively small body, its EQ of 25.4 was likewise slightly less than 40% of ours (Table 5.1).

Another primitive characteristic of naledi was the marked curvature of its fingers, approximating that of early hominins and nonhuman apes. The condition is diagnostic of lifelong arboreality, implying habitual climbing and suspension from branches, an extraordinary conclusion given the species' recent existence (Kivell et al., 2015).

Sapiens

Modern *Homo sapiens* (Figure 5.12) is held to have emerged about 195,000 years ago (Wood & Boyle, 2016), with Africa our point of origin. The earliest undisputed specimens are known from Ethiopia and Israel. Although an origin as early as 350,000 years ago has been claimed for fossils collected from a Moroccan site (Hublin et al., 2017), they have not yet been subjected to detailed criticism. They also appear to have some features that are not modern (i.e., the presence of a brow ridge and an undeveloped chin).

It should be noted that the boundary between *heidelbergensis* and *sapiens* is arbitrary, as there is a continual gradation. Nevertheless, the range of variation is sufficient to justify a division into two species (Wood & Constantino, 2004). The emergence of modern *sapiens* was accompanied by substantial cultural and cognitive developments, which will be outlined in subsequent chapters. The average brain size has been 1350 cc (nearly three pints).

Neanderthalensis

The middle Pleistocene is also known for the closely related *Homo neanderthalensis* (Figures 5.13–5.14). The earliest unambiguous Neanderthal fossils date from about 130,000 years, and the species survived until about 40,000 years ago (Wood & Boyle, 2016). Remains largely come from Europe, but some are from western and central Asia.

For many years there was a controversy regarding the Neanderthals' status, with some designating them a subspecies of modern humans (White, Gowlett, & Grove, 2014). Species status seemed confirmed by the earliest research that examined mitochondrial and chromosomal DNA from Neanderthal remains. It reported little affinity to living humans (Green et al., 2006; Orlando, Darlu, Toussaint, Bonjean, Otte, & Hanni, 2006). However, subsequent study of the full Neanderthal genome did find similarities. Specifically, Europeans and Asians were found to be more similar to Neanderthals than were Africans,

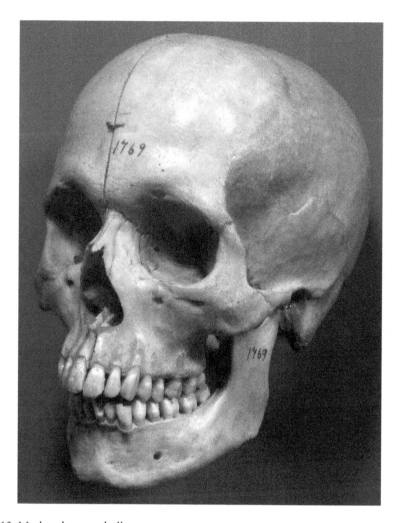

Figure 5.12 Modern human skull.

Source: Homo sapiens skull (Holocene) 3, CC BY 2.0, James St. John.

strongly implying that interbreeding occurred in areas of Neanderthal habitation. While the species designation remains, it may not represent the final word in classification given that up to 4% of the genome of non-Africans derives from Neanderthals (Green et al., 2010).

In that regard a distinction should be made between *coalescence time*, the latest date at which Neanderthals and humans had completely common ancestry (and thus the date at which there was no distinction at all between the species), and *split time*, the latest date at which there was any genetic interchange between the species. Coalescence time averages about 575,000 years across four studies using DNA from the Y chromosome (Mendez, Poznik, Castellano, & Bustamante, 2016), the *autosomes* (chromosomes other than X and Y; Green et al., 2010; Noonan et al., 2006), and mitochondria (Green et al., 2008).

It is unsurprising that the coalescence time substantially predates the earliest identified Neanderthal fossils, dating from about 130,000 years ago. First, as previously indicated,

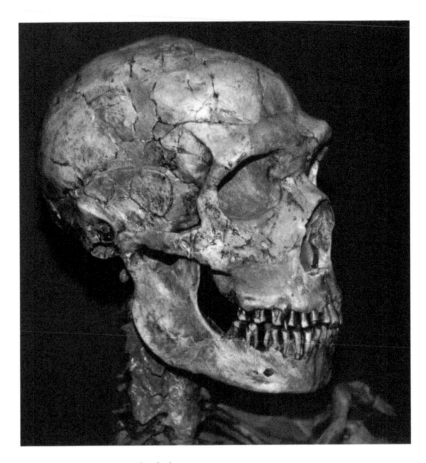

Figure 5.13 Skull of *Homo neanderthalensis*.

Source: Homo neanderthalensis fossil hominid (Pleistocene; Europe) 3, CC BY 2.0, James St. John.

fossils postdate the first appearance of a species for probabilistic reasons: The initial population is so tiny that finding the first fossils is practically an impossibility. Second, as noted in the discussion of Toumai, following their split, two species are so similar that it may not be possible to distinguish between them. Thus older fossils are not only difficult to find, but even if found, they might not appear unambiguously Neanderthal.

For its part, split time was initially estimated at 370,000 years ago, based on an incomplete Neanderthal genome (Noonan et al., 2006). However, that conclusion has yielded to recognition that late interbreeding must have occurred between *sapiens* and *neanderthalensis*, possibly as recently as 50,000 years ago (Simonti et al., 2016; Wall & Brandt, 2016).

Following the *sapiens-neanderthalensis* divergence, there was a further split off the Neanderthal line leading to the Denisovans, known from a finger bone and teeth recovered from a Siberian cave. This group survived as late as 29,000 years ago. Recovered DNA sequences demonstrate its affinity with Neanderthals and indicate that it contributed genes to modern human populations in Melanesia and Australia, and to a lesser extent in East Asia. Its species status is uncertain, and none has yet been assigned (Slatkin & Racimo, 2016; Wood & Boyle, 2016).

Figure 5.14 Reconstruction of *Homo neanderthalensis*.

Source: Neandertaler reconst.jpg, CC BY-SA 3.0, Stefan Scheer.

Neanderthals typically lived in cold, "marginal" environments. They had a larger brain than *sapiens* in absolute terms, but a slightly smaller EQ because of their substantially larger body mass (Table 5.1). They were powerfully built, and their skulls had thick, continuous brow ridges (Figure 5.13). They left cultural remains including evidence of intentional burials (Cowen, 2005; Wood & Constantino, 2004).

Floresiensis

The final hominin is also arguably the most bizarre, and not just because it was a late survival in a world populated by *sapiens*. A skeleton including a skull of *Homo floresiensis* was discovered in 2004 in a cave on the island of Flores, Indonesia, by a team of Australian and Indonesian researchers. The relatively recent remains were in a prefossilized, extremely fragile state. Remarkably, even though the specimen was an adult, it stood only slightly over one meter (3 ft) tall and would have weighed only about 22 kg (50 pounds). The nickname "The Hobbit" was accordingly bestowed because of its diminutive size. The skull was correspondingly small and had a low cranium (Figures 5.15–5.16), producing an estimated brain size of only 417 cc, a little less than a pint (Brown et al., 2004; Falk et al., 2006). The resulting EQ is only 24.2, the smallest of all Pleistocene species (Table 5.1).

Figure 5.15 Skull of *Homo floresiensis.*
Source: Homo floresiensis.jpg, CC BY 2.0, Ryan Somma.

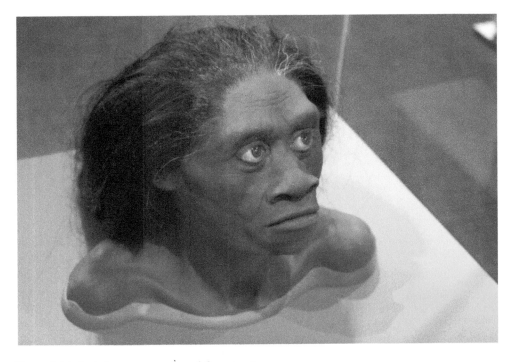

Figure 5.16 Artist's reconstruction of *floresiensis.*
Source: Homo floresiensis, CC BY 2.0, Ryan Somma via Flickr.

The find was initially greeted by skepticism and a counterproposal that the single individual might have been a *microcephalic* (abnormally small-brained) dwarf but nevertheless a *sapiens*. Its diminutive stature was subsequently confirmed, however, following the discovery of partial remains of eight other individuals. Height could be estimated for one of them, which turned out to be slightly *less* than that of the original find (Morwood et al., 2005). Furthermore, measurements based on the skull were found to much more closely match normal humans than a sample of microcephalics (Falk et al., 2007). Other analyses based on the wrist (Tocheri et al., 2007) and the shoulder, foot, pelvis, and femur (Aiello, 2010) likewise indicate that the Hobbit was not a pathological *sapiens*.

Initial dating estimates based on charcoal found in the cave found the youngest of the Flores fossils to be only 12,000 years old. However, this date was discovered to be false, the result of inadvertent sampling from a different geological section than the fossils. An array of dating methods now firmly date the youngest of the floresiensis-related deposits at 50,000 years old (Sutikna et al., 2016).

Recently, teeth and a jaw bone of *floresiensis* excavated in 2014 were described and dated to the much older period of 700,000 years ago. They were just as small as the later remains, and if anything, a bit smaller. Although they do not address the size or state of the cranium at that period, they do establish the antiquity of the species on the island of Flores (van den Bergh et al., 2016).

Hominin trends and connections

It is useful to highlight trends over the 7.5 million years of hominin evolution, because as we will see, they correspond in some degree to the evolution of cognitive capabilities. They also makes it possible to identify the most likely possibilities for ancestral connections among species.

Size trends

An examination of Table 5.1 highlights a strong trend in brain size. Across species and with few exceptions, brain size has increased, from a sub-400 cc level in the Miocene to the mid-400s in the Pliocene, as high as 987 cc in the early Pleistocene, and finally to around 1400 cc in the later Pleistocene species of *Homo neanderthalensis* and *sapiens*. All told, this was an increase from ¾ of a pint to three pints. The trend is so powerful that historically, it rightly assumed a major role in the reconstruction of hominin *phylogeny* (evolutionary relationships).

Body size has varied as a secondary trend, though in a more complex fashion. Most species through the early Pleistocene weighed less than 50 kg on average, and several less than 40 kg. However beginning with the first *Homo* species after *habilis*, most exceeded 50 kg, with some over 60 kg. Indeed, it can be argued that some of the gain in absolute brain size was accomplished simply by upscaling body size.

Considering the size trends jointly, however, indicates that growth of the brain outstripped growth of the body. Thus the ape and Miocene hominin EQ values of a little under 20 increased to 23–24 in the Pliocene, to over 30 in some early Pleistocene species, and then with some exceptions to 50 and even upwards of 60 in the middle and late Pleistocene. The trends reveal something else as well: A portion of the later increase in encephalization came about not just because absolute brain size increased, but because body size *decreased*. The early *Homo* species of *ergaster* and *erectus* had large brains but also large-body masses, resulting in EQs only modestly advanced over that of the first *Homo* species, the small-bodied *habilis*. Mean body mass then decreased by 10 kg in *heidelbergensis* and *sapiens*, which with a 200–400 cc increase in brain volume boosted their EQ values to high levels.

Notable exceptions to these trends are the late-surviving species *Homo naledi* and *Homo floresiensis*, whose small brains and bodies result in EQs comparable to the australopithecines. In fact, their EQs are so similar that they raise the question of a possible affinity between the species. However, a recent phylogenetic analysis shows them widely separated, with *naledi* most resembling *Homo antecessor* (i.e., early *Homo heidelbergensis*), and with *floresiensis* most resembling *Australopithecus africanus* and *Homo habilis* (undated Bayesian analysis in the supplementary online material of Dembo et al., 2016).

Naledi was too recently discovered to permit any firm speculation as to its origins. With respect to *floresiensis*, however, Falk et al. (2005) suggested that it might represent an example of *island dwarfism*. That is an evolutionary phenomenon in which isolation on an island causes large species to become reduced in size over time. Island dwarfism is specifically observed among primates (Montgomery, 2013). It probably occurs because islands prevent migration during famine, resulting in a selection pressure toward lower energy consumption and thus smaller body sizes.

After performing a cladistic analysis of 60 characters of the cranium, mandible, and postcranium, Argue, Morwood, Sutika, Jatmiko, and Saptomo (2009) concluded that the Hobbit may have split off the line leading to us, much earlier than its age implies. They suggested that Hobbit ancestors may have diverged from the *Homo* line around the time of *Homo habilis*, placing the split at about two million years ago. On this basis, the researchers suggested that the species may have started out and stayed small. Their position receives general support from the aforementioned phylogenetic analysis of Dembo et al. (2016), finding that the Hobbit is most similar to the 2-million-year-old *Australopithecus africanus*, and next most similar to *Homo habilis*.

However, a number of others report that *floresiensis* shows an affinity with African *Homo erectus* in several respects, including the skull and shoulder (e.g., Baab & McNulty, 2009; Lyras, Dermitzakis, Van der Geer, A. A. E., Van der Geer, S. B., & De Vos, 2009), and the teeth and mandible (van den Bergh et al., 2016). Zeitoun, Barriel, and Widianto (2016) even suggest that *floresiensis* should more properly be classified *as* erectus!

Skull trends

Additional trends over the course of hominin evolution are apparent from skulls. The brow ridges of *tchadensis* seven million years ago (Figure 5.1) waned, waxed, and waned again in size over successive species but were consistently present until the emergence of modern *sapiens*. At the same time, the face became flatter and more vertical, and the rear teeth generally decreased in size as diets included more meat and less fibrous vegetable matter. All of these characters, like brain size, have figured prominently in the reconstruction of our phylogeny.

Hominin phylogeny

What, then, was the structure of the human family tree? Before addressing this question, it is helpful to consider some influences on phylogenetic inferences. One already referred to is the occasional tendency of the discoverers of a fossil to overweight its importance. That can easily affect how that particular team reconstructs phylogeny. However, the beauty of science is that others are free to reconsider the evidence to potentially come to more balanced conclusions.

A second influence is that the age of a species is rarely precisely known even if it left accurately dated fossils. Earlier and later fossils may never have formed, or they may still be in the ground waiting to be discovered. Thus the species may have had a longer evolutionary lifespan than is apparent from fossil evidence.

Third, the assignment of fossils to particular species is error-prone. There are many historical examples of the reassignment of a misassigned fossil following the acquisition of additional information. That obviously affects the reconstruction of phylogeny.

Finally, a fourth influence on phylogenetic inferences should be inserted as a cautionary note. This is that it is generally impossible to determine whether *any* fossil is from a direct human ancestor in a strict genetic sense. The best we can say is that any given fossilized individual, and every human alive today, traces back to a common ancestor at some unknown point during or *prior* to the existence of the fossilized individual. This raises questions as to whether we can say that we descend from a particular species as defined by known fossils. However, in many cases, we can infer that we must descend from an individual *very much like* that species. With that proviso, it is nevertheless common practice, as a kind of shortcut, to refer to specific species defined by known fossils when reconstructing phylogeny.

We can now consider the likelihoods, tracing backwards. *Homo sapiens* (0–195,000 years ago) is almost universally recognized as a descendant of *Homo heidelbergensis* (0.1–1.0 million years ago) as broadly defined, although those recognizing the earliest examples of *heidelbergensis* as a separate species called *antecessor* sometimes draw the connection directly to that (e.g., Wood & Constantino, 2004). In turn, *heidelbergensis* is assumed to have descended from *Homo erectus* (0.3–1.8 million years ago), most likely the early African forms since later Asian ones are probably too derived to be ancestors of modern humans (Wood & Constantino, 2004). *Erectus* likely descended from the more primitive *ergaster*, even though at present its oldest fossils, dated at 1.8 million years ago, predate the oldest *ergaster* fossils, at 1.7 million years ago.

What path is traced backward from *ergaster* depends on which characters are emphasized. A strict chronology-and-encephalization approach traces the line back through *Homo habilis*. However, others emphasize the *postcranial* skeleton (the skeleton below the head) and suggest that *habilis* had limb proportions that were too apelike to make it a likely human ancestor (Richmond, Aiello, & Wood, 2002). Nevertheless increasing encephalization was clearly a major hominin theme, and that plus the cultural remains associated with *habilis* do seem to justify its placement in the ancestral tree.

There is also controversy with respect to the link between *habilis* and earlier forms. There is near-consensus that *Australopithecus afarensis* (3.0–3.7 million years ago) was ancestral, but less agreement that the 600,000-year gap between *habilis* and *afarensis* was filled by *Australopithecus africanus* (2.4–3.0 million years ago). Some nominate *Australopithecus garhi* (2.5 million years ago) for the intermediate role, but remains seem too fragmentary at this point to draw that conclusion with any certainty.

As indicated earlier, a descent of *afarensis* from *anamensis* (3.9–4.2 million years ago) is considered plausible, and in fact, the across-three-species ancestral connections between *habilis*, *afarensis*, and *anamensis* have been accepted by Donald Johanson, one of the discoverers of "Lucy", based on derived characters that seemingly passed from one to the next (Johanson, 2017).

Although a connection from *anamensis* to *ramidus* (4.3–6.3 million years ago) has been proposed, *ramidus'* widely splayed big toe is problematic. For that reason it seems wiser to place a speculative link from *anamensis* to *tugenensis*. What relationship *tugenensis* – or for that matter, *ramidus* – had to *Sahelanthropus tchadensis* remains to be seen. Wood and Boyle (2016) raised the possibility that all three should be assigned to the same genus, or even collapsed into one or two species instead of three, in recognition of their general similarity. But for now *tchadensis* (7 mya) can be tentatively placed at the head of the hominin ancestral line of descent, as the earliest of the three forms.

A well-pruned bush

What is most notable about Table 5.1 is that while many hominin species have existed across 7.5 million years, only one survives. Our portion of the tree of life more resembles a heavily pruned bush than it does a tree, with a single branch emerging from the crown and all others trimmed away.

Heavy pruning is not unusual in evolutionary history, another example being the survival of birds as the only representatives of a once-numerous suborder of dinosaurs. By its nature, evolution often takes a crooked path from past to present. The randomness of mutation combined with diverse and changing environments results in many unsuccessful "experiments", a useful term as long as we recognize that no intentionality is involved. One educated guess is that of all the species ever generated by evolution, 99.9% are now extinct (Raup, 1991, pp. 3–4).

Of course, one difference between birds and humans is that birds radiated to become thousands of species worldwide, while there is now only one human species. Not only can any population of humans interbreed with all others, but genetic analyses show close relatedness in evolutionary terms. For example, studies have examined the similarity of populations with respect to mitochondrial DNA, the genetic instructions accompanying the mitochondria in our cells that are not part of our 46 chromosomes. Because mitochondria are inherited as part of the egg rather than the sperm, their genetic content follows the maternal line of descent, mother to daughter to daughter. Taking a molecular-clock approach suggests that living human populations trace to a common female ancestor, a "mitochondrial Eve". She is estimated to have lived sometime between 125,000 and 200,000 years ago. Furthermore, Africans show the greatest variability in mitochondrial DNA and therefore the deepest convergence on that ancestor, strongly arguing that she lived in Africa (Cyran & Kimmel, 2010; Ingman, Kaessmann, Pääbo, & Gyllensten, 2000).

A second line of research has examined population similarities in the sequencing of the Y chromosome, which is inherited in the paternal line, father to son to son. Again, application of molecular-clock principles indicates that a common ancestor of all men lived in the relatively recent past, perhaps between 70,000 and 200,000 years ago. Furthermore, like Eve, this "Adam" is believed to have lived in Africa, where today's San hunter-gatherers represent the earliest known branch off the common ancestral tree (Poznik et al., 2013; Scozzari et al., 2014).

It should be stressed that African origins apply to all of our genetic material, not just mitochondrial and Y chromosomal DNA. The DNA from our autosomes similarly shows greatest variability in African populations, implying that it originated there and thus has had maximal time to develop diversity. Yet even the least diverse population globally – the Suruí, a Brazilian tribe – captures 60% of total worldwide diversity within its members (Hunley, Cabana, & Long, 2016), reflecting the relatedness of human populations.

Thus 7.5 million years of human evolution have come down to just one interrelated human species. This certainly was not always the case, as hominin ancestry has been bushy in the past. At least six species were alive two million years ago, and three species as little as 50,000 years ago (Table 5.1).

What accounts for such severe pruning? Undoubtedly, part of the answer is that all bipedal apes have occupied very similar environmental niches. They have been omnivorous, able to travel distances while carrying babies or tools, and large enough to defend against predators and doubtless other bipedal apes. Such similarity in the demands placed on the environment, and in responses to it, created direct competition for resources when different species came into contact.

Part of the answer, too, must be that the playing field implied by competition became radically tilted when human cultural innovations became commonplace. In turn, cultural innovations were produced by cognitive innovations, ensuring the selective survival of "wise man", or in Latin, *Homo sapiens*.

Conclusion

The term "hominin" refers to humans, their ancestors, and their immediate relatives, back to the chimpanzee-human split 7.5 million years ago. The *Sahelanthropus tchadensis* ("Toumai") fossil dates back seven million years, almost to the split itself. Signs of bipedalism in the skull are echoed in *Orrorin* fossils a million years later, showing the long femoral neck characteristic of early bipeds. *Ardipithecus* is a third Miocene hominin dating to 6.3 million years ago. However, the species had a splayed, grasping big toe, postdating a recently discovered trackway in Crete lacking that feature, and thereby suggesting that it may not have been directly ancestral to modern humans.

Hominin species emerging in the Pliocene included *Australopithecus anamensis, afarensis*, and *africanus*, and *Kenyapithecus platyops*. *Anamensis*, dating to 3.9–4.2 million years ago, represents a plausible transition from earlier forms. "Lucy" is the most famous example of *afarensis*, consisting of a partial skeleton including a long-necked, bipedal femur. This species likely made the trackway at Laetoli. *Afarensis* may also have been partially arboreal, judging from the shoulder blade and hand, but there is a good chance that either it or *Kenyapithecus* made the earliest known stone tools.

Australopithecus garhi (2.5 million years ago) was possibly also an early maker of stone tools. Its contemporary *africanus* was the first-discovered australopithecine and probably lived in a grassland or woodland environment in South Africa. The australopithecines *robustus, boisei*, and *aethiopicus* are sometimes assigned a separate genus, *Paranthropus*, in recognition of distinctive features including massive jaws and a sagittal crest.

The genus *Homo* emerged about 2.4 million years ago in the form of *Homo habilis*, a species marked by a noticeable increase in encephalization. It was probably the ancestor of *Homo ergaster*, itself giving rise after a very brief time to *Homo erectus*, with the latter surviving in Asia until relatively recently (about 310,000 years ago). *Ergaster* was an African hominin that for the first time showed an absence of arboreal adaptations, indicating full-time bipedalism.

Erectus was a likely ancestor of *Homo heidelbergensis*, or archaic *Homo sapiens*, dating as far back as one million years ago. Modern humans emerged about 195,000 years ago in Africa. Three other relatively late species were *Homo neanderthalensis, naledi*, and *floresiensis*, the former of which contributed to the biological ancestry of modern *sapiens*. *Naledi* and *floresiensis* were small-brained species with low EQs, and *floresiensis* also had a very small body, possibly a case of island dwarfism.

Throughout humankind's descent, brain size has generally increased both in absolute terms and in relation to body size. EQ values more than tripled between "Toumai" and today, a trend aided from the onset of archaic *sapiens* by a reduction in body size. Another trend has been a consistent trimming of the hominin "bush", from at least six coexistent species two million years ago to today's single species. There is strong evidence that all humans descend from a common female ancestor, "mitochondrial Eve", living 125,000–200,000 years ago; and all men from a Y-chromosome "Adam", living 70,000–200,000 years ago.

Determining the hominin line of descent is subject to biases, imprecise dates, and the impossibility of determining whether any fossil is a direct ancestor in a genetic sense. The likelihood is that modern *Homo sapiens* descends from *Homo heidelbergensis*, itself descending from *Homo erectus* and then in turn from *Homo ergaster* and *Homo habilis*. *Habilis* was likely descended from *Australopithecus afarensis* and *Australopithecus anamensis*. The remaining connections are more conjectural, but links to *Orrorin tugenensis* and, finally, *Sahelanthropus tchadensis* are possible.

Section II
Sensation and movement

6 The mechanical and chemical senses

A large brain serves as only a crude indicator of the sensory, motor, and cognitive abilities *Homo sapiens* acquired over seven million years of hominin evolution. Our ancestors used brain specializations, as do we, to collect inputs from the environment, to understand meaning, and to produce outputs in response. How sensation and movement evolved provides important context for understanding the evolution of the cognitive faculties that bridge them. In this chapter we primarily consider the mechanical and chemical senses of touch, balance, hearing, smell, and taste, and the evolutionary trajectory shared by some of them.

Early sensing

Even early single-cell organisms needed a way to interact with their environment. Life depended on it, for metabolism within the cell was contingent on obtaining raw materials from outside and on eliminating waste products from within. Both these needs required the ability to sense the presence of substances outside or inside the cell.

The simplest early sense mechanism was mechanical. When pressure inside increased past tolerable limits, large pores in the cell membrane opened, pouring out some of the cell's contents until pressure was lowered. This mechanical *gating* (the controlled passing of substances through a membrane) exists in similar form across all three domains of bacteria, archaea, and eukaryotes, testifying to an ancient evolutionary origin (Balleza, 2011).

At a very early evolutionary point, cell membranes acquired the ability to pass some molecules but not others. The ability to distinguish between what to pass and what to block from entering the cell was a crude form of external sensing. Early on, naturally occurring amino acids and phosphates were candidates for transportation into the cell (Trevors, 2003).

There is evidence that the cross-membrane transport of *ions* (electrically charged atoms) occurred early in the tree of life. Potassium ions (symbolized K+) form when potassium salts dissolve in water, and are actively transported across cell membranes in all three biological domains (Hänelt, Tholema, Kröning, Vor der Brüggen, & Wunnicke, 2011). Maintaining appropriate K+ levels within the cell is critical to phosphate transport, as well as to enzyme-mediated chemical reactions. Derst and Karschin (1998) constructed a phylogenetic tree of genes involved in K+ transport and tentatively concluded that some originated in prokaryotes, later descending to eukaryotes about two billion years ago.

A proton pump – a membrane pump of hydrogen ions (H+) – was probably an additional early development. Membrane proteins capable of pumping H+ out of a cell are believed to have existed in the last common ancestor of all life (Deamer & Weber, 2010;

Wei & Pohorille, 2015). Using an external source of energy such as thermal or light energy from the sun, H+ was pumped out of the cell. Then, when the energy source was no longer available, such as at night, H+ was let back into the cell, freeing energy to be used in chemical reactions such as phosphate synthesis.

Other chemical transport mechanisms believed to have had prokaryote precursors are those for the transport of sodium (Na+) and calcium (Ca+) ions, evolutionary descendants of the K+ mechanism. By about 600 million years ago, most of today's ion transport channels had evolved (Derst & Karschin, 1998).

Even at the unicellular level, the ability to sense compounds outside the cell can have behavioral consequences. Today's bacteria are capable of surprisingly sophisticated chemical signaling techniques to enhance mass action. *Quorum sensing* refers to the ability to detect increasing numbers of other bacteria by sensing a building concentration of chemical excretions. It is used to coordinate the release of toxins when attacking a host, and to establish *biofilms*, which are cooperative communities of bacteria, each contributing some component to the matrix (Bassler & Losick, 2006). An early origin for quorum sensing is suggested by the fact that rhomboid protein, a signaling chemical, is found across the life domains (Henke & Bassler, 2004).

Touch

Mechanical gating and ion transport probably served as sufficient sense mechanisms for many early single-cell organisms. However, metazoa, with complex body plans and in many cases improved mobility, required more complicated sensory systems. These included the presence of specialized touch receptors, and in most phyla a nervous system to carry touch information beyond the limits of the cell.

Mechanoreceptors

Mechanoreceptors (specialized sensory receptor cells that react to mechanical forces) predated nervous systems and possibly contributed to their evolution. The earliest mechanoreceptors are thought to have been in the form of thread-like cellular outgrowths. These were either small (*cilia*) or large (*flagella*), surrounded by *microvilli* (tiny projections of cell membrane) that were used in feeding. The outgrowths had both sensory and motor function, so that their stimulation caused movement. Their possession by choanoflagellates dates their origin to 1.0–1.5 billion years ago (Arendt, Benito-Gutierrez, Brunet, & Marlow, 2015; Fritzsch, Pauley, & Beisel, 2006).

Hair-based mechanoreceptors may descend from ancestral choanoflagellates, but more certainly from ancestral metazoans over 600 million years ago. These had a central hair surrounded by microvilli, and were probably the origin of later touch receptors (Fritzsch et al., 2006). The ancestors of early cnidarians probably also had hair-based mechanoreceptors in the form of stinging cells, possibly the origin of neurons (Box 6.1).

Early-emerging taxa with mechanoreceptors that synapsed onto neurons included tunicates (or urochordates, including sea squirts), with hair cells arranged in a *coronal organ* in the oral region that detected vibration in water and thus motion (Burgihel et al., 2003). They also included ancestral lancelets, whose descendant amphioxus has similar cells with microvilli, also located in the oral region (Kaltenbach, Yu, & Holland, 2009).

In humans, *somatosensation* (the sense of touch, or the *tactile* sense) begins when receptors in the skin are temporarily deformed due to mechanical pressure (Figure 6.2). *Merkel's cells* (or *discs*) and *Meissner's corpuscles* are two types of touch receptors, both lying just below the outer skin. They detect fine details and flutter, respectively. *Hair*

Box 6.1　The origin of neurons

Miljkovic-Licina, Gauchat, and Galliot (2004) proposed that *neurons* (nerve cells) had their origin as mechanoreceptors, specifically in the immediate ancestors of early cnidarians. Along with some *ctenophores* (comb jellies), cnidarians were the first metazoans to have identifiable neural nets. The earlier-emerging sponges did not.

But how could neurons have evolved from mechanoreceptors? Miljkovic-Licina and colleagues pointed out that cnidarian stinging cells sense mechanical disturbances. They respond by ejecting barbed tubes or threads that inject venom to immobilize both attackers and prey (Figure 6.1). Evolution might have caused these cells to differentiate, with one line of descending cells evolving into neurons. In turn, these might have connected with the stinging cells, possibly bidirectionally to allow signals to pass back and forth, improving coordination during attacks (Liebeskind, Hillis, Zakon, & Hofman, 2016). Partially supporting this supposition, stinging cells and neurons share the same regulatory genes (Miljkovic-Licina et al., 2004).

However, there is by no means a consensus that neurons evolved from stinging cells. Some believe they instead evolved in conjunction with muscles, quickly building into diffuse neural nets controlling movement. Certainly, cnidarians and ctenophores both have nets directly connected to muscle fibers. This arrangement could then have extended to control the muscles involved in feeding and digestion (Arendt, Tosches, & Marlow, 2016).

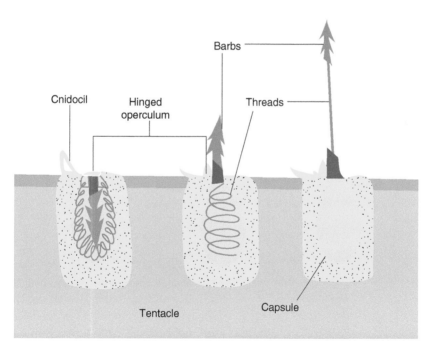

Figure 6.1 Representation of a cnidarian stinging cell firing sequence (left to right), with hair "trigger" (cnidocil) and coiled thread ready for ejection.

Source: Nematocyst discharge.png, public domain, Spaully.

Whatever view of neuronal origins is correct, continuity between cnidarians and the bilateria (including chordates and vertebrates) is indicated by the use of the same genes to specify cnidarian neural nets and bilaterian nervous systems (Arendt et al., 2016). Similarities between mechanical sensors found in cnidarians, chordates, and the vertebrate inner ear likewise suggest a common origin. Thus the nervous system in the crude form of a neural net, and its support of mechanical sensing, originated in shared ancestry a little over 605 million years ago (Burgihel et al., 2003).

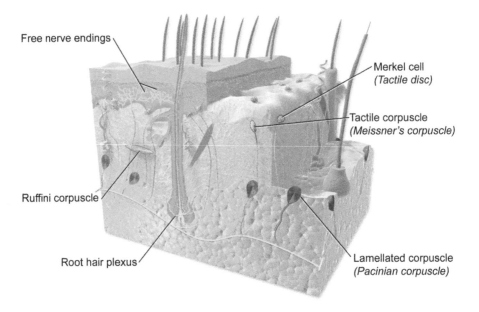

Figure 6.2 The skin receptors.

Source: Blausen 0809 Skin TactileReceptors.png, CC BY-SA 3.0, BruceBlaus.

follicle receptors are a third type, consisting of a *root hair plexus* (nerve endings wrapped around the base of a hair) that respond to bending. More deeply embedded in the skin are *Ruffini's corpuscles* and *Pacinian corpuscles*, responding to skin stretching and vibration, respectively. Each receptor is thus a sensory cell having a specialized function.

Little is known about the evolution of hairless touch receptors, with the exception of Meissner's corpuscles. Found in both marsupials and primates (Hoffman, Montag, & Dominy, 2004), they presumably originated at least 170 million years ago in the common ancestor of both (Drummond et al., 2006). In primates they are related to frugivory, with more frugivorous species having a higher density of Meissner's corpuscles, perhaps to better judge the texture of fruit (Hoffman et al., 2004).

Regardless of type, a skin receptor is stimulated when mechanical pressure impinges on it either through the skin or by hair movement. That stretches its cell membrane, resulting in the opening of ion channels that permit sodium ions (Na+) to enter. As a result, a net positive charge is created inside the cell. With sufficient stimulation, the increase in positive charge, called a *generator potential*, increases past a threshold value at which point other Na+ channels open. Even more ions flood into the cell, creating a *spike potential* (a sudden increase in positivity), which travels up an *axon* (a long thread-like projection) on the other end of the cell.

Evolution produced this Na+ gating mechanism by building on the one present in single-cell organisms. Thus, a method initially used to detect and admit to the cell ions critical to metabolism became an important link in signaling touch.

Tactile neural pathways

In humans, somatosensory cell axons first synapse on neurons in the *cutaneous nerves*, which enter the spinal cord. From there they synapse onto neurons of two different systems, called the *medial lemniscal* and *extralemniscal* systems. Axons of the medial lemniscal system run upward to the medulla of the brain, where there is a synapse onto another set of neurons. These cross over the body's midline, travel up through the thalamus, and end in the primary somatosensory cortex (designated SI), occupying nearly all the postcentral gyrus, the most forward part of the parietal lobe (Figure 6.3). Because of the crossover, the medial lemniscal system is *contralateral* (opposite-sided), with the left side of the body represented in the right hemisphere of the brain and vice versa.

In contrast, the extralemniscal system is largely unlateralized. Some of the axons in this system cross over, but others do not, so that each side of the body is represented in each hemisphere. The final connections of the extralemniscal pathway are to the parietal

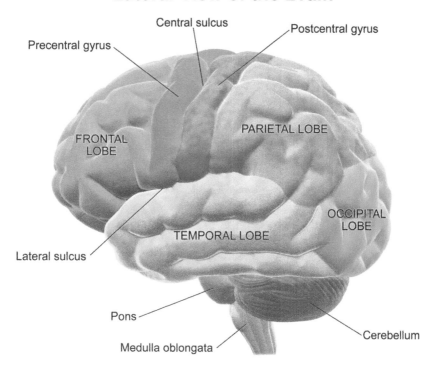

Lateral View of the Brain

Figure 6.3 The lobes of the brain, and the location of the postcentral gyrus.

Source: Blausen 0101 Brain LateralView.png, CC BY-SA 3.0, BruceBlaus.

lobe in an area designated SII (located in the ceiling of the lateral sulcus adjacent to the postcentral gyrus; see Figure 6.3), as well as to the rear of the parietal lobe and the frontal lobe.

It is known that SI and SII exist across the mammalian phyla, including marsupials (Kaas, 2004a). Thus, like Meissner's corpuscles, their origin dates back at least 170 million years. Their size follows a "principle of proper mass" enunciated by Harry Jerison (1973, p. 8). This states that there is a relationship between the neural mass devoted to a function, and the amount of processing involved. Accordingly, the most sensitive parts of the human body, such as the lips and hands, have the greatest cortical representation. In contrast, naked mole-rats, which live underground in the dark and depend heavily on somatosensation, have greatly expanded cortical areas representing touch to their incisors and head. Overall, SI is 50% larger in naked mole-rats than in laboratory rats. It is also enlarged in other somatosensory-dependent species such as star-nosed moles and nocturnal owl monkeys (Catania & Henry, 2006).

Balance

Like touch, the *vestibular* sense (balance) also involves hair cells. Some are located at the base of each of three semicircular canals (ducts) in the inner ear, in structures called ampullae (Figure 6.4). The canals are filled with fluid and are set nearly (but not quite) at right angles to one another. Because of the slightly off-kilter angles, any rotational acceleration results in fluid movement in at least two of the canals. That causes hair bending and thus sensing in at least two planes, enabling the precise neural computation of acceleration information. Other hair cells are located in fluid-filled cavities, called the *utricle* and *saccule*, and respond to vertical and horizontal acceleration (Lundberg, Xu, Thiessen, & Kramer, 2015).

The base of the hair cells synapse onto neurons, which coalesce into the vestibular nerve, in turn connecting to nuclei in the brainstem (Cronin, Arshad, & Seemungal, 2017). This pathway produces the vestibular reflex that allows us to keep our eyes fixated on an object while our heads move. Another pathway terminates, like somatosensation, in SI of the parietal lobe (D. Schneider, L. Schneider, Claussen, & Kolchev, 2001).

It was once believed that vestibular hair cells originated in the *lateral line system* of fish. This system uses hair cells arranged in a line from gills to tail, detecting water currents, the presence of prey, and possibly the location of sounds. However, the lateral line system and the inner ear differ in form, development, and function. Therefore, the current view is that both evolved in parallel from loosely arrayed somatosensory hair cells, rather than one from the other (Duncan & Fritzsch, 2012).

The earliest-emerging taxa with an identifiable inner ear were the hagfish and lampreys, placing its origin about 520 million years ago. However, those cartilaginous fish have only one and two semicircular canals, respectively. The third canal emerged about 430 million years ago with the evolution of bony fishes (Beisel, Wang-Lundberg, Maklad, & Fritzsch, 2005).

A striking aspect of the vestibular apparatus is its extraordinarily restricted size range across mammalian species. The size ratio of the smallest to largest is less than 1:10, in contrast to a ratio of possible body lengths of nearly 1:1000. Possibly its size is constrained by physical properties such as the inertia of its fluids and the range of hair displacement in the receptors. When these and other properties are mathematically modeled, a narrow range of sizes is indeed possible, and the human apparatus is neatly positioned in midrange (Squires, 2004).

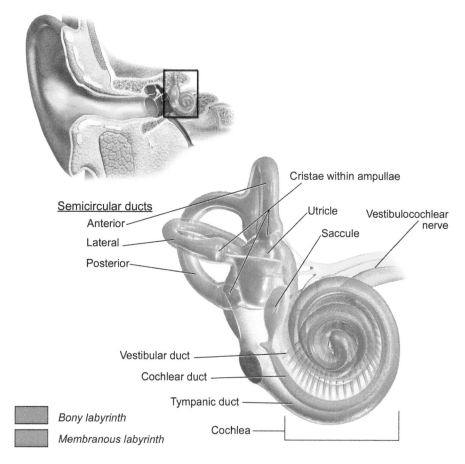

Figure 6.4 Structures of the inner ear, including the semicircular ducts (canals), utricle, and saccule.
Source: Blausen 0329 EarAnatomy InternalEar.png, CC BY-SA 3.0, BruceBlaus.

Nevertheless, some size variation is present across species, although its meaning is unclear. We might expect highly active species like gibbons to have large semicircular canals to provide more precise acceleration information. However, there is no consistent relationship between activity/agility and canal size (Kemp & Kirk, 2014; Malinzak, Kay, & Hullar, 2012). Larger canals may instead benefit species with more acute vision, in order to better stabilize vision during movement via the vestibular reflex (Kemp & Kirk, 2014).

In agile species, large canals may be less useful than canals oriented 90° to one another. Across mammalian species, the more the canals are at right angles to each other, the greater the estimated sensitivity to angular acceleration (Berlin, Kirk, & Rowe, 2013). Even within a specific group of primates, i.e., strepsirrhines, actual measurements of the speed of head rotation in a species is correlated to the perpendicularity of its semicircular canals (Malinzak et al., 2012). The implication is that evolution has reoriented the canals to be more perpendicular to one another, in species emphasizing speed and agility. However, the limited species examined so far do not permit a phylogenetic analysis across primates, let alone animals more generally.

Hearing

Hearing, or the *auditory* sense, likewise depends on hair cells to detect mechanical disturbances, in this case in the air or water medium surrounding the organism. Sound propagates pressure waves through the medium, and their amplification and detection result in hearing.

However, even at the basic mechanical level, auditory processes are complex. A sound entering the ear is transformed by the physical structure of the *pinna* (outer ear), enhancing middle frequencies in the range of 2000–5000 Hz. It then impinges on and causes vibration of the eardrum (tympanic membrane), which transmits the vibration through three connected *ossicles* (small bones in the middle ear) to the oval window of the inner ear's *cochlea* (Figure 6.5).

The sound causes the *basilar membrane* inside the cochlea to vibrate, and hair cells located on the membrane detect the vibration (Varin & Petrov, 2009). Thus, even before the first neural response to sound, a series of complex mechanical events takes place.

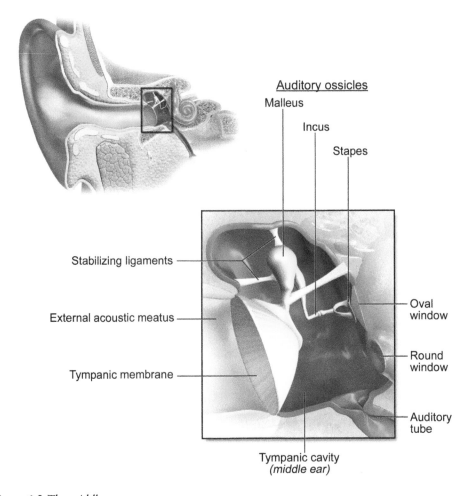

Figure 6.5 The middle ear.

Source: Blausen 0330 EarAnatomy MiddleEar.png, CC BY-SA 3.0, BruceBlaus.

As we have seen, hair-based mechanoreceptors are very ancient. As with the sense of balance, auditory hair cells of the inner ear likely evolved from loosely arrayed receptors of that type. While nonvertebrate chordates have hair cells, they are not located in an ear but rather in the oral region. Thus the co-option of hair cells for hearing must have postdated the chordate emergence 535 million years ago. Probably it was long after, as will become apparent.

As the primary receptor surface, the basilar membrane is of central interest in evolutionary analyses. Its initial form was as *basilar papillae* (a papilla being a rounded projection), one in each ear, each containing a few hair cells. Coelacanth fish have them, as presumably did their ancestors about 395 million years ago as a vestibular system modification (Fritzsch, 1987). However, the lungfish lineage, not the coelacanth lineage, is believed ancestral to tetrapods, including humans – and lungfish lack basilar papillae. Why do tetrapods share them with their more distant coelacanth relations, but not with their closer lungfish cousins?

The likely answer is that basilar papillae were inherited from the last common ancestor of all three groups but were subsequently lost in lungfish. It has been suggested that the common ancestor was a lobe-finned fish whose papillae were used to assess pressure differences between its primitive lungs and its membrane-covered middle ears, allowing the monitoring of depth (Fritzsch et al., 2013). The shallows-dwelling lungfish presumably had little need for depth information and so eventually lost the papillae, but not before giving rise to tetrapod descendants who incorporated them into a more elaborate hearing mechanism. In tetrapods, a few hair cells became many hair cells. Ultimately, the "carpet" of hair cells became continuous over a lengthened basilar membrane, achieving the mammalian form that humans have today.

Another evolutionary step coiled the basilar membrane inside a cochlea, packing a maximum of membrane surface into minimum space (Figure 6.4). Comparative analyses indicate that full coiling evolved after monotremes emerged about 195 million years ago, but before marsupial and placental mammals diverged about 170 million years ago (Manley, 2000; Vater, Meng, & Fox, 2004). Nevertheless, the earliest known fossil showing coiling dates back only about 80 million years (Vater et al., 2004).

Frequency selectivity

With these modifications came frequency selectivity, the ability to distinguish between sounds of differing pitch. Turtles and lizards have little, and birds an intermediate amount, but mammals including humans have substantial selectivity (Manley, 2000). In mammals, selectivity is mostly achieved through organization of the basilar membrane. Hair cells at its base (nearer to the bones of the middle ear) detect high frequencies, while those near the membrane's apex (on the opposite end) detect low frequencies (Breedlove, Watson, & Rosenzweig, 2010). Frequencies can then be distinguished based on the location at which hair cells are stimulated.

Lizards lack basilar membrane selectivity but have evolved frequency-related modifications of the hair cells themselves so that particular cells are more responsive to certain frequencies than others. Birds also use hair cell modifications, as even mammals do to supplement the use of location on the basilar membrane (Manley, 2000).

In humans, the modification of vestibular structures for auditory function has been accompanied by a substantial shift in the ratio of hair cells to connecting neurons. Vestibular hair cells show an approximately 5:1 ratio to neurons, indicating some funneling of information, but auditory hair cells show a remarkable 1:8 ratio – in effect a reverse funneling or elaboration of processing at the first neural juncture. The elaboration is particularly pronounced in the case of high-frequency hair cells toward the base of the basilar membrane (Fritzsch et al., 2002). Indeed, "high end" expansion has occurred over some 350 million years. Turtles, which are reptiles, may be limited to hearing frequencies under 1000 Hz, while birds are capable of detecting frequencies up to about 10,000 Hz. Humans, at least when young, can hear up to about 20,000 Hz, and some other mammals can do so up to 100,000 Hz! (Manley, 2000).

Within primates there is evidence that high-frequency sound detection is related to the amount of social calling. Species living in larger groups, and therefore employing a wider variety of social calls, have greater sound sensitivity overall but especially greater high-frequency sensitivity. This exploits the fact that in the natural environment, background noise tends to be pitched low. Thus having social calls pitched at high frequencies makes them easier to distinguish from the background (Ramsier, Cunningham, Finneran, & Dominy, 2012).

For many species, another advantage of a higher frequency range is improved sound localization (Masterton, Heffner, & Ravizza, 1969). Blockage of sound by the head is an important location cue, and high frequencies (short wavelengths) experience much greater blockage than low frequencies (long wavelengths). The greater blockage of high frequencies thus allows a computation of location using differences between the ears in perceived frequency (Breedlove et al., 2010). The same differential blockage is true of other obstacles as well. That is why we are usually more annoyed by a neighbor's bass, which tends to pass through walls, than by their treble, which doesn't.

However, if an animal's head is large, there is a more powerful cue than differential blockage, based on timing. Sound reaches the near ear earlier than the far ear, and the larger the head, the larger the timing difference. Timing differences tend to be less affected than frequency differences by other objects in the environment, posing an advantage for animals large enough to use them. Thus we should look to large mammals to have a reduced "high end", because they have less need for high-frequency hearing. That is certainly true of elephants, which top out at only 11,000 Hz. It is also true of humans, topping out at 20,000 Hz. Other primates, which tend to be smaller, typically have higher "high ends" than humans, with nearly all being able to hear over 32,000 Hz. Within land mammals generally (Heffner, 2004), and primates specifically (Ramsier et al., 2012), evolution appears to have optimized frequency response to match body size. The range shifts toward higher frequencies in small-bodied species, and toward lower frequencies in large-bodied species.

Although humans have reduced high-frequency hearing compared to our fellow primates, we do have other unusual auditory abilities. Specifically, we have lower "low ends" than most primates, being able to routinely detect frequencies close to 30 Hz, or even 20 Hz under ideal conditions. At our most sensitive frequency, we can also hear fainter sounds than most (Coleman & Ross, 2004; Heffner, 2004).

Origin of the ossicles

Finally, a remarkable evolutionary story concerns the ossicles, the three small bones that transmit sound through the middle ear. Amphibians, reptiles, and birds have only one ossicle. Some 385 million years ago, it and the cavity in which it rested were part of the tetrapod *Panderichthys*' breathing apparatus, serviced by breathing holes toward the rear of the skull. But in *Acanthostega* 20 million years later, these structures had evolved into a middle ear with true auditory function (Brazeau & Ahlberg, 2006). Yet with only one ossicle, the "high end" of hearing was limited.

Mammals, on the other hand, have three ossicles. Where did the other two come from? Astoundingly, instead of evolving anew, they were adapted from bones that detached from the lower jaw, moved away, and reduced in size. As a result, mammals have only one bone in the lower jaw while other groups have several.

The evolution of the three-ossicle assemblage improved the detection of high frequencies by matching air's low *impedance* (the amount of pressure required to produce a given displacement) to the high impedance of fluid in the inner ear (Manley, 2000; Masterton et al., 1969). This occurred after mammals appeared, sometime after 225 million years ago, and is first known from 195-million-year-old fossils of the tiny mammal *Hadrocodium* (Luo, 2001). At some point an eardrum evolved, substantially increasing sound sensitivity. That development occurred independently, through convergent evolution, in several groups of amniotes including lepidosaurs, archosaurs, and mammals (Manley & Clack, 2004).

From this discussion, it is clear that hearing was a relatively late arrival at the sensory party. It probably emerged in its most primitive form shortly before basilar papillae were inherited by coelacanths about 395 million years ago. Subsequently, it further evolved in land-based tetrapods for reasons not the least bit mysterious: Simply put, the thin medium of air constituted a selective pressure for improved vibration detection. Evolution's answer was the fully developed ear, with an external part capturing and amplifying airborne sound, a middle portion transmitting it while matching its impedance to a fluid medium, and an inner portion containing hair cells to transduce the energy into a form detectable by the nervous system. Its highest expression is in mammals, which have used hair cell modifications, a greatly expanded basilar membrane surface, and an elaboration of neural processing to extend frequency perception upward.

Smell

As with mechanical sensing, chemical sensing began with simple cellular transport mechanisms. Yet today, we enjoy a complex ability to smell and to taste. While the *olfactory* (smell) and *gustatory* (taste) sensory systems are closely related in function, more is currently known about the evolution of the former.

In humans, olfaction begins when a molecule of an odorant binds to the membrane of a cell in the *olfactory epithelium*, a layer of neurons lining the upper portion of the nasal passageway. When binding occurs, a complex chain of reactions is triggered that opens some of the Na+ channels in the cell membrane. The same scenario plays out as with somatosensation, with a spike potential propagating throughout the cell and into the axon on its far end.

Taken together, the axons from millions of receptor neurons make up the olfactory nerve, which travels a very short distance to an olfactory bulb of the brain (Figure 6.6). There is one bulb on the left and another on the right of the brain's midline, with an olfactory nerve connecting to each in *ipsilateral* (same-sided) fashion. This is the only sensory system which does not predominantly cross over with each side of the body linked to the opposite side of the brain.

From the olfactory bulbs, the olfactory tract courses backward in the brain to connect with cortex on the *medial* (inner) surface of the temporal lobes. Pathways then run through the midbrain to connect with the *prefontal cortex*, the forward portion of the frontal lobes (Figure 6.3). Together, these comprise the brain's olfactory sensory system.

During prenatal development, the olfactory epithelium develops from a *placode* (a specialized cell cluster), visible as a thickening of the *ectoderm* (the cells giving rise to external features like the skin, nails, and hair, and counterintuitively, the brain and nervous system). Placodes exist for other senses as well, developing into structures of the eye and ear among others (Steventon, Mayor, & Streit, 2016).

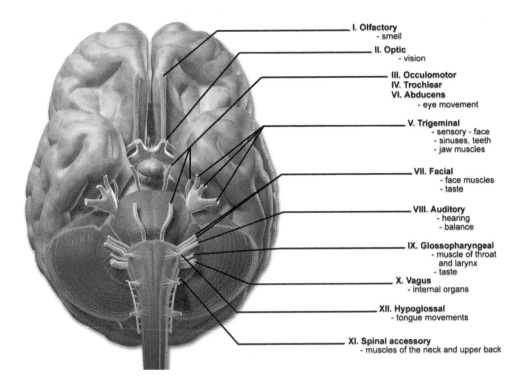

Figure 6.6 The olfactory bulbs (upper part of I), and olfactory tract (lower part of I) on the underneath side of the human brain. Also shown is the optic nerve (II) among other numbered cranial nerves.

Source: Blausen 0284 CranialNerves.png, CC BY-SA 3.0, BruceBlaus.

Comparing species suggests that the olfactory placode originated shortly after the emergence of chordates, but before the emergence of vertebrates. Thus, while it exists across the full spectrum of vertebrates, its existence in chordates depends on the species. Amphioxus, considered representative of early ancestral chordates, shows either no evidence at all for placodes (Schlosser, 2017), or the possibility of tiny ones consisting of only one or two cells (Lacalli, 2004). Furthermore, evidence for either smell or taste in amphioxus is ambiguous at best (Lacalli, 2004). Yet descendants of early-emerging vertebrate lineages such as lampreys do have identifiable olfactory bulbs (Nieuwenhuys, 2002). *Haikouella*, a transitional form between early chordates and vertebrates, may have had nostrils and olfactory nerves 530 million years ago, but the evidence is ambiguous (Mallatt & Chen, 2003). In any case, it seems likely that the neural system underlying olfaction traces its origin to a point after the splitting off of amphioxus's ancestors, but before the splitting off of those of lampreys, perhaps 530 million years ago.

Olfactory genes

Research on the genes underlying olfaction has yielded new insight into the evolutionary rise, and in our case fall, of this chemical sense. A few olfactory genes trace back to the early ancestors of chordates (Niimura, 2012), just before evolution of the olfactory placode.

The evolution of tetrapods was accompanied by a tremendous expansion of olfactory genes, so that amphibians and terrestrial vertebrates now have several hundreds to thousands of these. The expansion greatly increased the receptors sensitive to airborne odors. In contrast, coelacanth fish lack airborne odor receptors entirely, and in some whales and seals whose ancestors shifted from land to sea, their numbers are dramatically reduced. Thus there is good reason to believe that the transition to land some 370 million years ago, and the availability of airborne odors, provided the impetus for olfactory expansion (Bear, Lassance, Hoekstra, & Datta, 2016).

Today's terrestrial mammals have an average of about 1500 olfactory receptor genes. They are believed to be directly related to the development of specialized odor detectors in the olfactory epithelium, so that the more genes there are, the larger the range of detectable odors. Actual behavioral data demonstrating such a relationship are quite limited, however (Kay, Campbell, Rossie, Colbert, & Rowe, 2004).

A mammalian expansion in olfactory sensitivity is also observable in the form of larger olfactory bulbs. Larger bulbs service larger numbers of receptors (Heritage, 2014). Because the bulbs sometimes make impressions on the inside of fossil skulls, their volume can be measured in extinct species. It is found that a three-stage increase occurred during the evolution of mammals, all within a 30-million-year span beginning about 225 million years ago (Rowe, Macrini, & Luo, 2011). Thus there has been a strong, general mammalian trend toward evolving larger olfactory bulbs.

It is all the more striking, then, that the more recent primate trend has been the exact opposite, toward olfactory bulb reduction. Controlling for overall brain size, the olfactory bulbs of strepsirrhines are sized about the same as those of close relatives like tree shrews. But from there on down the line, it was all reduction: The olfactory bulbs got smaller and smaller, relative to brain size, from haplorhines (splitting off 73 million years ago) to anthropoids (72 million years ago) to catarrhines (45 million years ago) to Hominoidea (32 million years ago; Heritage, 2014).

The formation of olfactory pseudogenes

The downgrading of the olfactory system has continued to the present day in modern humans. Of our olfactory receptor genes, 50–60% are nonfunctional *pseudogenes*, i.e., genes that have become deactivated because of disruptive mutations (Gilad, Man, & Glusman, 2005; Go & Niimura, 2008). Deactivation appears to have accelerated within the last four million years of human evolution (Gilad et al., 2005). In contrast, only about 20% of the olfactory receptor genes in mice and dogs are pseudogenes.

What can account for the loss of human olfactory genes? Gilad and colleagues suggested that it is related to increased color vision. They found that pseudogene formation has been over 10% greater in nonhuman apes and Old World monkeys than in most New World monkeys. However, howler monkeys, a New World species, share the higher rate of pseudogene formation. The significance is that among primates, only the apes, Old World monkeys, and howler monkeys have true *trichromatic* (three-color) vision. Other New World monkeys do not (Melin, Khetpal, et al., 2017).

An association between olfactory pseudogene formation and increased color vision does not prove a causal connection, but it is suggestive. If it is causal, it could be because improved color vision reduced selection pressure for odor recognition. For example, if ancestral Old World primates were able to detect ripe fruit more accurately or at a greater distance using trichromatic vision in place of the sense of smell, that could have rendered some olfactory receptors unnecessary and led to their loss by the transformation of genes to pseudogenes.

While that explanation can account for increased pseudogene formation in frugivorous Old World monkeys, it is unclear how it might account for it in non-frugivorous howler monkeys. An alternative hypothesis, based on leaves rather than fruit, was advanced by Dominy and Lucas (2001; Dominy, 2004). They proposed that increased color vision allowed perception of subtle differences between high-protein young leaves and less nutritious older ones. As the most folivorous of New World monkeys, howler monkeys would have found that a useful ability, as would many Old World primates during times of fruit shortage. Indeed, howler monkeys have been found to feed predominantly on young leaves that are distinguishable from mature leaves on the basis of greater red coloration (Melin, Khetpal, et al., 2017). Thus in both Old World primates and howler monkeys, the evolution of trichromacy presumably allowed the deactivation of many olfactory receptor genes.

But, then, what accounts for the accelerated deactivation in hominins, who were neither folivorous nor heavily frugivorous? Gilad and colleagues suggested that human pseudogene formation was less connected to trichromacy than to the cooking of food, which deactivates many toxins. The fewer toxins, the fewer olfactory receptors that were needed, and so olfactory receptor genes were converted to pseudogenes (Gilad, Bustamante, Lancet, & Pääbo, 2003).

The cooking hypothesis might account for recent pseudogene formation in hominins. However, it cannot explain an acceleration in pseudogene formation as far back as four million years ago, for there is no evidence of the use of fire at that time. A simpler explanation may be that in hominins of that period, increased raw meat consumption was linked to a reduced dietary reliance on plants and fungi. That also would have reduced the range of toxins requiring detection, in turn allowing deactivation of the corresponding olfactory receptor genes. However, after turning to gustation, we will see that toxin-based explanations run into another problem that is not so easily solved.

Taste

Gustation, or the sense of taste, is a second chemical sense. The receptors are in the form of cells clustered inside taste buds, in turn located in papillae on the tongue's surface. The tastes of sweet, sour, and *umami* (meaty or savory) are detected much as odors are detected, with complex molecular binding on the membranes of the cells. Salty and sour tastes, however, involve simpler mechanisms of ion transport through receptor membranes, with the sodium ($Na+$) ions of salt directly transported and with the hydrogen ($H+$) ions of sour acids blocking the transport of potassium ($K+$) ions.

A receptor cell responds to taste by releasing a neurotransmitter that crosses a synapse and activates a neuron in one of three *cranial nerves* (sensory and motor nerves completely located in the head), specifically those designated VII, IX, and X (Figure 6.6). These nerves run to the brainstem, forming a tract that, after a couple of synapses, reaches the somatosensory cortex in the postcentral gyrus of the parietal lobe (Figure 6.3).

Comparative analyses indicate that all vertebrates including lampreys, but not hagfish, have taste buds, and that even hagfish have chemically sensitive cells in the mouth area that are sensitive to tastes (Kirino, Parnes, Hansen, Kiyohara, & Finger, 2013). Thus the gustatory system must date from about 520 million years ago with the emergence of ancestral hagfish.

Much of the recent research on gustation's evolution has focused on bitter tastes, in part because bitterness characterizes many toxins, and in part because of the discovery of a family of bitter taste receptor genes designated T2R. The common ancestor of all vertebrates is reconstructed as having fewer than 10 T2R genes. Today, the number varies widely over species (Li & Zhang, 2013). Mammals have many more, between 19 and 41 in the species that have been studied (humans, mice, dogs, cattle, and opossums), while zebrafish, pufferfish, and chickens have only three or four (Go, 2006).

As in the case of olfaction, there has also been notable transformation to pseudogenes in some lineages. Although under 10% have been transformed in nonmammalian species, 20–40% have been in mammals. Humans occupy the midpoint of the mammalian range (Go, 2006).

An average level of gustatory pseudogene formation in humans raises questions about toxin-based explanations, both in gustation and olfaction. If either cooking or raw meat-eating reduced the toxicity of food, for instance, and that led to olfactory pseudogene formation, there should have been at least as much impetus to convert gustatory genes to pseudogenes. Thus the failure to find a higher level of gustatory pseudogenes in humans than in other mammals casts doubt on the toxin explanation. Unfortunately, it is not yet clear what alternative can take its place.

Complicating the issue is that when assessed widely across mammalian species, there *is* a positive correlation between the amount of vegetable matter in the diet and the number of T2R bitter taste genes (Li & Zhang, 2013). This is exactly what would be expected of a toxin-based explanation of pseudogene formation, because carnivores wouldn't need sensitivity to plant toxins and thus should have fewer functional T2R genes. Herbivores should have a greatly increased need and thus more functional genes. However, the size of the correlation is quite modest, $r = +0.43$, indicating that other unknown factors are in play. Perhaps if they were known, they would account for the failure to find increased gustatory pseudogene formation in humans relative to other mammals.

Common trajectories in mechanical and chemical sensing

The near-simultaneous emergence of the vestibular, olfactory, and gustatory systems 520–530 million years ago is surely significant. One of the constraints preventing their earlier appearance was undoubtedly the lack of either a neural net or nervous system prior to the evolution of cnidarians about 605 million years ago. Nevertheless, the chronological gap of about 80 million years between the emergence of the neural net and of sensory systems suggests that an additional constraint was at work, one requiring a long time for evolution to work around.

What could that constraint have been? A good guess is that it was the limited neural architecture permitted by early sensory cells. Specifically, the exclusive use of *primary sensory cells*, cells with sensors at one end and an axon out the other end, limited the kinds of neural connections that could be made because the axon prevented the processing of sensory information prior to its terminus in the central nervous system (CNS). In contrast, the later-evolving *secondary sensory cells*, in which receptors have a sensory end and a broad synaptic surface on the other end, but no axon (Lacalli, 2004), allowed receptors to form any number of more immediate connections with other neurons. Such secondary sensory cells in the form of hair cells emerged with the urochordates and are present in amphioxus, suggesting an origin of somewhat over 535 million years ago.

Although that narrows the gap a bit, it still follows the emergence of neural nets by about 70 million years. Perhaps such a long span of time was required because the evolution and exploitation of secondary sensory cells involved three interacting components: (a) the loss of sensory cell axons in at least some receptor types, (b) the substitution of connecting neurons for those axons, and (c) the organization of the connecting neurons as a system of inputs into the CNS. A major restructuring may have taken a major interval of time. However, once it was in place, it was then exploited more or less simultaneously by the emerging vestibular, olfactory, and gustatory sensory systems.

The vertebrate plan differed from the invertebrate one in important ways that highlight the extent of the restructuring. Vertebrates came to rely on a large number of relatively undifferentiated neurons, forming ganglia with the cell bodies inside and axons outside, leading to a CNS organized inside the skull and vertebrae. In contrast, invertebrates had relied on a small number of complex neurons, forming ganglia with the cell bodies outside and axons inside, and a CNS organized around the digestive tract (Rosenzweig, Breedlove, & Watson, 2005). Thus vertebrates achieve sensation by adopting a generalized nervous system structure in which interchangeable sensors plug into a complex network of simple cells, arranged hierarchically, leading to a central processor. Previously, invertebrates had achieved sensation by elaborating a few complex neurons using localized processing.

There is a general understanding that the vertebrate CNS must have evolved from the simple neural nets possessed by cnidarians, although the intermediate steps remain murky. This may be because the transitional forms are extinct and their fossils have not been recognized. In any case, given the scope of the restructuring necessary to overcome limited invertebrate architecture, the wonder may be that it took "only" 70–80 million years to create the vertebrate computational apparatus.

The other mechanical and chemical senses

Touch, balance, and hearing are not the only mechanical senses. We also use mechanical sensing to assess joint position and muscle tension, the *kinesthetic* sense. However, there appears to be little understanding of its evolution.

Similarly, smell and taste are not the only chemical senses. Receptors in our arteries engage in chemosensation through sensitivity to changing oxygen and carbon dioxide levels. The *vomeronasal* system senses pheromones using receptors in the nasal passages. The system is vestigial in Old World monkeys and apes, to the point that humans lose their vomeronasal nerves during fetal development, but it is functional in strepsirrhines, tarsiers, and New World monkeys. This suggests that the system was largely lost in a common ancestor of Old World monkeys and apes, in round numbers about 30 million years ago. In species with vestigial vomeronasal systems, the olfactory system has taken over the role of pheromone detection (Smith, Rossie, & Bhatnagar, 2007).

Conclusion

Our senses have their roots in simple membrane gating mechanisms that relieved excess pressure and allowed needed nutrients to enter the cell and waste products to leave. These mechanisms must have existed since the dawn of cells close to four billion years ago. Today, bacteria are capable of sophisticated chemical signaling techniques in order to attack a host or establish biofilms.

Hair-based mechanoreceptors may have appeared 1.0–1.5 billion years ago in choanoflagellates, later becoming incorporated into cnidarians as stinging cells. Neurons possibly evolved from such cells, allowing more coordinated attacks and avoidance of the creature's own stings. But however they emerged, neural nets and their support of mechanical sensing originated a little over 605 million years ago. Hair cells that synapsed onto neurons are believed to have appeared in early chordates.

Humans have a number of skin-based mechanoreceptors as part of the somatosensory system. It has two divisions, the medial lemniscal system and the extralemniscal system, with the former crossing over to the opposite side of the brain and the latter only partially doing so. Collectively, they project to cortical areas SI and SII, among others. The size of the cortical areas serving different body parts varies with the sensitivity of the part.

About 520 million years ago, the vestibular sense appeared in the form of an inner ear, building on the existence of mechanoreceptors and improving balance and visual tracking during body movement. Three semicircular canals set nearly at right angles provide our sense of balance. Two of these were present in ancestral lampreys, but the third only emerged 430 million years ago in bony fishes.

Hearing also emerged as an inner ear faculty, shortly before 395 million years ago. Basilar papillae, each with a few hair cells, exist in coelacanths. Tetrapods also have them, but not lungfish, so it appears that while a mutual ancestor of the three groups had them, they were later lost in lungfish. Eventually, in tetrapods, a carpet of hair cells became continuous over a lengthened basilar membrane. Full coiling of the membrane inside a cochlea occurred about 180 million years ago. Humans largely distinguish sounds of differing frequency by the locations stimulated on the basilar membrane.

Sounds are localized differently by small-headed and large-headed primate species. Small-headed species use the blockage of high-frequency sounds by the head, allowing a computation of location by comparing the frequencies perceived by the two ears. Large-headed species, in contrast, take advantage of a more powerful cue, which is the difference in the time that a sound arrives at the ears. This is believed to account for the observation that in general, small-headed animals can hear higher frequencies than large-headed mammals.

Early tetrapods had only one of the ossicles now found in the mammalian middle ear. The other two were adapted from bones in the lower jaw. The three together improved the detection of high frequencies by matching the impedance of air to that of fluid in the inner ear, and they were functioning together in that way by 195 million years ago.

The neural systems involved in chemoreception seem to have evolved about 520–530 million years ago after the emergence of the chordates but before that of the hagfish and lampreys. This is true both of olfaction and gustation. The number of olfactory genes increased rapidly after the emergence of tetrapods, who needed to be sensitive to airborne odors. Mammalian olfactory bulbs also expanded, but in primates more specifically, the olfactory bulbs progressively shrank. Also in humans, many chemoreception genes have transformed into nonfunctional pseudogenes. Probably this was at least partially diet-related, although whether it was in compensation for greater reliance on visual food cues or a decreased need to detect toxins is subject to debate. In humans, pseudogene formation in gustation has not been nearly as pronounced as that in olfaction, raising a problem for explanations based on toxins.

The near-simultaneous emergence of the vestibular, olfactory, and gustatory systems 520–530 million years ago is surely significant. It was probably due to nervous system changes allowing unspecialized neurons to connect to synaptic surfaces on sensory receptors, as opposed to combining neuronal and sensory function in the same cell. Such hierarchical structure became characteristic of vertebrates.

7 Vision

One of our most important senses does not depend on either mechanical or chemical stimulation. Although *vision* involves a chemical reaction, the chemistry involved is internal to the cell, and stimulation involves light-based rather than mechanical energy transfer. In this chapter we will first consider how vision began and led to the evolution of eyes. We will examine how day vision came to sense colors, and how it gave rise to highly sensitive night vision. We will also see how visual acuity increased in the human ancestral line, producing improvement in vision that went hand in hand with changes in the brain.

Vision began with single-cell organisms that were neither animal nor plant. Among single-cell organisms, *photomotility* (movement in response to light) is widespread and involves light-absorbing pigments from several chemical classes. These *photopigments* are embedded in the cell membrane, in cytoplasm, or even at the base of a flagellum, as in the single-celled *Euglena*. In *Euglena*, rotational movement of the cell causes the light-sensitive pigment to be shaded by a nearby red eyespot, allowing the direction of light to be determined. This alters the flagellum's activity, keeping the organism in an area where light levels are suitable for photosynthesis (Ogawa et al., 2016).

The evolution of eyes

Metazoan vision involves much greater complexity than that of single-cell organisms. Initially it may have evolved from a symbiotic relationship, with unicellular organisms providing primitive vision to multicellular ones. Presumably this was in exchange for protection or nutrition, much like the relationship leading to the incorporation of mitochondria into eukaryote cells (Zachar, Szilágyi, Számado, & Szathmáry, 2018). An observation supporting this is that some single-cell *dinoflagellates* (ocean-dwelling unicellular organisms with two flagella) possess primitive vision and are common symbionts in cnidarians. Thus, dinoflagellate genes may have transferred to some ancient cnidarian, making metazoan vision permanent (Gehring, 2005).

Whether or not that supposition is true, a watershed in the metazoan acquisition of vision must have been the evolution of multicellular eyes feeding into a nerve net or nervous system. The first eyes of this type may have appeared among cnidarians, some of which respond to light by combining pigment cells with neurons. The *cubozoans*, or box jellyfish, have fully developed eyes. In fact, each individual has 24 eyes of various types, including complex lensed eyes arranged in pairs of small and large members, complete with corneas and retinas. These are connected by a nerve ring with ganglia (Garm, Oskarsson, & Nilsson, 2011).

The purpose of this impressive sensory array is not entirely clear, but it is known that box jellyfish avoid obstacles and orient toward light. Unlike most jellyfish, they live near the shore, often in or near mangrove swamps, coral reefs, and beaches. Thus, one major function of the eye array is to avoid potential obstacles that could abrade or destroy delicate tissues (Garm et al., 2011).

Until recently, a debate raged over whether or not all metazoan eyes descend from a common origin. One argument in favor of multiple origins (i.e., convergent evolution) is that eye designs differ radically across species. There are compound eyes with many lenses, like those of house flies, but there are also homogeneous lens eyes, lensed eyes, mirror eyes, multiple lens eyes, pinhole eyes, pit eyes, refractive index lens eyes, and two-lens eyes (Fernald, 2000). That suggests that eyes emerged independently many times in different lineages, with different designs emerging in response to different needs. For example the Cambrian species *Anomalocaris* – a name meaning "strange shrimp", and a huge one – some 1–2 meters (3–6 feet) in length – had large compound eyes well suited to a highly mobile predator operating in well-lit conditions and equipped with a visual brain (Paterson, García-Bellido, Lee, Brock, Jago, & Edgecombe, 2011). But judging from living examples, scallops, also originating in the Cambrian, had mirror eyes that served merely as shadow detectors. These were appropriate to a prey species needing to escape from predators but possessing little neural architecture (Schwab, 2018). Other kinds of eyes are believed to have evolved in the human ancestral line, each again meeting a different need (Box 7.1).

Box 7.1   Intermediary eyes in the human ancestral line

The stages of eye evolution leading up to humans are believed to be understood to a first approximation (Figure 7.1), even though supportive evidence is incomplete. In each instance, an intermediary form improved vision compared to the form immediately preceding. This satisfied a key requirement – for evolution to proceed, each product had to leave descendants with a selective advantage over its immediate predecessor.

In the first stage, a small set of photoreceptors, or even a single receptor, appeared on the skin and was connected to a nerve net (Figure 7.1.a). Probably first appearing in cnidarians, the advantage of vision in this instance lay simply in signaling the presence or absence of light, a distinct advantage over complete blindness.

In the second stage, a depression or folding of the skin created a cavity (Figure 7.1.b). Because the sides of the cavity produced shading of those receptors not in the direct line of light, this form improved vision by indicating the direction of a light source, while still being able to signal the presence or absence of light. Eyes of this type are believed to have evolved in the immediate ancestors of bilateria (Arendt & Wittbrodt, 2001).

Further contraction of the cavity's opening, in the third stage, resulted in a pinhole eye (Figure 7.1.c). Due to optical physics, this created a sharper image, improving vision by allowing objects to be pictured for the first time. Of course, the direction of light and its presence or absence were still signaled. If you have corrected vision, try removing your contact lenses or glasses, and peer through a pinhole in a piece of foil to experience the improved, though dimmer, vision a pinhole provides.

Closure of the eye, so that it was covered by a transparent membrane (Figure 7.1.d), eliminated entry of water-borne debris, improving vision by

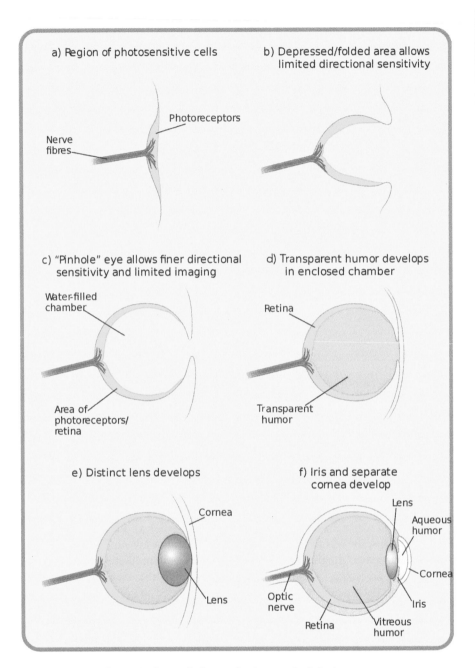

Figure 7.1 Intermediaries in the evolutionary background of the human eye.

Source: Diagram of eye evolution.svg, CC BY-SA 3.0, Matticus78.

minimizing the obstruction of light. Imaging, directionality, and presence/ absence information could still be extracted. That remained true when a distinct lens evolved (Figure 7.1.e). Initially this may have been little more than a water-filled sac, but whatever the structure, it allowed images to come to a better focus on the back of the eye. Simultaneously, the entry for the light pathway became larger, allowing more light to be gathered that was possible with a pinhole eye. Early vertebrates are believed to have had paired, lensed eyes (Bardack, 1991).

In the final form of the eye (Figure 7.1.f), in which a bulging cornea emerged along with an iris, the image was brought to a precise focus, and the amount of light entering the eye could be regulated. Pattern vision was much improved, and over a wider range of light levels. The cornea probably emerged first in early vertebrates as a protective covering. It provided little optical advantage while its possessors lived in water, but increasingly did so as the ancestors to tetrapods ventured onto land. Increasing stiffening of the cornea occurred later in mammals and continued in humans (Koudouna et al., 2018).

Thus, in a series of changes, over many millions of years, vision was nearly perfected in our ancestors. It would require other changes – especially the invention of external artificial lenses – to perfect it further.

Other indications of the multiple origins of eyes include differences in (a) the organization of photopigments in the receptors, (b) the receptors' ion transport channels, and (c) the development of the eye in embryos. On the other hand, arguing for a common origin is that all eyes use variations of vitamin A-based photopigments even though other light-sensitive pigments exist (Fernald, 2000).

The genetic basis of eyes

Fortunately, the acquisition of genetic data has brought greater clarity to the debate over common origins in vision. It has been found that a set of genes called Pax genes operates across vertebrate species to direct the development of eyes. One of these, Pax6, appears to be a "master gene" that controls eye development across its many varieties. Three others (Pax2, Pax5, and Pax8) appear in mammals, where they influence organ and nervous system development. Humans have an additional five Pax genes beyond these (Pichaud & Desplan, 2002).

Significantly given their early evolution, sponges and cnidarians have yet another member of this gene family, designated PaxB. It was likely the ancestral gene from which other Pax genes evolved (Kozmik, 2008). Its appearance can be viewed as the first of a number of important developments in the evolution of vision leading up to primates (Table 7.1).

These observations suggest that the basic control mechanism for eye development was present 665 million years ago in sponges. In that sense, eyes have a common origin. However, whether eyes actually evolved in a particular lineage depended on the evolution of other Pax genes from PaxB. Thus, there has been a substantial amount of convergent evolution even though all eyes have some genetic commonality.

Many sponges, which have only PaxB, sense light with rings of pigment-containing receptor cells (Rivera et al., 2012). Some are known to withdraw their *oscula* (mouth-like openings) when exposed to light, and their larvae swim directionally in response to light. They are most sensitive to violet light with a wavelength of 440 nanometers (nm), a response that is consistent with a class of photopigments called *flavins* or *cryptochromes* (called "crypto" because for many years their identity was unknown; Leys, Cronin, Degnan, & Marshall, 2002).

Table 7.1 Some important developments in the evolution of vision in the primate ancestral line. mya = millions of years ago.

Characteristic	Earliest example	Approximate date
Pax genes	sponges	by 665 mya
flavin photopigments	sponges?	by 665 mya
opsin photopigments	cnidarians	by 605 mya
cup eyes	bilaterians	by 580 mya
receptor hyperpolarization	chordates	535 mya
multiple cone types	chordates?	535 mya?
a single, optically reversing eye	chordates	530 mya
paired, lensed eyes	vertebrates	520 mya
rhodopsin (rods)	vertebrates	520 mya
reduction to two cone types	early mammals	180 mya?
large corneas	primates	73–83 mya
enlarged optic nerve	haplorhines	73 mya
well-developed fovea	anthropoids	72 mya
corneal reduction	anthropoids	70 mya?
recovery of three cone types	Old World monkeys	21–45 mya

Perhaps as our legacy from such ancient ancestors, flavin-based photopigment today underlies the human *circadian rhythm*. This rhythm governs our sensitivity to the 24-hour day, raising respiration, temperature, and activity levels during the day and lowering them at night. Light helps set this rhythm and, at least in mammals, does so through visual sensing by some of the retina's ganglion cells. Output from the ganglion cells then projects to the *suprachiasmatic nucleus*, which controls the circadian rhythm (Do & Yau, 2010).

However, in cnidarians, Pax genes are involved in the production of a different family of photopigments called *opsins*. Such genes have been found in cnidarians, but are absent in sponges (Plachetzki, Degnan, & Oakley, 2007). Consistent with this, peak visual sensitivity in the cnidarian genus *Hydra* falls within the range of wavelengths to which opsins respond. Specifically, hydra are most sensitive to blue-green light, at wavelengths between 480 and 530 nm (Leys et al., 2002).

Putting these observations together, it seems that sponges' photosensitivity is based on flavin-based photoreceptors, using a receptor chemistry that is repeated in our regulation of the circadian rhythm. Receptors based on opsins, which led evolutionarily to complex eyes, emerged later in cnidarians.

The bilateria appeared soon after cnidarians (see Figure 2.2), about 580 million years ago. These are animals showing bilateral symmetry, and they include humans. Arendt and Wittbrodt (2001) concluded that the immediate ancestors of bilateria probably had paired, pigmented cup eyes as larvae, each possibly as simple as a single photoreceptor containing a light-sensitive pigment, with a second pigment close by. The second pigment functioned like the eyespot in *Euglena*, providing shading and thus an indication of the directionality of light. The Pax6 gene and opsin figured into the chemistry of these receptors (Arendt & Wittbrodt, 2001; Nilsson, 2013).

Chordate and early vertebrate eyes

Early chordates emerged after cnidarians and bilaterians about 535 million years ago, and may have been similar to today's amphioxus. Amphioxus larvae have a single Pax-expressed frontal eye involving a multicellular pigment cup backed by rows of neurons. It is suited to determining the direction of light, perhaps sufficient to escape predators

(Lacalli, 2004, 2018). The salp, a primitive jelly-like chordate, likewise has a single eye during the solitary phase of its life (Braun & Stach, 2017). Accordingly, it seems possible that even though bilaterian ancestors had paired eyes, only a single eye passed from chordates to the ancestors of vertebrates (Lacalli, 2018).

While amphioxus is important in suggesting the nature of a single ancestral eye, hagfish, which emerged shortly after, are equally important in implying the reappearance of paired eyes. Living hagfish have two eyes hidden under skin. They are lensless even though lens placodes are present in embryos, suggesting that the eyes may have degenerated during evolution. Consistent with this supposition, the sole example of a hagfish fossil, dating to about 300 million years ago, shows a pair of dark eye spots including small white dots that may represent lenses (Bardack, 1991). If so, the lenses were subsequently lost. It is possible that hagfish eyes are now used not for vision, but for circadian rhythm regulation (Sun et al., 2014).

It is important to note that the ancestral hagfish eye reversed images left-to-right (and up-to-down). That is because of the geometry involved in its pit structure, as illustrated in Figure 7.2 (Lamb, Collin, & Pugh, 2007). Whether or not it had a lens did not impact

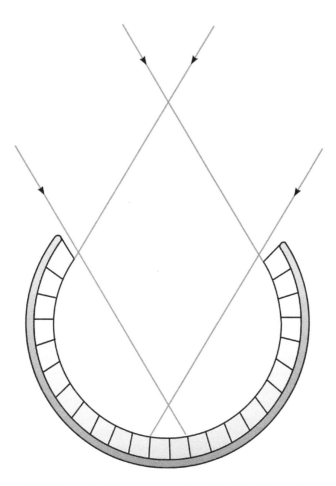

Figure 7.2 A pit eye. Light from the left projects to the right side of the retina, and vice versa.

Source: Adaptation of Pit eye.svg, CC BY-SA 3.0, Gagea.

this, and images would have reversed in either case. However, if present, a lens would have strongly improved image focusing.

Lampreys, the next group to diverge, definitely have paired, reversing eyes, with lenses that in the adult produce focused images (Kröger, Gustafsson, & Tuminaite, 2014). This argues that lenses originated at least 520 million years ago, although they evolved convergently in multiple taxa. Some of the transparent proteins forming the lens are specific to vertebrates and presumably derive from a common ancestor (Bloemendal et al., 2004).

The earliest *fossil* evidence of eyes among vertebrate ancestors is in *Haikouella*, a transitional form. A Cambrian species dating to about 530 million years ago, several of its fossils unambiguously show paired eyes (Mallatt & Chen, 2003). The basic structure of the human eye may thus have been set a half billion years ago during the chordate-vertebrate transition.

Color vision

Although the abilities to sense light and to form images from it are fundamental to vision, the ability to see color is also important to the recognition and evaluation of objects. Some objects, such as fruit, carry color cues signaling suitability for consumption. Others carry warnings, as in distinctively colored poisonous snakes, or advertising, as in the colorful plumage of male birds seeking mates.

In humans, and in all apes and Old World monkeys, normal daytime color vision is accomplished with a system of three retinal receptor types called *cones*. Each type of cone responds to a broad range of wavelengths and thus colors, but each shows a *peak* response at a particular wavelength. Our short-wavelength cones show a peak at 420 nm, in the violet portion of the spectrum; medium-wavelength cones at 530 nm, a green color; and long-wavelength cones at 560 nm, a yellow-green. However, these three cone types are often referred to as "blue", "green", and "red" cones, respectively, because each is sensitive to longer wavelengths as well that include those primary colors. An opsin photopigment is involved in each.

It may seem curious to you that we see only a narrow range of wavelengths in an electromagnetic spectrum running from gamma rays (wavelengths on the order of a thousandth of a nm), up to the emissions of AC circuits (on the order of 10 to the 15th power nm). Why is our vision limited to the tiny range of about 400 to 700 nm?

A large part of the answer is that our sensitivity reflects the predominant wavelengths available on Earth's surface in daytime (Figure 7.3). Above and below visible wavelengths, there is relatively less to detect. That can only mean that evolution tuned the response of our cones to the radiation characteristics of our environment.

However, the greater availability of some wavelengths over others only tells part of the story. While our maximum daytime sensitivity is at 555 nm, fairly close to the availability peak (Delgado-Bonal & Martín-Torres, 2016), this may not be the case for animals generally. Bees, for example, show maximum sensitivity at 344 nm, in the ultraviolet range of light (de Ibarra, Vorobyev, & Menzel, 2014). The reason for this is unclear. Although it is often attributed to the reflection of ultraviolet wavelengths by flowers, in fact most flowers absorb ultraviolet light and do not reflect it (de Ibarra, Vorobyev, Brandt, & Giurfa, 2000).

The daytime system is not very sensitive to low levels of light, and so there are additional receptors called *rods* which we rely on at night to see dim objects. These receptors are exquisitely sensitive: A single photon or quantum of light is sufficient to trigger a rod response, and 5–7 responding rods are sufficient to result in the perception of light

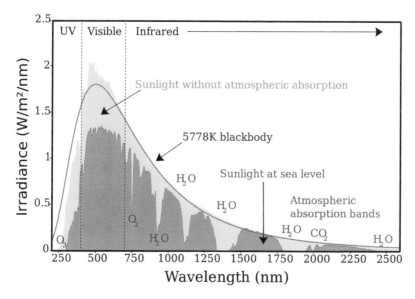

Figure 7.3 Solar radiation above the atmosphere (yellow) and at sea level (red).

Source: Solar spectrum en.svg, CC BY-SA 3.0, Nick84.

(Field & Sampath, 2017). However, because only one receptor type is involved in dim nighttime vision, there is virtually no ability to perceive colors unless an object is bright enough to be sensed by the cones.

The rods and cones are located in the retina, shown in Figure 7.4. The receptors in the retina operate in backward fashion compared with other senses in two different ways, the first involving their position and the second involving their effect on neurotransmitters.

First, rods and cones are literally backward in position: The light-capturing ends of the receptors are oriented *away* from the light, so that light has to pass through neural machinery to be captured by photopigment. In fact, as Figure 7.4 illustrates, light must pass through the *ganglion cells*, the connecting *bipolar cells*, and the rod and cone nuclei and cell bodies before it reaches the light-sensitive disks of the receptors. Fortunately, neurons are nearly transparent, limiting the absorption and scattering of light. Also, the problem is minimized in the *fovea* (the central area of acute vision) because the connecting neurons bend away from it, taking many of their cell bodies out of the direct light path (Figure 7.5).

Why the system is arranged this way is an interesting evolutionary question in its own right. One possibility is that the tight packing of the receptors so severely limits space that the connecting neurons and major blood vessels cannot both be on the side away from the light. If so, they must be separated. In that case, it is better to have the mostly transparent connecting neurons in the light path and the dark, occluding blood vessels away from it, rather than the other way around. Even so, except in the fovea, some capillaries are located in front of the receptors. Normally, we adapt to their presence and are unaware of them. But under artificial circumstances such as the beaming of a narrow light shaft at the corner of the eye (a cell phone light works well with some fiddling), we may suddenly see the shadow cast by the network of capillaries.

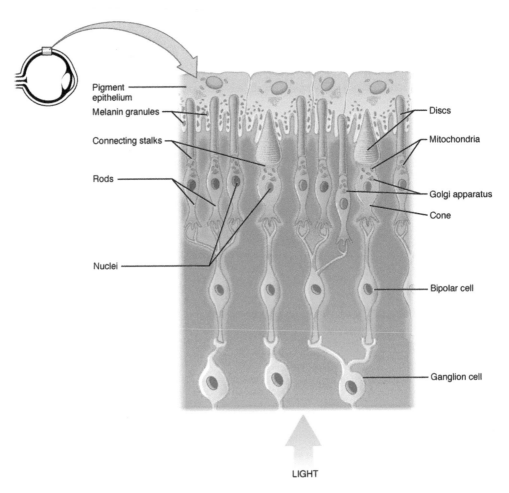

Figure 7.4 The retina. Light from the bottom must pass through several layers of cells before reaching the rods and cones.

Source: Adaptation of 1414 Rods and Cones.jpg, CC BY-SA 3.0, OpenStax College.

The second backward aspect of the visual system is that light stimulation slows the flow of neurotransmitters from the receptors instead of initiating it. When light strikes a receptor, Na+ channels close, and the receptor *hyperpolarizes*, becoming negatively charged. This slows the release of neurotransmitters in a graded manner related to the intensity of the light: The brighter the light, the less transmitter that is released (Breedlove et al., 2010). This use of receptor hyperpolarization appears to date back to ancestral urochordates, now exemplified by sea squirts, some 540 million years ago (Lamb et al., 2007).

Trichromacy

Our possession of three cone types – "blue", "green", and "red" – is the origin of the term *trichromacy* (the condition of possessing three types of cones). Oddly, trichromacy is not an advanced evolutionary feature. Genetic analyses indicate that the ancestral state

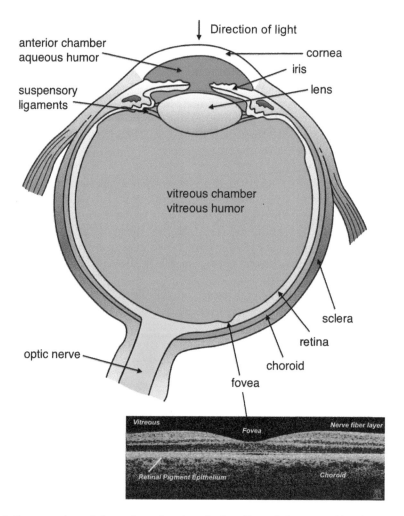

Figure 7.5 Cross-section of the retina, showing the bending of the nerve fiber layer away from the fovea.

Credits: Adaptation of Three Internal chambers of the Eye.png, CC BY-SA 3.0, Holly Fischer, and Retina-OCT800.png, CC BY 2.0, origin medOCT-group.

in vertebrates was *tetrachromacy*, the possession of four cone types. The "extra" type was tuned to very short wavelengths, allowing sensitivity to ultraviolet light. Consistent with this, some lampreys have four cone types. Many reptiles, birds, and bony fish do as well, and some species have cone subtypes, resulting in more than four spectral peaks (Bowmaker & Hunt, 2006).

What happened to reduce the range of color vision in apes and Old World monkeys? The story is even more dramatic than it appears, because our ancestors passed through an even tighter bottleneck of only two cone types (*dichromacy*), meaning that we lost not just one but two of the original four. This is believed to have occurred in early mammals, although its exact dating is unclear (Davies, Collin, & Hunt, 2012). Today, most mammals continue to have only two cone types.

These losses were connected to the nocturnal nature of early mammals. Nocturnal animals generally have fewer cone types than diurnal ones, as well as a smaller cone population relative to rods. Early mammals lost first one, and then a second cone type as their lives became increasingly dependent on rod-based night vision, following a well-known "use it or lose it" evolutionary principle. Biological structures require metabolic expenditures to develop and maintain, and so if they fail to convey a compensating selective advantage in behavior they are selected against. In the case of cones, losing two types reduced metabolic costs while maintaining visual sensitivity to a range of wavelengths beyond those detected by rods. The loss was accomplished by converting cone genes to pseudogenes, rendering them inactive (Arrese, Hart, Thomas, Beazley, & Shand, 2002; Bowmaker & Hunt, 2006).

As explained in Chapter 6, Gilad et al. (2004) identified a link between the re-evolution of trichromacy and olfactory pseudogene formation, with Old World monkeys and the howler monkey developing three-cone color vision along with a large percentage of olfactory pseudogenes. With these monkeys' diurnal lifestyle came the opportunity to judge food sources at a distance by color cues, whether fruit or leaves (Melin, Khetpal, et al., 2017). Re-evolving trichromacy presumably rendered some olfactory receptors unnecessary and led to their loss.

The re-creation of trichromacy in primates beautifully illustrates the role of chance in evolution, because different primate groups accomplished it in varying ways with a varying quality of outcome. Old World monkeys and apes accomplished it by duplicating the gene for the long-wave receptor, and then modifying the duplicate to become sensitive to middle wavelengths (Bowmaker & Hunt, 2006; Davies et al., 2012). The duplication created different gene sites, so all individuals normally inherit genes for all three cone types. In Old World monkeys and apes, this solution must have occurred sometime between 21 and 45 million years ago.

New World monkeys, on the other hand, generally accomplished trichromacy by evolving different alleles of the same gene and shifting the sensitivity of one toward middle wavelengths. Furthermore, the gene involved is located on the X chromosome. This means that *females* often show trichromacy, if they inherit differing alleles on their two X chromosomes. In that case, one X chromosome is randomly inactivated in every cell, resulting in a mosaic of photoreceptors, with half being the middle-wavelength type and half the long-wavelength type. That mosaic, together with the third cone type, creates three-color vision. But a price is paid: Females who inherit the *same* allele on both X chromosomes are dichromatic. So are all males, because they have only one X chromosome and therefore only one of the alleles. Thus, most New World monkeys are only trichromatic as a species, not as all individuals within the species (Bowmaker & Hunt, 2006; Davies et al., 2012).

The howler monkey, as noted earlier, is exceptional among New World monkeys. After the evolution of the alternate alleles, it underwent an actual gene duplication, allowing the two cone variations to become fixed in the population (Surridge, Osorio, & Mundy, 2003). Molecular-clock estimates place the divergence of howler monkeys from other New World monkeys at about 7 million years ago, so the duplication was a fairly recent event (Cortés-Ortiz et al., 2003; Perelman et al., 2011).

Night vision

There is strong evidence that the rods used in night vision evolved after cones. Hoke and Fernald (1997) cited analyses of DNA sequences from humans, chickens, and fish indicating that rhodopsin, the specific opsin found in rods, emerged prior to divergences among these groups, but after the appearance of color-sensitive opsins. Furthermore, these researchers pointed out that rod output converges on cone pathways for transmission from the retina, not the other way around, and that during gestation, rods develop later than cones. All of these considerations suggest that rods evolved after cones (Kim et al., 2016).

This conclusion seems surprising because as the superficially less specialized system, involving only one type of receptor, the rod system appears more primitive than the cone system. Yet in truth, the rod system is highly specialized, having evolved to allow an exquisite sensitivity to low levels of light. As we have seen, a single photon is sufficient to cause a response in a rod containing rhodopsin.

The cone-then-rod order makes sense from a broader evolutionary perspective as well. The largest selective pressure in the history of vision was no doubt toward evolving *any* sensitivity to light. Because light is mostly present in the daytime, selective pressure was maximal for daytime sensitivity, and so it emerged first. Later, more subtle selective pressures fine-tuned it so that many creatures became sensitive to the low nighttime levels of light.

More recent work on gene sequences agrees with the conclusion that cones came first, and it dates the cone-rod divergence to a vertebrate common ancestor about 520 million years ago. Furthermore, it appears that the first of the cone types to evolve was the long-wave ("red") type, followed by short-wavelength ("blue") cones, then medium-wavelength ("green") cones, and finally rods (Pisani, Mohun, Harris, McInerney, & Wilkinson, 2006; Wilson, 2007).

Mechanisms for increasing light sensitivity

Recall that, in humans, a single photon can produce a rod response. Since it is difficult to improve on perfection, how, then, can we account for the fact that many animals, such as owls, cats, and tarsiers, have better night vision than humans and, in fact, better night vision than anthropoids generally? The answer lies in species differences in (a) the light-gathering apparatus that delivers photons to the retina, (b) the density of rods in the retina, and (c) the neural machinery collecting information from the retina. The first of these, the light-gathering apparatus, refers largely to the size of the cornea, the transparent bulge in front of the pupil that is one of the eye's optical components. Just as a larger telescope lens gathers more light to image faint details of the cosmos, a larger cornea gathers more light to detect fainter variations in brightness at night. More photons are delivered to the retina, increasing the number captured by rods.

Large corneas are the rule in primates generally. It is therefore likely that they originated with early nocturnal primates prior to the emergence of ancestral strepsirrhines about 73 million years ago (Ross & Kirk, 2007). Tarsiers, whose ancestors diverged from our ancestral line about 72 mya, provide a graphic example of enlarged corneas (see Figure 4.5). Across living primates, there is a strong empirical relationship between nocturnality on the one hand and corneal size relative to eye diameter on the other. Anthropoids, including humans, have smaller corneas relative to other primates, attesting to the existence of later diurnal ancestors (Ross & Kirk, 2007).

A second light-gathering solution is the use of a reflective backing to the retina. The *tapetum lucidum* ("silvery carpet"; Land & Nilsson, 2012, p. 139) is the reflective layer behind the retinas of cats and dogs, responsible for making their eyes shine at night when a beam of light is directed at them. Photons passing through the retina without being captured are reflected back, affording a second opportunity for capture. The result is increased night vision, though at the expense of a slight blurring of the image.

Both of these solutions – large corneas and a tapetum – appear in tawny owls, a nocturnal species. Significantly, a tapetum is frequently found in the mostly nocturnal strepsirrhines, but not in the later-evolving tarsiers and anthropoids. Because strepsirrhines emerged early in primate history, these characteristics provide additional arguments that early primates were nocturnal (Ross, 2000).

Another improvement on perfection is for nocturnal species to pack a higher density of rods into the retina. Some owls have close to a million rods per square mm, compared to not quite 200,000 per square mm in humans (Goodchild, Ghosh, & Martin, 1996; Ross, 2000). Such tight spacing increases the chances that a photon reaching the retina will be absorbed by a rod. The tradeoff is that nocturnal species typically have fewer cones, although enough to support a moderate quality of day vision.

A final improvement lies in how information is collected from the receptors. In the human fovea – the central area of acute vision in the retina – the ratio of receptors (exclusively cones) to ganglion cells is 1:1, providing sharp, detailed day vision. But in the periphery of the visual field (which contains mostly rods), the convergence onto ganglion cells is as high as 1000:1 (Goodchild et al., 1996). Although it makes night vision blurrier, the massive funneling of rod output increases light sensitivity, since several stimulated rods can signal the same ganglion cell. Nocturnal animals have even better low-light signaling because they funnel rods into ganglion cells at even higher ratios. For example, in the nocturnal owl monkey, rod convergence onto ganglion cells is as much as 16,200:1. By way of comparison, in the diurnal capuchin monkey (see Figure 4.8), the maximum rod convergence is 6500:1 (Yamada, Silveira, Perry, & Franco, 2001).

In the final analysis, the reason why humans do not have better night vision is because our ancestors did not need it. At night, our ancestors were at rest and hidden, minimizing their vulnerability to predators. In contrast, as diurnal animals, they needed excellent day vision, and natural selection pushed them – and us – in that direction instead.

Acuity

The funneling of rods and cones has important consequences for *acuity* (the ability to resolve fine visual details). The human fovea's 1:1 ratio of cones to ganglion cells preserves high acuity, while the 1000:1 ratio for human rods does not. Day vision is therefore more acute than night vision. Similarly, whether a primate species is diurnal or nocturnal has much to do with their visual acuity. Among New World monkeys, diurnal species have been found to have two to five times better resolution than nocturnal species (Ross, 2000).

The fovea's contribution to acuity

The fovea, a relatively recent innovation, is responsible for the high acuity most primates enjoy relative to other mammals. It consists of a tiny pit in central vision, part of the macula, a general vertebrate innovation. Among mammals, only anthropoids have well-defined foveas that are not occluded by blood vessels, a state presumably reached soon after their emergence about 72 million years ago. The earlier-emerging Strepsirrhini and tarsiers exemplify a transitional state, with galagos and tarsiers having a shallow or variably expressed fovea with occluding blood vessels.

Given these observations, it seems safe to conclude that the fovea emerged early in primate evolution, was partially present in ancestral Strepsirrhini, and was perfected in anthropoids. Ultimately, visual acuity in the Old World primates and apes improved to the point of being two to three times greater than that in their near relatives, the tree shrews (Ross, 2000).

As you might infer from these differences, substantial disparities exist in the width of the area of best vision. In the gerbil, it is a bit over 180° – i.e., the visual field shows uniform, relatively low acuity. In strepsirrhines, it is about 30°, in macaques 4°, and in humans 1.5°, about the width of our thumbnail held at arm's length. Thus, as acuity improves across these species, the area of highest acuity decreases.

Furthermore, the width of the area of best vision has influenced auditory localization thresholds, with narrower areas of best vision accompanying better ability to localize sound. Thus, within the horizontal plane, gerbils can localize a sound only to within about 28°, macaques to about 5°, and humans to about 1.5°. Summarizing these results, Heffner (2004) suggested that selection pressure tuned sound localization to better direct visual attention to objects. Species with a narrow field of acute vision, like humans, need to judge sound location accurately to bring vision to bear on a sound source. Species with a wide field have less need for this and make do with poorer sound localization.

Eye and brain

Evolutionary changes to the eye might be expected to cause changes in the brain. To examine this issue, it is first necessary to describe the neural projections from the retina to the visual cortex.

As we have seen, the rods and cones constitute the visual receptors of the retina. These synapse onto bipolar cells, which in turn synapse onto ganglion cells. The axons of the ganglion cells, taken collectively, become the optic nerve, which exits the eye socket through a small opening in the bone called the *optic foramen*. The bundle of axons splits at the optic chiasm, with half crossing the midline and the other half not, in a way that ensures that the left half of the visual scene (called the left visual field) projects to the right side of the brain, and vice versa (Figure 7.6). From there, most of the axons travel

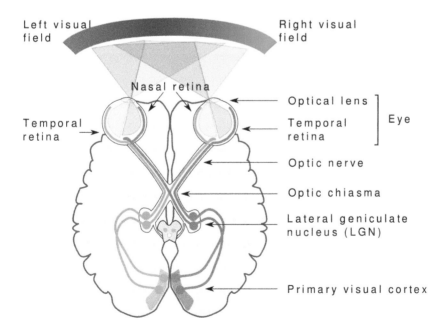

Figure 7.6 The primary visual pathway, from eye to brain.

Source: Human visual pathway.svg, CC BY-SA 4.0, Miquel Perello Nieto.

to the lateral geniculate nucleus (LGN) in the thalamus, where they synapse upon neurons that in turn project to the primary visual cortex at the rear of the occipital lobe. The primary visual cortex is also known by a number of alternative names, including V1 (i.e., the first visual area in the brain).

The suggestion that sensory input to the brain might influence its evolution is an old one, dating at least as far back as a proposal by Grafton Elliot Smith in 1927 (Kirk, 2006). The idea is simple and logical: If evolution produces changes to the eye, then changes to the brain should occur as well to accommodate them. In particular, evolution of the fovea and the re-evolution of trichromacy should have been accompanied by a corresponding increase in the cortical areas devoted to vision.

Kirk (2006) addressed this issue by examining the size relationship between the optic nerve and brain in living as well as extinct primates. The size of the optic nerve was measured using the area of the optic foramen, under the empirically supported assumption that the opening's size reflects the size of the nerve bundle passing through it. Because the sizes of the optic nerve and brain are both highly correlated to body size, the effect of body mass was statistically removed when examining size relationships. These relationships proved substantial: Correlations (r) of optic foramen area to brain volume were +.66 in living primates and +.67 in fossil primates, indicating that about 45% of the variation in the brain size of living and extinct primates is explained by variation in optic nerve size.

Although correlation is not causation, the robustness of these relationships is suggestive. As Kirk noted, they are consistent with Jerison's "principle of proper mass". This principle, introduced in Chapter 6, holds that there is a relationship between the neural mass devoted to a function and the amount of processing involved. Those primates relying more on cone receptors, with increased processing needs, have expanded cortical areas devoted to vision and thus larger brains.

The most interesting results of the study emerged from the fossil data. Two groups of fossil primates were included, adapiformes and haplorhines. Of the two, the adapiformes were more ancient in terms of divergences from the human ancestral line (see Figure 4.2). They were found to have small optic nerves and small brains – in fact, outside the range of living primates. In contrast, the haplorhines had relatively large optic nerves and large brains, well within that range (Kirk, 2006).

These results are telling, for haplorhines with their large brains emerged about 73 million years ago, while anthropoids with their perfected foveas emerged about 72 million years ago (Table 4.2). In other words, the evolutionary threshold for larger optic nerves – and bigger brains – appears to be located at about the same place in primate phylogeny as the threshold for a well-developed fovea. As such, the results support a link between acute vision and brain expansion in primate evolution.

A link between eye and brain continues to the present day. Pearce and Dunbar (2012) reasoned that in modern populations, eye size (and thus brain size) should be affected by geographic latitude. At higher latitudes, low-light days are more frequent than at lower latitudes. Thus larger eyes should be selected, because larger eyes have larger corneas that collect more light. In turn, larger eyes have more ganglion cells, requiring a larger brain to support visual processing. Pearce and Dunbar therefore examined modern human skulls curated in a British museum. They found, as predicted, that both eye size and brain size increased with increases in the geographic latitude of the population that a skull was drawn from.

Of course, acuity-related brain expansion, starting in the Paleocene, was only the first of several brain expansions in the primate line leading to modern humans. Later ones were to have other causes.

Conclusion

Vision originated in single-cell organisms in the form of photopigments allowing the organisms to orient to light. Possibly it was introduced into multicellular animals through symbiosis, with an initial cooperative relationship with a light-sensitive unicellular organism eventually becoming genetic.

The first true eyes feeding into neurons were probably in cnidarians, specifically jellyfish. Today's jellyfish commonly use pigment cells as a navigation and migration aid, and box jellyfish have complex lensed eyes. Along with the earlier-emerging sponges, cnidarians have a PaxB gene known to play a role in eye development. It was probably ancestral to a number of other Pax genes involved in expression of the eyes, nervous system, and other organs. Thus, the basic control mechanism for eye development was likely present by 665 million years ago, although the later appearance of photoreceptor cells and true eyes involved independent evolution in multiple taxa.

Opsin photopigments may originally have been regulated by PaxB. Flavins (cryptochromes) are another class of photopigments involved in setting the circadian rhythm via retinal ganglion cells and the suprachiasmatic nucleus. Ancestral eyes were probably simple light sensors in early cnidarians, cup eyes in bilaterians and chordates, and paired lensed eyes in 520-million-year-old vertebrates.

The human visual system is "backward" in two respects. First, the light-capturing end of the receptors is located away from the light, behind several layers of cells. This design may actually minimize the blocking of light, since the cells are nearly transparent while blood vessels behind the retina are not. Second, when light is detected, receptors hyperpolarize instead of depolarize. That slows the flow of neurotransmitters, a feature tracing back to ancestral chordates.

Old World monkeys and apes, including humans, have three retinal cone receptors with differing wavelength sensitivities, which are used in daylight conditions. The cones are tuned to the predominant wavelengths available at the Earth's surface. The ancestral state in vertebrates was four cone types, only two of which were retained in our nocturnal mammalian ancestors. A third type, sensitive to middle wavelengths, was regained independently in New World and Old World monkeys, though in different manners. This independent development led to individual-level trichromacy in the Old World branch but in only species-level trichromacy in most New World monkeys. An exception is the New World howler monkey, which recently evolved species-level trichromacy.

The cones of the retina are the eye's original receptors, and rods later evolved from cones. Rods are sensitive to low levels of light, with a single rod capable of responding to a single photon, and with sensitivity of the system as a whole increased by substantial funneling of information from the rods to the cortex. Nocturnal species show more funneling than diurnal ones and often have other low-light adaptations, such as a large corneal light-gathering area, a reflective tapetum lucidum, and a dense packing of rods in the retina.

The difference between the rods and cones in the amount of funneling affects acuity; the rod system in the periphery of vision has low acuity while the cone system at the macula has very high acuity. The area of the visual field with the highest acuity has decreased throughout primate evolution and is only 1.5° in extent in humans. There is a strong correlation between optic nerve size and brain size in primates, even after correcting for body size, suggesting that part of the primate brain expansion may be accounted for by specialization in vision.

8   The origins of motion

As we have seen, the earliest form of sensation allowed substances to pass through cell membranes. However, early organisms had only a limited ability to exploit such resources, because they depended on whatever was brought to them by water. Acquiring movement was a quantum leap in evolution, freeing single-cell life from relative immobility and at least to some extent from the vagaries of currents. Motion meant that by trial and error, organisms could shift from areas of scarcity to areas of plenty.

Movement would prove to be a key building block in cognitive evolution. The ability to shift between locations imposed navigational requirements calling for highly developed spatial processing abilities. Favored locations needed to be remembered so that movement could be directed back to them, imposing a need for improved memory. Pattern recognition processes were required to recognize known locations, and to direct movement toward new ones affording familiar types of food as well as water and shelter. In humans, movement also led to bipedalism and a resulting freeing of the hands, allowing the evolution of more dexterous reaching and grasping – as well as what those imply for tool use and manufacture.

In this chapter we lay the groundwork for those developments. First, we will look briefly at the movement of single-cell organisms, before examining early metazoan motion. We then consider movement innovations introduced in the human line of descent by chordates, tetrapods, amniotes, mammals, and primates. The neural mechanisms of movement are considered, as is the origin of contralateral organization in both the motor and sensory systems.

Single-cell organisms

Locomotion (movement from place to place) originated in early flagellated bacteria. Rotation of the flagella propelled a bacterium on a straight course. In a beneficial microenvironment, repeated momentary reversals kept the bacterium in place. Entering a harmful microenvironment, however, caused abrupt reversal of the rotation, introducing a tumbling motion and a random change of direction (Machemer, 2001).

As cell sizes increased, gravity posed an increasing challenge, because cells were denser than water and thus subject to the deadly consequences of sedimentation. In response, eukaryote membranes evolved mechanical sensitivity to the force exerted by water displacement on their undersides. This triggered upward swimming, supplemented by photomotility in light-sensitive organisms (Machemer, 2001).

Metazoa

Early metazoans encountered similar challenges to those faced by single cells, but magnified by increased mass. Sponge larvae swam, using the beating of cilia on their surface and end (Leys et al., 2002). Thus, since sponges were the earliest-emerging metazoan phylum with living representatives, movement likely appeared near the outset of animal history. Ctenophores, next to emerge, reached a similar solution. They used columns of cilia ("combs") beating in unison, propelling the animal through water.

Many early metazoans adopted a benthic lifestyle, surrendering to gravity instead of counteracting it. However, a benthic lifestyle alone could not cope with local variations in nutrient concentrations. In times of plenty, a stationary animal might thrive, but in times of scarcity, there was a strong selective pressure favoring mobility and a seeking of new resources.

Trace fossils (fossils of behavior as opposed to physical form) depicting trails about 1 mm in width implicate a surface-grazing form of locomotion in some benthic metazoans of the Ediacaran (also known as Vendian) period 570 million years ago. Beginning 540 million years ago, at the onset of the Cambrian period, trails increased to about 3–5 cm in width, and burrowing became common at depths up to 5 cm (Collins, Lipps, & Valentine, 2000; Fedonkin, 2003).

Judging from living animals, the trails were caused by creeping movements. Secreted mucus provided lubrication, and cilia propelled the animal forward. Some sea anemones, classified as cnidarians, move in this way. They minimize their "footprint" by keeping only a tiny fraction of their body surface in contact with the seabed, thus leaving trails only 2–3 mm wide from bodies up to 10 times wider. The secreted mucus binds the substrate material, leaving a trail that can fossilize. Some flatworms likewise creep using mucus secretions and cilia, leaving ridged trails resembling some Precambrian trace fossils (Collins et al., 2000).

Cilia-based benthic movement achieves speeds of 1–2 mm/sec, or .002–.004 miles per hour. Movement accomplished by peristaltic waves, similar to that of earthworms, evolved a little later, providing increased speed as bodies became larger.

Flatworms are notable not just for creeping, but for swimming, involving full-body undulation using muscles. The later-emerging chordates also undulated to swim, but in a more fish-like fashion due to their stiff notochord. Thus one possibility is that the cilia used in swimming by early sponge larvae and ctenophores were adapted in benthic species for creeping movements, in turn followed by the evolution of muscle tissue that allowed a reacquisition of swimming.

However, a complication is that muscles may have first evolved for spawning rather than locomotion purposes. Seipel and Schmid (2005) proposed that an early swimming metazoan (using flagella or cilia) evolved muscle cells for reproductive purposes. Its descendants then included benthic bilateria as well as swimming ctenophores and cnidarians. All inherited muscles, which began to be used for movement. This inheritance became particularly significant to the bilateria, whose bottom-dwelling lifestyle led to differentiation between *dorsal* (top) and *ventral* (bottom) sides. That set the stage for the evolution of body segments, allowing for more complex muscle arrangements capable of more complicated movements (Finnerty, 2005).

It is indisputable that body segments permitted more complicated movements. In segmented chordates like amphioxus, swimming involves head-to-tail waves of body contraction controlled by the nerve cord. A swimming rhythm is maintained by alternating cord segments, a movement pattern that descended to later fish, and from them to tetrapods and eventually to mammals (Viala, 2006).

Tetrapods

A major feature of the fish-to-tetrapod transition was a shift in emphasis from power supplied by the pectoral region (the forward fins or limbs) to the pelvic region (the rear ones). The *Panderichthys-Acanthostega-Ichthyostega* evolutionary sequence (Figure 3.3) illustrates the changes. The early lobe-finned *Panderichthys*, in the words of Boisvert (2005), was a "front-wheel drive" model, pairing fish-like undulation of the front part of the body with anchoring pelvic fins.

In contrast, to varying degrees, *Acanthostega* and *Ichthyostega* were "rear-wheel drive" models. Their appendages shifted toward the side, both front and back; their muscles and bones enlarged for body support; and the "fin rays" became digits. All limbs lengthened, but the increase was greater in the rear, where the pelvis also enlarged for increased support. Thus the transformation of fins to limbs began in the front, spread to the rear, and ultimately resulted in transferring power to the rear (Boisvert, 2005).

This shift toward "rear-wheel drive" is significant in light of later locomotor developments in dinosaurs including birds. But what caused it? The initial "front-wheel drive" state sensibly used anchoring rear fins to stabilize an undulating body. But once tetrapod ancestors were free of the mud, what caused the shift toward rear propulsion?

The explanation is that in land animals, the tail became critically important to propulsion. This is visible in the archosaurs. Ancestors of the crocodilians and dinosaurs, they left extensive fossils allowing visualization of muscle attachments to bone. All three groups had a massive muscle, the *caudofemoralis longus*, located in the tail but connected to the rear limbs. Although the muscle helped move the tail, biomechanical analysis reveals a more important opposite function. Specifically, the muscle provided power to the rear limbs, using the heavy tail as an anchoring point (Gatesy, 1990). Figure 8.1 illustrates this aspect of its use.

The relevance to tetrapod evolution is that the caudofemoralis longus is a tetrapod character. It is found even in the early-diverging salamanders, where it does help power strides (Ashley-Ross, 1994).

Thus the shift to "rear-wheel drive" involved propulsory use of the tail. It is a beautiful example of evolution using an existing structure for new purposes, much as jaw bones were used in the auditory evolution of mammals. The tail, finned and directly propulsive in water, could no longer be used that way on land. But it could be used indirectly, by anchoring muscles powering newly evolved rear limbs.

Another tetrapod characteristic was a "step cycle" that moved the limbs in a particular order. The sides alternated, specifically right hind–right front–left hind–left front. It

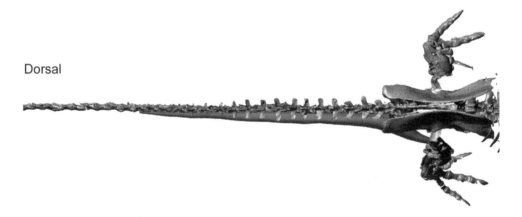

Dorsal

Figure 8.1 The tail as anchoring point, dorsal view. The caudofemoralis longus is in reddish brown, connecting the tail to the femur of a *Tyrannosaurus*.

is noteworthy that the same step cycle is observed in lobe-finned fishes, indicating that *gait* (the cyclical means by which the limbs move the body) traces back to the immediate ancestors of tetrapods (Prothero, 2007, p. 220).

As a result of these changes, tetrapods enjoyed a leap forward in traveling speed over a solid surface. One species of salamander, an early-emerging tetrapod lineage, has been clocked trotting at 0.6 miles per hour (Ashley-Ross, 1994), a 200-fold increase over cilia-based movement. However, early tetrapods sprawled severely. Limbs were held outward and the belly was supported by the ground.

Amniotes and mammals

Several changes occurred en route from the quadrupedal movement of early tetrapods to that of amniotes and mammals. An ability to partially erect the legs evolved, lifting the body free of the substrate (Blob, 2001). That ability, similar to what crocodiles do when walking, existed in *Orobates pabsti* about 260 million years ago. A 4 kg (9 pound) tetrapod believed to be very similar to early amniotes, it had a center of gravity nearer to the hind legs than to the forelegs, making it a mostly rear-wheel drive model. Yet its limbs, though sprawling, held the body off the ground. Fossilized trackways known to have been made by the creature show no evidence of a dragging belly (Nyakatura et al., 2015).

Eventually in mammalian ancestors, limbs further evolved so that body weight was placed more directly over the legs, reducing mechanical stress. Thus an erect quadrupedal posture became the mammalian norm (Blob, 2001). There is some evidence that the forelimbs could erect earlier than the hindlimbs. But in any case, mammals were soon brought to an "all-wheel drive" stage, perhaps by 125–150 million years ago. *Eomaia*, the early eutherian dating from 125 million years ago, had a fully upright posture (Kielan-Jaworowska & Hurum, 2006).

While these changes were occurring, the limbs evolved from two segments to three, arranged in a zigzag pattern (see Figure 8.2). Propulsion was provided mostly by the segments closest to the body, aided by spinal movement. Movement transformed from side-to-side undulation to up-and-down oscillation. The shoulders became moveable, while the hindlimbs evolved ankle joints to incorporate lengthened feet. These changes were probably adaptations for movement over uneven ground, a challenge for small mammals (Fischer, Schilling, Schmidt, Haarhaus, & Witte, 2002). Once established, they then carried over to later, larger mammals.

Limb changes were accompanied by gait changes. The side alternation of early tetrapods was a *symmetrical gait* in that a movement cycle took the same length of time for the forelimbs as for the hindlimbs (Ashley-Ross, 1994; Gasc, 2001). Mammals, however, evolved an *asymmetrical gait* with different timing cycles for the forelimbs and hindlimbs, resulting in higher speeds. (Note that in this sense, asymmetry concerns front-back differences, not left-right differences.) Often both gaits are available, as in the horse's symmetrical trot but asymmetrical gallop (Gasc, 2001).

The absence of asymmetrical gaits in early tetrapods was due to their sprawling posture. Without support directly under the body, sprawling compromises the balance needed for an asymmetric gait. Even today, sprawling species like salamanders use a symmetrical gait, nearly always with two or three limbs in contact with the ground even when trotting. In contrast, horses do not sprawl, and their asymmetric gallop can take all four limbs off the ground simultaneously. Although salamanders can support themselves momentarily on a single leg, this is maintained for only one or two strides before they flop on their bellies (Ashley-Ross, 1994). Thus the evolution of a fully upright

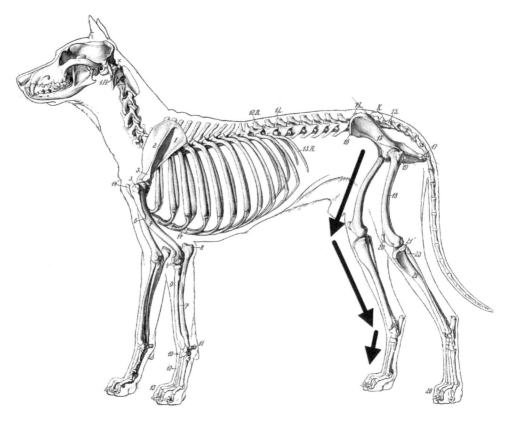

Figure 8.2 Three zigzag segments in the hindlimb of a dog.

Source: Adaptation of Dog anatomy lateral skeleton view.jpg, public domain, Wilhelm Ellenberger and Hermann Baum.

quadrupedal posture in mammals, putting weight directly over the legs, was an important milestone in making fast movement possible.

Primates and hominins

Primates, whether scurrying mouse lemurs, climbing and leaping sikafas, or crawling humans (try it!), have a characteristic quadrupedal stride in which the forward placement of limbs follows a right hind–left front–left hind–right front series. This diagonal sequencing, of hind and front limbs on opposite sides, differs from that of nearly all other mammals, which use the tetrapod's side-alternating step cycle of right hind–right front–left hind–left front. Also unlike other mammals, most primates experience less force on their forelimbs than on their hindlimbs when walking. Counterintuitively, however, they extend their forelimbs more than other mammals (Granatosky, Tripp, Fabre, & Schmitt, 2016; Schmitt & Lemelin, 2002).

The fine-branch environment

The reason for these gait modifications is that primates evolved to occupy an arboreal, fine-branch environment. Moving along fine branches emphasizes accurately placing grasping hands. Diagonal sequencing improves balance on narrow supports, and shifting force to the hindlimbs allows the front limbs to better grasp food. Increased extension of the forelimbs both increases the range of grasping and reduces branch oscillations. Linking primate gait to fine-branch environments is confirmed by the similar gait of woolly opossums, also a grasping fine-branch quadruped (Granatosky et al., 2016; Schmitt & Lemelin, 2002).

Other adaptations in many quadrupedal, arboreal primates are unusually mobile limbs and large, heavy tails. When an Old World monkey walks on a branch, it uses forearm rotation to produce slight side-to-side turns of the body to maintain balance. It is probably easier to do this when there is reduced force over the front limbs relative to the back limbs, again helping to account for the rearward shifting of force in primates. A second mechanism for maintaining balance is rotation of the tail. For example, if the animal starts to fall to the right, the tail rapidly rotates in that direction. In turn that causes the rear of the body to make a reactive leftward shift (Larson & Stern, 2006). A mobile tail also facilitates leaping. Merely lifting it during leaps does much to stabilize the body, reducing pitch error (Libby et al., 2012).

Diagonal sequencing of stride is widely spread among primates, but tails less so. They are not just an arboreal feature because baboons, which are terrestrial, have them. Baboons, however, are subject to open-space predation and often have to run, making the balancing function of a tail useful even on land. Somewhat more puzzling is why apes do not have tails, as they are all at least partially arboreal. Indeed, *Proconsul* lacked a tail even though it is believed to have been an arboreal quadruped (Larson & Stern, 2006).

The effect of large bodies?

It is quite probable that body size was involved in the loss of tails on apes. If early apes had large bodies, tail movement might have had little effect on balance. Once they proved ineffective, tails would then have been lost as useless appendages (for a similar suggestion, see Williams & Russo, 2015). Supporting this notion, *Proconsul* is estimated to have weighed up to 90 kg (200 pounds), and the contemporaneous *Morotopithecus* 40 kg (90 pounds). Furthermore, modeling indicates that as tails become smaller fractions of body weight, pitch correction becomes less effective (Libby et al., 2012). It therefore seems quite plausible that a large-body stage accounts for tail loss.

Primates also differ from most mammals in the nature of their middle-speed gait. Most mammals trot with an off-the-ground phase, but primates have an "amble", allowing continuous engagement with the substrate while minimizing vertical oscillations. It has obvious advantages for moving along branches, although ambling is also available to some large terrestrial animals such as horses and elephants. An advantage to large animals is reduced peak forces on the limbs (Schmitt, Cartmill, Griffin, Hanna, & Lemelin, 2006).

Thus far the description of primate locomotion has emphasized quadrupedalism. That should be interpreted as including the knuckle walking of gorillas and chimpanzees, involving folding four fingers into the palm and contacting the ground with the second row of knuckles (Simpson, Latimer, & Lovejoy, 2018). However, quadrupedalism is deemphasized by the other nonhuman apes, namely the gibbons, siamangs, and orangutans. All are primarily arboreal. Gibbons and siamangs are considered true brachiators, while orangutans use a combination of brachiation, climbing, and clambering, and are sometimes called semi-brachiators.

At this point it is unclear whether ancestral apes were arboreal quadrupeds or brachiators. *Morotopithecus*, a 20-million-year-old ape, had a mobile shoulder consistent with brachiation (MacLatchy, 2004). However, *Proconsul*, contemporary to *Morotopithecus*, appears to have been quadrupedal. Both are plausible candidates for the common ape ancestor.

The hominin background

Only a little clearer is the nature of the last common ancestor of the African apes including humans. Was that ancestor mostly arboreal, or more of a terrestrial quadruped in the knuckle-walking style of gorillas and chimpanzees? While knuckle walking as a human ancestral trait has had its proponents (Young, Capellini, Roach, & Alemseged, 2015), the preponderance of evidence is against it. For one thing, the detailed knuckle-walking anatomy of chimpanzees and gorillas differs sufficiently that it likely had independent origins postdating their (and our) last common ancestor (Kivell & Schmitt, 2009; Lovejoy, Simpson, White, Asfaw, & Suwa, 2009). Features of the fossil hominin wrist and *humerus* (upper arm bone) have also been interpreted as inconsistent with knuckle walking (Arias-Martorell, Potau, Bello-Hellegouarch, Pastor, & Pérez-Pérez, 2012; Selby, Simpson, & Lovejoy, 2016).

This conclusion is strengthened by recent analysis of the australopithecine shoulder joint. The joint shows closer affinity to arboreal species, such as the orangutan and New World woolly monkeys, than to gorillas and chimpanzees. The curved fingers of australopithecines likewise attest to a partially tree-based lifestyle. That has been described as "generalized arboreality" and presumably included both above-branch and below-branch (suspensory) locomotion. It would not, however, have included the kind of agile brachiation shown by the small-bodied gibbons and siamangs (Arias-Martorell et al., 2015).

Neural mechanisms of movement

In humans, striking effects on movement are often observed after strokes. Specifically, damage to one hemisphere of the brain often impairs movement on the *opposite* side of the body. Why does this happen, and how did such a crossover evolve? Before tackling these questions, we first need a basic grounding in the neural mechanisms of movement.

Muscles move because motor neurons synapse directly onto muscle fibers at what is called the *neuromuscular junction*. Muscle fibers are actually elongated cells lying parallel to one another, composed of *myofibrils* (chains of protein) that contract when stimulated. When enough muscle fibers are stimulated, the muscle as a whole contracts.

Motor neurons originate either in the spinal cord or in cranial nerve nuclei. A number of motor actions are initiated as reflexes at these levels, without brain input. For example, stepping on a sharp object in bare feet elicits a rapid foot withdrawal because the sensation triggers a multisynaptic pathway looping from the foot to the spinal cord and back to the foot.

Most complex movements, however, require direct intervention by the brain. These are mediated by two major sets of pathways extending down the spinal cord from various levels of the brain including the cerebral cortex. One is the *corticospinal tract*, also known as the *pyramidal* or *dorsolateral tract*. This system is a contralateral one: In other words, it shows complete crossover from one side of the brain to the opposite side of the body. The tract originates in the precentral gyrus, or primary motor cortex, the rearmost strip of the frontal lobe (see Figure 6.3). Moving downward, most of its crossover occurs at the level of the medulla. The tract continues to descend to a segment of the spinal cord, where it synapses with motor neurons that travel to the muscles.

Damage to the primary motor cortex therefore affects movement on the opposite side of the body due to the crossover. Because the tract largely controls movement of the *distal* (farther from the core) musculature, damage to it is more likely to affect the hands and lower limbs than it is the trunk muscles.

The other tract is called the *extrapyramidal tract*, although there are actually several distinct pathways. What they have in common is that crossover is only partial. Thus the pathways from the brain divide, with part going down the same side, and part crossing over to the opposite side. This allows either side of the brain to control both sides of the body. However, unlike the corticospinal tract, the extrapyramidal tract largely innervates *proximal* (nearer to the core) musculature such as trunk muscles that maintain posture.

With this background, we can easily address the question of why damage to one hemisphere often impacts movement on the opposite side of the body. What is observed in such cases is the effect of the corticospinal system, which crosses over. With one hemisphere damaged, cortical control of the opposite distal musculature may be disrupted. Proximal musculature is much less affected because the *bilateral* (two-sided) extrapyramidal system allows those muscles to be controlled by the undamaged hemisphere.

That answers the "why" question of opposite-side effects of hemispheric damage. But how did such a crossover evolve? That is both more interesting and more difficult to answer.

The origin of contralateral organization

To begin with, it is important to recognize how fundamental crossover is in the nervous system. It is not merely a motor phenomenon. As we saw in Chapter 7, when we fixate vision on a point in space, everything to the left of that point projects to the right hemisphere of the brain, and vice versa. Auditory input and a touch to the skin largely project to the contralateral hemisphere as well, though not as completely.

At first glance, this arrangement seems nonsensical, producing an unnecessarily complicated architecture requiring neurons to pass from one side of the body to the other on the way to or from the cortex. What accounts for this crossover of sensory and motor projections?

Cajal's proposal

The famous neurophysiologist Santiago Ramón y Cajal (1852–1934) argued that the origin of crossover was the reversal of images by the eyes, a simple optical property (Vulliemoz, Raineteau, & Jabaudon, 2005). Given that images reverse, the problem faced by the nervous system is how to achieve a continuous representation in the brain. According to Cajal, an uncrossed system results in discontinuity, as at the bottom of Figure 8.3 where the word "BEAR" is represented across the cerebral hemispheres as "ARBE". A crossed system, in contrast, achieves continuity and reproduces the two halves of the visual scene in the correct order (top of the figure). When the two sides of the midline are bound together using the between-hemisphere connections of the corpus callosum, a fully coherent representation is achieved and "BEAR" is represented as "BEAR".

Cajal assumed that with the left side of the visual world represented in the right hemisphere, and vice versa, the other senses would need to be arranged similarly. That would allow a coherent global representation of the world in which all input from one side would be integrated in the same hemisphere. The motor system would also need to follow suit, to control responses on the same side on which stimulation is received (Vulliemoz et al., 2005).

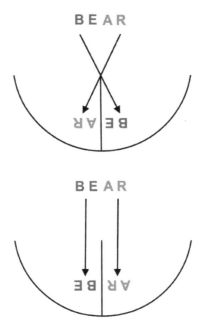

Figure 8.3 Image continuity in a crossed system (top), and discontinuity in an uncrossed system (bottom). The two curved segments at the bottom of each diagram represent the left and right cerebral hemispheres.

Source: Original figure.

A problem with Cajal's proposal

Cajal's explanation is elegant, and it captures something important about crossover: Potentially, the simple optical properties of an image-reversing eye provide the seeds for contralateral organization. However, his proposal depends on interconnected hemispheres to make visual continuity relevant by knitting together the two sides of the midline. That did not occur until long after reversing eyes evolved. Indeed, the corpus callosum, the primary cross-hemisphere structure connecting the hemispheres, was a relatively late innovation. It first appeared in eutherians, dating its origin at about 170 million years ago (Suárez, Gobius, & Richards, 2014).

In contrast, the reversing eye is quite ancient, tracing at least as far back as chordates in our ancestral line and therefore well over 500 million years ago. Amphioxus, the living representative of primitive chordates, has a visual system that has been interpreted as *homologous* (traceable from common origins) to that of vertebrates. Thus its single frontal eye has *interneurons* (small laterally connecting neurons) as well as what appear to be the equivalents of bipolar and ganglion cells, all found later in vertebrates (Butler, 2000; Lacalli, 1996).

On this basis it seems likely that vertebrates descend from an ancestral condition similar to that of amphioxus larvae. The startling aspect of that condition, of course, is that there was only one eye. Indeed, it has been suggested that the condition may possibly represent "a primitive undivided stage in eye evolution" (Lacalli, 1996, p. 244).

In recent decades the genetics of eye development has come to be better understood, in particular the means by which one eye becomes two in vertebrates. During development a single eye patch first emerges under control of the Pax6 gene, but a genetically controlled signaling molecule whimsically named "sonic hedgehog" causes it to split, leading to the development of paired eyes. Interrupting sonic hedgehog can lead to the failure of this process and the development of a single "cyclopic" eye in minnows, tadpoles, and mice (Butler, 2000; Chow & Lang, 2001). Amphioxus seems stuck at this developmental stage.

The seed of contralateral representation that was proposed by Cajal is present in the single amphioxus eye. The visual receptors are arranged in cup form and are shaded by the cup's sides (Lacalli, 1996). The eye is therefore a reversing eye, as in Figure 7.1. Neural output from the receptors largely project back along the same side (although receptors at the midline of the eye project to both sides; Lacalli, 1996). As a result, while the neural pathways themselves do not cross over, representation of the visual world is mostly contralateral because of image reversal (Figure 8.4, left).

A modified proposal

There have been other proposals for the origin of crossover, e.g., that it resulted from a physical "twist" that occurred in a proposed flatfish-like ancestor of vertebrates (de Lussanet & Osse, 2010). However, Cajal's reversing eye seems compelling as a proposed point of origin. It is proposed here that the emergence of paired eyes, which occurred no later than ancestral hagfish (Chapter 7), is what resulted in the full crossover of neural representation illustrated on the right side of Figure 8.4.

Evolutionarily, what is proposed is the following. First, a single reversing eye appeared. At roughly the same time, a partial motor crossover evolved in order to coordinate the two sides of an undulating tail. Contraction on one side of that midline structure needed to be paired with relaxation on the other, creating a need for motor commands to affect

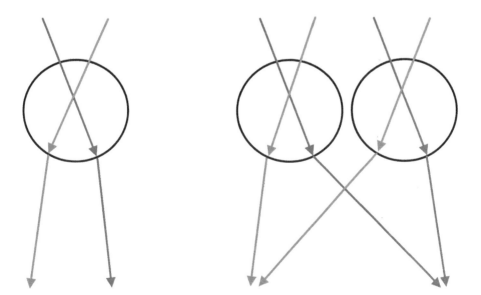

Figure 8.4 Illustration of the proposed origin of the visual crossover in paired reversing eyes. Left: Single eye, with same-side projection of each retinal half. Right: Paired eyes, following the same projection pattern.

Source: Original figure.

muscles on both sides. Partial crossover probably also allowed the use of timed opposite contractions, allowing the tail to cyclically flip back and forth (Lacalli & Kelly, 2003). Notice, though, that the effect of a partial motor crossover is to allow both sides of the body to be moved by either side of the nervous system. This is the arrangement of the extrapyramidal tract.

Next, the eye duplicated, an evolutionary success because the eyes provided somewhat differing views and thus improved surveillance. The duplicated eyes, it is proposed, simply followed the same projection pattern of the single eye: The left side of each eye back along the left side, and the right side of each back along the right side. The result was essentially the modern crossover of projections (Figure 8.4, right).

Given this simple beginning, the remaining puzzle pieces fall into place much as Cajal proposed. Because each half of the visual world was mapped onto the opposite side of the nervous system, selection also favored crossover in the auditory and tactile sensory modalities. That kept the global view of the world consistent. As motor responses came to be made not just by midline structures but by distal ones away from the midline (e.g., fins and limbs), increasing crossover continued to be favored in order to bring the motor and sensory worlds into congruence. Therefore, the corticospinal tract is contralateral.

Finally, paired eyes and crossover led to a divided brain with left and right cerebral hemispheres. Both were present in vertebrates as an early evolutionary emergence after chordates (Kapoor & Khanna, 2004). Thus it is proposed that a cascade of mutually reinforcing evolutionary changes caused crossover to become a primary organizing principle of the vertebrate world, all of it traceable to a reversing eye.

Conclusion

Although movement originated with flagellum-propelled bacteria, early free-swimming metazoa used cilia for thrust. Cilia were probably also used by benthic species for grazing, made possible by lubricating mucus secretions. Trace fossils indicate this type of locomotion existed about 570 million years ago.

Later, peristaltic waves from muscular contraction pulled animals forward. However, it is possible that muscles were initially used for spawning rather than locomotion. Hox gene-based segmentation of the body beginning with chordates allowed swimming by way of head-to-tail waves of body contraction.

The transformation from fish to tetrapods produced a shift in power from the pectoral fins to the rear limbs. Once bodies could clear the substrate, the rear limbs continued to power strides because of their anchoring to the tail. One tetrapod characteristic was a "step cycle" that moved the limbs in a particular order, the cycle having its origin in lobe-finned fishes.

Early amniotes had a sprawling posture that nevertheless held their bellies off the ground. About 125–150 million years ago, the sprawl gave way to an erect posture in mammals, spreading power more evenly between the limbs. Limbs evolved a third segment, forming a zigzag pattern that produced an up-and-down rather than side-to-side oscillation, an adaptation to uneven ground. A symmetric gait characterized by the simple alternation of sides, transformed to asymmetrical walking and trotting in which forelimbs and hindlimbs had different timing cycles. An example is the horse's gallop, which can leave all four limbs off the ground.

Primates, however, have a quadrupedal stride with diagonal sequencing and forearm extension, both of which are adaptations for an arboreal, fine-branch environment. In many species, heavy tails help maintain balance by producing a counterrotational shift of the body when moved to one side. Lifting the tail also helps stabilize the body during leaps. However, apes lost their tails, probably because they passed through a large-body phase of evolution in which a counterbalancing tail was of little help. Primates including apes also have an "amble" that allows continuous engagement with the substrate.

Knuckle walking, characteristic of gorillas and chimpanzees, has been thought by some a human ancestral trait. More recent evidence, however, has found features of the hominin wrist and humerus to be inconsistent with knuckle walking. More likely our hominin ancestors engaged in "generalized arboreality" with both above-branch and below-branch locomotion.

One prominent feature of the neurology of movement is that control of distal movement crosses over, with the left hemisphere controlling the right limbs and vice versa. This may have occurred to make motor control congruent with sensory representation. That also largely crosses over, an arrangement that probably traces to optical image reversal by the eyes. It is proposed that neuronal arrangement in the single-eyed amphioxus, in which each half of the retina projects straight back along the left and right sides, was simply duplicated when the eyes duplicated, producing true crossover of neural representation. Other senses followed crossover schemes to produce consistency in the global view of the world.

9   Bipedalism

At the onset of the hominin clade, primate quadrupedalism gave way to bipedalism. How and why this occurred have always been central issues in human paleontology. Many subsequent developments depended on a bipedal stance, for it freed the hands. In turn, our hands manufactured tools as well as the other multivaried artifacts of culture, giving form to human cognition.

As we have seen, the cranium of "Toumai", the 7-million-year-old hominin close to the chimpanzee-human divergence, shows a sign of bipedalism in a downward-opening foramen magnum. However, the lack of postcranial remains leaves residual doubt that it walked on two feet. In contrast, *Orrorin tugenensis*, a million years younger, leaves little doubt because of its long femoral neck among other bipedal features (Almécija et al., 2013). Certainly by 3.4 million years ago, the strongly bipedal characteristics of *Australopithecus afarensis* can leave no doubt that full bipedalism had evolved (Ward, Kimbel, & Johanson, 2011).

The most parsimonious conclusion is that bipedalism was present in some form since the onset of the human clade. Indeed, it may very well define it. If so, we need to know why it began, how it evolved, and what consequences came of it.

The causes of bipedalism

When considering the possible causes of hominin bipedalism, the influence of earlier forms of primate locomotion should not be overlooked. The primate extension of arms and rearward shift of weight can be viewed as preadaptations for later bipedal postures. In effect, with the hands freed of a major support role and body weight predominantly on the legs, only limited adaptations were needed to stand up. Of course, the question is why we took up habitual bipedalism, while other apes did not. This will be a good question to keep in mind as we consider several possible causes of our bipedalism.

Efficiency

Could it have evolved because it was a more efficient means of locomotion? For many years it was assumed, based on running, that it is energetically *inferior* to quadrupedalism. For example, Alexander (1992) reported that when running, humans are less efficient than a typical mammal of similar mass. It was concluded, therefore, that increased efficiency could not be bipedalism's cause.

Analyses since then have modified that conclusion. Steudel-Numbers (2001) found that at three miles per hour, human walking saves 20% of the energy used by a typical similar-sized mammal. Nevertheless, Steudel-Numbers discounted this difference because a third of similar-sized quadrupedal mammals proved more efficient than humans, and because early bipedal hominins would have walked less efficiently than modern ones. Halsey and White (2012) took a similar stance. They reported that walking is more efficient (and running less efficient) in humans than in the average similarly sized mammal, but deemphasized the outcome because humans fell within the overall range of comparison species.

However, Leonard and Robertson (1997) correctly pointed out that the proper comparison group is not mammals generally, but apes specifically. Thus the walking efficiency of the human ape need only have exceeded that of its ancestors for increased efficiency to help explain bipedalism's appearance. As it happens, apes are notably inefficient walkers: The human energy savings over their quadrupedalism falls in the astounding range of 50–75% (Leonard & Robertson, 1997).

Also under reconsideration is the assumption that early bipeds were notably less efficient walkers than modern ones. Initially *Australopithecus afarensis* was thought to have moved with a shuffling bipedal gait (Hunt, 1994), or even quadrupedally like a baboon but with occasional bipedalism (Wood, 1992). Further research soon indicated, however, that *afarensis* was a habitual biped (Ward, 2002). A notable *afarensis* feature is a long femoral neck (Figure 9.1), permitting an angling-in of the thighs to place body weight directly over the knees in a bipedal stance (Lovejoy, Meindl, Ohman, Heiple, & White, 2002).

Biomechanical simulations indicate that *afarensis* was capable of walking efficiently in an upright position (Wang & Crompton, 2004). Furthermore, it was most efficient at a speed of 1.0 m/sec (2.2 mi/hr), matching the speed inferred from adult footprint trails at

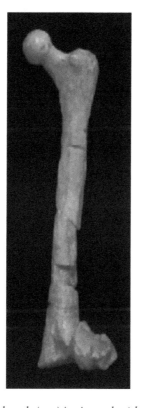

Figure 9.1 Femur with long femoral neck (top) in *Australopithecus afarensis*.

Source: Adaptation of Lucy blackbg.jpg, CC BY-SA 3.0, 120.

Laetoli (see Figure 5.6). Although *afarensis* was slower than modern humans, who walk most efficiently at 1.5 m/sec (3.4 mi/hr), it was also shorter. But in any case, its gait was notably more efficient than chimpanzee quadrupedalism, allowing the conclusion that it was "a fully competent biped" (Sellers, Cain, Wang, & Crompton, 2005, p. 439).

This conclusion is reinforced by close analysis of *Australopithecus africanus* femurs. The un-apelike greater thickness of cortex on the inferior as opposed to superior surface of the femoral neck is consistent with bipedalism. However, the femoral neck angle was reduced compared to modern humans, probably implying some side-to-side rocking, and thus a somewhat less efficient gait (Ruff & Higgins, 2013).

One problem with australopithecine efficiency analyses, however, is that they focus on a period postdating the appearance of bipedalism by some three to five million years. The limited fossil remains do not allow a determination of whether bipedalism in its *earliest* form was notably more efficient than ape quadrupedalism. Thus the efficiency issue cannot be fully addressed until more extensive fossils are found of the earliest hominin species.

Climate change, foraging, and thermoregulation

Several other reasons for adopting bipedalism have been proposed. One possibility is that hominins were pressured toward it by climate change that resulted in more widely spaced food trees. With longer traveling distances, bipedal apes may have been advantaged over quadrupedal ones (Stanford, 2003).

An alternative explanation emphasizes foraging behaviors. Perhaps bipedalism improved the ability to reach fruit on overhanging branches. In woodland and forest, chimpanzees naturally adopt bipedal postures when foraging from tree limbs (Hunt, 1994; Stanford, 2006). Furthermore, experimental presentation of elevated food sacks to chimpanzees increases bipedal postures, though of an "assisted" type in which the legs support most of the body weight with minor assistance from other body parts (Videan & McGrew, 2002). It thus seems plausible that bipedal stances improve foraging.

Another possibility is that upright posture improved thermoregulation by exposing less of the body to direct sun (Dávid-Barrett & Dunbar, 2016). Modeling shows a favorable thermal balance for bipedalism compared to quadrupedalism. It even suggests that thermal considerations may have played a role in out-of-Africa migration. Unable to cope with the heat of low coastal altitudes, australopithecines might not have been able to leave at all, making *Homo ergaster*, a more efficient biped, the first migrating species (Dávid-Barrett & Dunbar, 2016).

Carrying

A final possibility is that a need to carry objects produced a selective pressure toward bipedalism. One version of the hypothesis is that bipedalism allowed mothers to carry their infants (Hodges, 2017). Others, however, emphasize the role of inanimate objects. For example, chimpanzees carry rocks to nut trees, at distances up to several hundred meters, where they are used as hammers and anvils to crack tough shells (Mercader, Panger, & Boesch, 2002).

Even food itself is carried. Providing captive chimpanzees with piles of fruit results in upper limb carries using unassisted bipedality, i.e., with full weight on the legs (Videan & McGrew, 2002). If nuts are provided requiring a hammerstone, both are carried bipedally (Carvalho et al., 2012). Thus prized food items are carried sufficiently far that they can be consumed without interference.

Although the carrying explanation of bipedalism is intuitively appealing, it is not entirely problem-free. First, early hand-carried loads must have been very restricted in weight. Simulations suggest that loads were limited to 50% of upper limb mass in *Australopithecus afarensis*, in contrast to 200–300% in *Homo ergaster* and nearly 800% in modern humans (Wang & Crompton, 2004). In species preceding *afarensis*, hand-carriable loads may have been even more limited. A second problem with the carrying explanation is that it seems most applicable to open savanna, where widely spaced trees might have required the transport of stones from one to another. Yet the earliest known hominin, "Toumai", lived in a mosaic environment consisting not only of savanna, but also woodland, grassland, and forest (Brunet et al., 2004).

A closer look, however, suggests these may not be problems for the carrying explanation so much as they are important features of it. First, perhaps only light loads were involved, well within the severe weight restriction on hand carrying. Such light loads might have included simple vegetation-based tools and hunting weapons. Second, a mosaic environment may have necessitated carrying items into areas where they were otherwise unavailable. For example, a spear might be difficult to acquire in grassland. Thus a mosaic environment may itself have been an important influence on the adoption of bipedalism.

Chimpanzees certainly hunt small animals for meat, so it seems reasonable to assume that the last chimpanzee-human common ancestor did as well. Chimpanzees encountering a tree with red colobus monkeys follow a strategy in which several individuals drive monkeys into part of the tree crown or to the ground where they can be caught. Once caught, the monkey is killed either with a bite to the cranium using the canine teeth or by flailing it against the ground or a limb. Adult males are the primary hunters, and they succeed the majority of the time (Stanford, 1996; Watts & Mitani, 2002). Other encounters are more spontaneous, as when a young pig or antelope is encountered and quickly grabbed. The annual caloric intake from hunting is not insubstantial, by one estimate approaching 8–9% of total calories (Stanford, 1996).

Of course this style of hunting is mostly uninformative with respect to a carrying explanation of bipedalism, since no weapons are involved. Recently, however, a second type of hunt has been observed that is more directly relevant. Chimpanzees in Senegal have been found to prepare branches for use as spears, involving breaking off a living branch, stripping it of leaves and twigs, and modifying one end to sharpen it. It is then used to forcefully probe tree holes for galagos sleeping during the day (Pruetz et al., 2015). It seems quite possible that early hominins engaged in similar behavior, advantaging the carrying of a prepared spear from one place to another.

Hominin vegetation-based weapons would be consistent with weight-limited carries in that neither spears nor clubs need be heavy. They are also consistent with an observed shrinkage of canines, the fang-like teeth used by chimpanzees in aggressive encounters including hunting. Canine reduction occurred as far back as seven million years ago in "Toumai" (Brunet et al., 2004). If weapons were used, they may have eliminated the need for large canines, leading to their reduction according to the "use it or lose it" principle.

A hunting-carrying contribution to the evolution of bipedalism is thus consistent with a wide range of evidence, including observations of hunting in chimpanzees; a variable, mosaic habitat; limitations in upper limb strength; and reductions in canine size. However, it remains controversial. Some resistance to the idea can be accounted for by the absence of stone tools, including weapon points, until later when bipedalism was highly advanced. The rejoinder, of course, is that the earliest weapons would have been crafted as chimpanzees craft them, from perishable materials such as sticks. These would likely not be preserved in the fossil record.

Why aren't other apes bipedal?

Combining all of these considerations, it seems likely that bipedalism originated in behavior already present in the chimpanzee-human common ancestor. That ancestor was probably capable of bipedal postures, useful for foraging in trees, and of limited bipedal locomotion, useful for carrying. Indeed, carrying seems especially promising as a cause of habitual bipedalism. But there is a final potential problem: What differentiated early hominins from ancestral chimpanzees, leading us and not them to become bipedal? Indeed, this is a critically important question that all proposals for the origin of bipedalism must address.

In this regard it is useful to first ask whether the hominin-chimpanzee split was allopatric or sympatric, i.e., whether the separation between the lineages was physical or behavioral. Traditionally the East African Rift (Figure 9.2) is cited as a physical barrier, with chimpanzees isolated to the west and hominins to the east of the valley (e.g., Stringer & McKie, 1996, p. 24). However, this interpretation became less plausible following the discovery of "Toumai" in Chad, a location far west and north of the rift. More recent interpretations tentatively suggest sympatric speciation but don't identify the specific behavioral differences that might have led to the cessation of interbreeding (Elton, 2008; McBrearty & Jablonski, 2005).

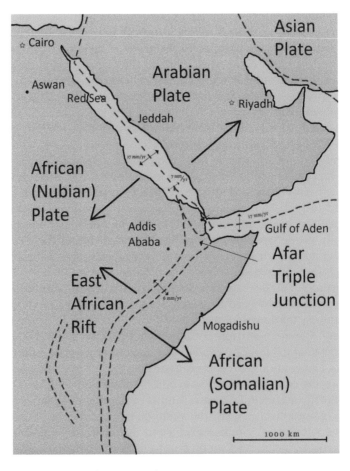

Figure 9.2 East African Rift, with tectonic plate movements.

Source: ATJ map (color).jpg, CC BY-SA 3.0, Razashah1.

It is proposed here that a critically important factor may have been the range of environments exploited by the two lineages. Although today's chimpanzees occupy a number of environments (McBrearty & Jablonski, 2005), there is presumptive evidence that this was not the case when their ancestors split from hominins. Specifically, there are no known chimpanzee fossils older than 550,000 years. As pointed out in Chapter 2, this is probably because most ancestral chimpanzees lived in hot, damp forest environments with acidic soils, promoting rapid decay (Orwant, 2005).

One must always be heedful of the adage, "The absence of evidence is not evidence of absence". However, in this case the complete absence of ancient chimpanzee fossils in places where hominins are (relatively) abundantly found argues that at the split, ancestral chimpanzees were restricted to forest environments.

To be sure, hominins also exploited forests – but they used other, more fossil-friendly environments as well. As noted, "Toumai" exploited a diverse mosaic environment that also included savanna, woodland, and grassland (Brunet et al., 2004; Elton, 2008). It seems possible that this explains why hominins, and not other apes, adopted bipedalism. Hominins needed to carry tools and weapons because a particular environment might not have appropriate source materials. Ancestral chimpanzees, in contrast, could depend on an unchanging environment to provide needed materials, substantially reducing their carrying needs.

Thus while the hominin-chimpanzee split likely involved sympatric speciation, the greater diversity of hominin environments may have pressured toward bipedalism in a way that did not apply to ancestral chimpanzees. Over time, behavioral differences then emerged that created reproductive barriers between the groups. While this hypothesis of course requires corroboration, it represents a potential path forward in understanding the causes of hominin-specific bipedalism.

Consequences of bipedalism

Once bipedalism became reasonably efficient, hominins could afford to embark on hunts that included the opportunistic collection of fruit and nuts as a secondary activity. This reversed the quadrupedal ape practice of foraging for fruit and nuts accompanied by opportunistic hunting. To the chimpanzee-human common ancestor, fruit was an imperative and meat a luxury; to later hominins, meat was an imperative and fruit a luxury. The long-distance hunt may have perfected human bipedalism (Stanford, 2003).

The achievement of habitual bipedalism was accompanied by significant changes in anatomy. Virtually all body segments were affected, head to toe. As we have already seen, the head moved to a more balanced position with respect to the trunk, visible in fossil skulls as a better-centered foramen magnum that opened straight downward rather than angling slightly backward (Figure 9.3). At the same time, the snout shortened, which improved head stabilization (Bramble & Lieberman, 2004).

The vertebral column and thorax

At least in the australopithecines (Ward, 2002), and probably even earlier in *Ardipithecus* (Lovejoy et al., 2009), bipedalism led to the vertebral column becoming sinusoidally curved, allowing the trunk to be carried vertically instead of hunched over. The shape acts as a spring (Martin, 1992). It also helps keep the body centered over shifting feet, which has been said (perhaps a bit apocryphally) to require "balance as fine as that required to spin a dinner plate on a broomstick" (Stanford, 2003, p. 47).

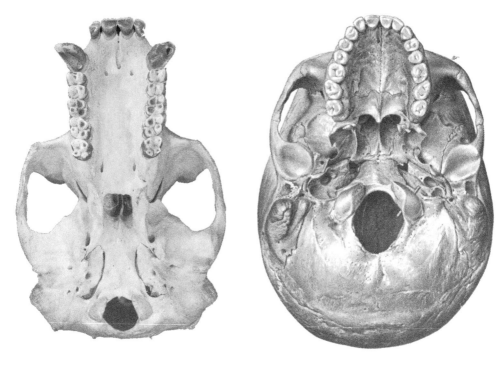

Figure 9.3 Underneath side of the gorilla and modern human skulls. The foramen magnum is pos-
teriorly shifted in the gorilla relative to the more central one in the human. Note that
the two skulls are not equally to scale.

Credits: Adaptation of Gorilla Male skull base.png, CC BY-SA 3.0, Didier Descouens, and Sobo 1909 41.png,
public domain, Dr. Johannes Sobotta.

The ribcage also changed, from the inverted-cone shape of apes to a more cylinder-like
shape in modern humans (Figure 9.4). The upper narrowing in apes allows their shoulders to more closely overhang the center of gravity, an aid to habitual arm-hanging from
tree limbs (Hunt, 1994). This narrowing was still present in *Australopithecus afarensis*
and *sediba* two to three million years ago (Schmid et al., 2013). It is often interpreted
as indicating preserved arm-hanging behavior in those species, despite their advanced
bipedalism on the ground.

However, it could simply be that the cone shape was retained until some additional
factor favored a cylinder. The Neanderthal thorax, for example, had a cone shape even
though Neanderthals were fully terrestrial. Possibly the additional factor favoring a cylinder was endurance running (Bramble & Lieberman, 2004). That is more efficient with
expansion of the top of the ribcage, a widening of the shoulders, and a generalized narrowing of the thorax (McHenry & Coffing, 2000). The changes allow greater counter-
rotation of the trunk relative to the hips (Bramble & Lieberman, 2004).

The pelvis

One of the most substantial areas of change, long-recognized and carrying high significance for childbearing, was in the pelvis. Our pelvis is lower and broader than that of a
nonhuman ape. The pelvic opening opens more distinctly downward, and the *ilium* (each
of the bladelike extensions at the back of the pelvis) wraps further forward around the

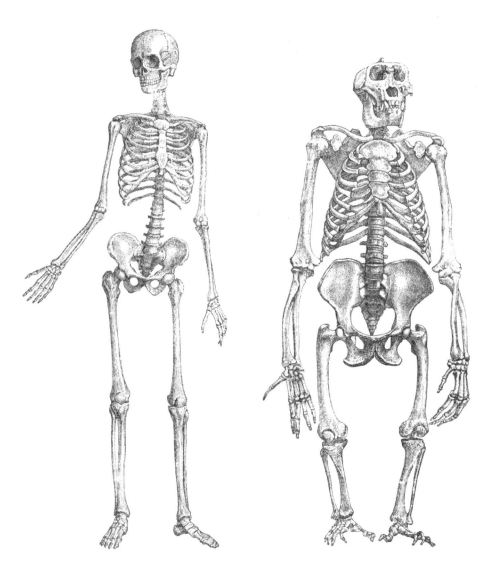

Figure 9.4 Human vs. gorilla skeleton, showing cylindrical vs. cone-shaped ribcages.

Source: Primatenskelett-drawing.jpg, public domain, unknown.

sides, providing a broad attachment surface for the hugely expanded *gluteus maximus* (butt) muscles. These muscles allow us to walk stably, with only one foot periodically in contact with the surface, unlike the side-to-side rocking of apes walking bipedally. In general the shape of our pelvis is more cup-like, cradling our internal organs (Martin, 1992; Stanford, 2003). These changes were well advanced by the time of the australopithecines (Figure 9.5, top). By the time of *Homo erectus* about a million years ago, the pelvis' birth canal had enlarged sufficiently to pass a baby's head sized within the modern range (Simpson et al., 2008). Modern humans have a pelvic opening that is more circular than that of australopithecines, better accommodating the shape of the newborn skull (Figure 9.5, bottom; Claxton, Hammond, Romano, Oleinik, & DeSilva, 2016).

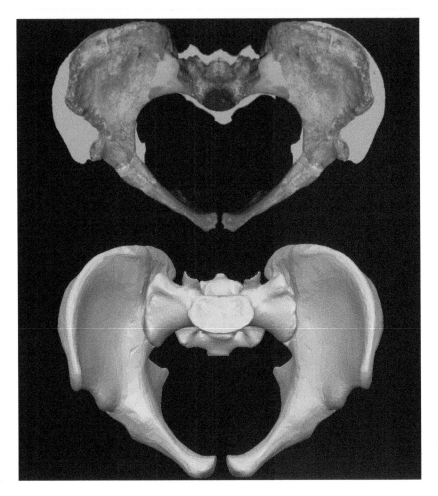

Figure 9.5 Pelvis in *Australopithecus sediba* (top) and a modern human (bottom). Note: Not to scale.

Credits: Adaptation of Pelvis MH2 Australopithecus sediba.jpg, CC BY-SA 3.0, Profberger, and Pelvis (male) 03-superior view.png, CC BY-SA 2.1 JP, BodyParts3D.

The arms and hands

Also in response to bipedalism, the arms shortened, as did the fingers, which straightened. The noticeable curvature of the fingers in *Australopithecus anamenis* and *afarensis* is a further argument, beyond that of the ribcage's shape, that these species remained partially arboreal. The fingers had straightened a bit by the time of *africanus* (Harcourt-Smith & Aiello, 2004; McHenry & Coffing, 2000; Ward, 2002), and continued doing so afterward.

The legs and feet

For their part the legs lengthened, allowing longer walking and running strides, and because of their weight, reducing the center of gravity for a more stable stance (Bramble & Lieberman, 2004; Martin, 1992). Reference has already been made to the longer femoral neck in early hominins, allowing the legs to angle in so that weight was distributed more directly over the feet (Lovejoy et al., 2002). Muscle distribution also changed, with hominins predominantly emphasizing the *quadriceps* (muscles on the front of the thigh that extend it). Apes emphasize the *hamstrings* and *adductors* (muscles on the back and inside of the thigh that move it backward and inward, respectively).

The hominin change in emphasis improved strides by powering the forward movements of the knee, allowing fuller extension. It also limited hip extension, which was less needed in walking than in tree climbing (Lovejoy et al., 2002). The Achilles tendon lengthened, improving the energetics of running through its spring-like storage of energy (Bramble & Lieberman, 2004).

As the only portion of the body in direct contact with the ground, the foot changed dramatically to promote both balance and propulsion. The arch of the foot became more prominent sometime after *Australopithecus afarensis* and allowed more efficient weight transfer while walking. In apes the *hallux* (big toe) is opposable and can be used to grasp limbs when climbing. In our ancestors this function was lost, and the hallux moved into the same plane as the other toes (Harcourt-Smith & Aiello, 2004).

Better documented is the angle of the hallux to the other toes, within that plane. In quadrupedally walking bonobos, the mean angle is about 32°, and one measured while walking bipedally showed an angle of 28° (Vereecke, D'Août, De Clercq, Van Elsacker, & Aerts, 2003). In the 3.66-million-year-old Laetoli footprints, believed to have been made by *Australopithecus afarensis*, the hallux was at a bonobo-like 25° angle – which, as noted in Chapter 5, is indistinguishable from that in a 5.7-million-year-old hominin trackway. In marked contrast, a 4.4-million-year-old *Ardipithecus ramidus* specimen had the hallux splayed away from the other toes at a whopping 68° angle, as approximated in Figure 5.3 (Lovejoy et al., 2009).

It is interesting, and relevant to the evolution of bipedalism, that even a 25° angle could produce dragging of the hallux, as in some prints from the Laetoli trackway (Figure 9.6). In 1.5-million-year-old fossilized footprints in Kenya, thought to have been

Figure 9.6 Detail of newly uncovered prints at Laetoli. The middle print reveals a dragging hallux as the individual stepped forward, evidenced as a trail behind the footprint.

made by *Homo ergaster* or *Homo erectus*, the hallux was at about a 15° angle. Now, in *Homo sapiens*, it is only about half that (Bennett et al., 2009), and we certainly do not drag our hallux.

The substantial departure of the *ramidus* measurement from the bonobo and *afarensis* results gives strong reason to doubt that *ramidus* was in the direct human ancestral line. Otherwise, the last four million years of hominin evolution has seen a smoothly progressing realignment of the big toe, increasing its efficiency in working with the other toes to promote bipedal locomotion.

Late anatomical developments in *Homo* helped improve bipedal performance when running. These included shortened toes commencing with *habilis* and lengthened legs beginning with *erectus*. Today, while we are relatively poor short-distance runners, we are fairly impressive at long distances. The speeds obtainable in endurance running, about 2.3–6.5 m/sec (5.1–14.5 miles/hr), manage to bracket the average 5.8 m/sec (12.9 miles/hr) speed of long-distance postal horses. However, this type of running is still metabolically more expensive, pound for pound, than it is for typical quadrupeds (Bramble & Lieberman, 2004).

Nevertheless endurance running may tie in with long-distance hunting. Although modern hunter-gatherers rarely engage in endurance running, it may have been an obligatory aspect of hunting before the invention of at-a-distance weapons like spearthrowers and arrows. Even if metabolically expensive, endurance running may have allowed animals to be run down through exhaustion, permitting hunters to approach and make a kill (Bramble & Lieberman, 2004).

Conservation and change in motor patterning

The emergence of bipedalism produced incremental change not just in anatomy but in motor patterning. However initially, bipedalism likely conserved already existing patterns.

D'Août and colleagues concluded that there are strong similarities in stance and gait between quadrupedal and bipedal modes of locomotion in bonobos. The angular displacement of joints is similar in the two modes, and the distribution of forces in the foot while striding are also similar although the heel tends to make greater impact during a quadrupedal stride. The roll-off of the foot at the end of the stride shows little difference between the two modes. The authors concluded that early bipedalism, assuming it mimicked these results, was likely a "free bonus" of movement patterns established with quadrupedalism (D'Août et al., 2004).

Conservation of gait during evolution is also stressed by proponents of a less terrestrial, more arboreal origin for bipedalism (Schmitt, 2003). Regardless of whether they are walking quadrupedally or bipedally, primate species in general use a "compliant" gait that absorbs shocks via *flexed* (bent) joints (Figure 9.7). The gait also prevents excessive vertical movement of the center of gravity and allows long strides. That is particularly beneficial to species walking in trees, because it minimizes branch shaking. Indeed, arboreal marsupials have compliant gaits similar to those of primates. Presumably a compliant style of walking was characteristic of early hominins, and only later did the stiff, noncompliant characteristics of modern bipedalism emerge (Larney & Larson, 2004; Schmitt, 2003).

The modern form was approximated by *Australopithecus afarensis* 3.4 million years ago. Analysis of its lower limbs indicates that the heel struck first while striding with the knee extended, which then extended further at mid-stride. This corresponds to modern, stiff bipedalism (Ward, 2002). Others analyzing the Laetoli footprints of 3.66 million years ago conclude that the *afarensis* stride was either virtually identical to that of

Figure 9.7 Flexed joints in a tripedal walk by a young chimpanzee.

Source: Adaptation of untitled, public domain, sarangib.

modern humans (Bennett, Reynolds, Morse, & Budka, 2016), or involved slightly more flexed limbs (Hatala, Demes, & Richmond, 2016). Nevertheless, all parties agree that compared to other primates, *afarensis* possessed a relatively stiff and less compliant gait, one close to the modern form. Even if it remained partially arboreal, it was well adapted to terrestrial bipedalism.

Our stiff gait's advantage is that it conserves about 70% of expended energy from one stride to the next, a figure nonhuman apes fall far short of achieving (Sellers et al., 2005). Thus it appears that a gait that minimized branch movement gave way to one that recovered energy between successive strides.

Conclusion

All indications are that bipedalism has been a hominin feature for about as long as there have been hominins. It was definitely present in *Orrorin* six million years ago, and very likely present in "Toumai" seven million years ago. Nevertheless, early hominins were simultaneously specialized for more than one mode of locomotion. Besides being well adapted for bipedal walking using a remarkably modern stiff, extended stride, *Australopithecus afarensis* also showed clear signs of arboreal adaptation in its cone-shaped thorax and curved fingers.

Both modes of locomotion make sense in terms of our evolutionary antecedents. The last common ancestor of chimpanzees and humans could probably stand and take a few steps bipedally, most often while foraging. Yet like chimpanzees, that ancestor was also comfortable in trees. Human bipedal walking is more efficient than ape quadrupedal walking, providing one reason for the evolution of bipedalism, but other factors must have been at work.

The factors pushing us into habitual bipedalism have become clearer, although they are not yet a matter of consensus. Naturalistic observation of our closest relatives indicates some capacity for bipedally carrying objects, a conclusion confirmed by experimental observations of food carrying. A reduction in the size of the canines six to seven million years ago in human ancestors likely indicates a reduced need for them in aggressive encounters. Together these observations suggest that carrying, including the carrying of simple weapons and other tools, provided impetus for us to walk upright. Other possible influences include improvements in upright foraging and in thermoregulation.

It appears that the split between hominins and chimpanzees was sympatric in that the environments they occupied overlapped. However, the absence of chimpanzee fossils of any significant age suggests that ancient chimpanzees were limited to hot, damp forests while hominins exploited a mosaic of woodland, grassland, forest, and savanna. If so, hominins may have had to carry tools and weapons to ensure that they had them in shifting environments, providing a possible reason why humans, but not chimpanzees, became bipedal.

An initially compliant bipedalism yielded over three million years ago to a stiffened gait that recovered energy from one step to the next. Then, further changes beginning about two million years ago adapted us for endurance running.

Throughout this evolutionary process, changes occurred in the hominin skeleton. The foramen magnum became more centered in the skull, allowing the head to balance during bipedal locomotion. The vertebral column became a curved spring. As hanging from branches became less common, the ribcage became more cylindrical than the cone shape found in nonhuman apes and early hominins, and curved fingers straightened. The pelvis became lower and broader, and its opening more circular to aid childbirth. The legs lengthened, and a longer femoral neck allowed them to angle in to distribute weight more directly over the feet (though this was to shorten again following changes to the pelvis). The feet developed a prominent arch, and the hallux came into the same plane as the other toes. The hallux also became much less divergent over several million years.

So much for our legs and feet. The next question was, what would we do with our freed-up hands?

Section III

Perception and cognition

10 Praxis and handedness

Whether or not carrying objects led to bipedalism, bipedalism certainly had profound effects on carrying objects. Standing upright freed the hands from any direct role in locomotion. From that point onward they could either wither away as an unnecessary metabolic expense, like the forelimbs of some theropod dinosaurs, or adopt new roles in the hominin lifestyle. That they took the latter course profoundly affected the history of cognition as it relates to the creation of objects and language.

In this chapter we first consider the nature of praxis and its underpinnings in the brain, before examining handedness and its primate origins and human components. The genetic foundations of handedness are considered, as is archaeological evidence attesting to its antiquity. Finally, we ponder why handedness exists, in both a right-handed majority and a left-handed minority.

Praxis

Consideration of how we voluntarily move, especially the hands but not limited to them, begins with *praxis* – the ability to conceive, initiate, and complete movement. Although this ability partially depends on motor pathways described earlier, brain mechanisms play a substantial role by allowing us to conceptualize and initiate movement.

Movement-related brain areas

Several movement-related brain areas were identified by early research on the effects of brain *lesions* (areas of physical damage, usually resulting from strokes or penetrating wounds). Lesions of the primary motor cortex, the rearmost strip of the frontal lobe (see Figure 10.1), can lead to paralysis. As we saw in Chapter 8, this strip originates the corticospinal tract that crosses to the opposite side. The body is systematically represented within it, with the tongue and face represented in the lowermost part, the legs and feet in the uppermost part, and the trunk in the middle. Thus a circumscribed lesion can paralyze a single part of the opposite side of the body. Observations like these indicate that the main function of the primary motor cortex is to initiate movement, passing signals down the motor pathways to produce muscular contraction.

Other brain areas, such as the premotor cortex and the supplementary motor area (Figure 10.1) also play roles in movement. For its part the supplementary motor cortex is activated during mental rehearsal of movement sequences. Lesions of it disrupt the voluntary control of movement, in part (if the left hemisphere is involved) by disrupting memories of sequential movements and their timing (Halsband & Lange, 2006).

A third area, the premotor cortex (Figure 10.1), plays a role in integrating spatial information with movement. The area activates when participants move using visual

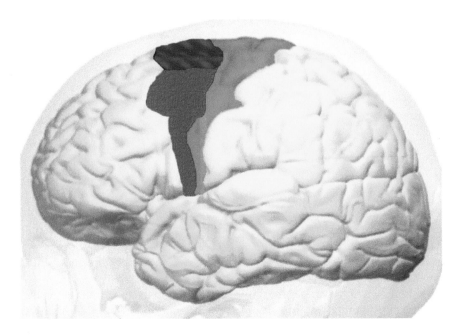

Figure 10.1 Motor areas of the frontal lobe. The primary motor cortex is in red, the premotor cortex is in green, and the supplementary motor area is in blue. The red and blue areas also extend onto the medial surface of the cortex.

Source: Adaptation of Brodmann area 4 lateral.jpg, CC BY-SA 2.1 JP, BodyParts3D.

guidance, as in reaching, writing, drawing, or in the laboratory, tracking a moving slot with a stylus. In the right hemisphere particularly, the area is involved when the integration of spatial information is demanding, as in the early phases of motor learning using visual guidance. Activation may switch to the left hemisphere later in learning (Halsband & Lange, 2006).

A notable aspect of the ventral portion of the premotor cortex is that it contains cells that fire rapidly when their possessor observes the actions of others. They also fire when he or she makes the same motions. Such *mirror cells* undoubtedly play a major role in skill learning through the observation and mimicry of others (Rizzolatti, Fogassi, & Gallese, 2002). In effect, they serve as a kind of preverbal communication system, with a teacher triggering neural activity in the learner that aids in copying behaviors.

The parietal lobe (Figure 10.2) plays an indirect role in praxis through spatial computation and the evaluation of sensory feedback. Although these could be viewed as perceptual, there is increasing recognition that the parietal lobe can play a more direct role in movement. Specifically, the intraparietal sulcus (Figure 10.3) helps direct eye movement, and in the left hemisphere helps produce gestures. It becomes active when hand motions are targeted to particular locations. Also, lesions in or near it interfere with reaching movements (Grefkes & Fink, 2005; Mühlau et al., 2005).

The functions served by the premotor cortex in the frontal lobe and the intraparietal sulcus in the parietal lobe are sufficiently similar that they are probably linked as parts of a larger system for the control of movement. Damage to either can result in disordered gestures due to spatial and timing errors. Although their functional differences are somewhat unclear, it may be that the premotor area is more involved in memory aspects

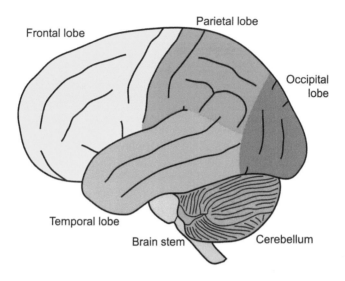

Figure 10.2 The cerebellum in relation to the lobes of the cortex and brainstem.
Source: Gehirn, lateral - Lobi + Stammhirn + Cerebellum eng.svg, CC BY-SA 3.0, NEUROtiker.

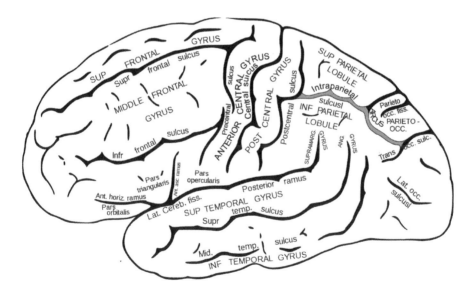

Figure 10.3 The intraparietal sulcus.
Source: Gray726 intraparietal sulcus.svg, public domain, Gray.

of movement, and the parietal area in spatial orienting and the positioning of body parts (Haaland, Harrington, & Knight, 2000).

A final brain area involved in praxis is the cerebellum. This large structure at the rear and base of the brain (Figure 10.2) plays a role in motor programming, specifically in "packaging" movements as a timed sequence of components. The cerebellum also directs real-time correction of movement due to its possession of overlaid sensory and motor maps of the body, permitting it to detect discrepancies and issue error messages (Kolb & Whishaw, 2003).

The evolution of movement-related brain areas

Cross-species comparisons indicate that movement-related brain areas evolved at varying times in the human ancestral line. The primary motor cortex traces at least as far back as the origin of eutherian mammals over 160 million years ago. In contrast, interconnected supplementary, premotor, and parietal motor areas date approximately from the origin of primates 83 million years ago (Stepniewska, Preuss, & Kaas, 2006). Tree shrews, close relatives of primates, do not have them (Kaas, 2004).

There have also been changes within the primate lineage. Even before the divergence of New World from Old World monkeys about 45 million years ago, control of the forelimbs by the premotor cortex had expanded from proximal to distal musculature, and stronger connections had formed between the premotor cortex and surrounding brain areas. These adaptations are thought to have been important to the evolution of reaching and grasping by primate forelimbs (Stepniewska et al., 2006; Sustaita et al., 2013).

In contrast, the cerebellum is a truly ancient structure. Its first clear appearance was in jawed fish, a group including the sharks and rays (see Figure 2–2). This places the origin of the cerebellum over 400 million years ago. It may be even older, consisting of a few cells in lampreys (Sugahara, Murakami, Pascual-Anaya, & Kuratani, 2017). If this *homology* (a similarity in traits traceable to common ancestry) is accepted, the origin of the cerebellum is pushed back to over 500 million years ago.

Evolutionary investigation of the cerebellum in primates has focused on its size relative to the rest of the brain. An enlargement in apes relative to monkeys suggests a size shift beginning about 32 million years ago. The enlargement may have been related to increases in movement complexity and motor planning (Rilling, 2006).

The primate origins of handedness

The importance of the comparative study of handedness is that *population right-handedness* (PRH) may serve as a behavioral marker for aspects of cognition that otherwise would be difficult to uncover. More particularly, cross-species comparisons of handedness may allow tentative conclusions regarding evolutionary changes in praxis.

Although species in which individuals show either a left or right limb preference are common, species in which most individuals prefer the same side are rare. *Population left-handedness* (PLH), in which a majority favor the left side, is sometimes reported in primates that diverged earliest from the human line. A 2005 review found that strepsirrhines tend to show PLH, an observation owing more to lemurs than to galagos or sikafas (Papademetriou, Sheu, & Michel, 2005). However, subsequent studies of lemurs have mostly failed to find significant population differences, although one reported PLH (Nelson, O'Karma, Ruperti, & Novak, 2009) and another PRH (Regaiolli, Spiezio, & Hopkins, 2016). Overall the results suggest that either the primate primitive character was PLH, or there was no population tendency at all.

Modern humans, of course, show a strong PRH, a derived character. Identifying the point of switchover to PRH is therefore of evolutionary interest. Following a massive review of great ape handedness, involving over 1500 individuals, Hopkins (2006) concluded that in tasks taken as a whole, orangutans show no sidedness, gorillas a significantly greater tendency than orangutans toward PRH, and the two chimpanzee species a significant PRH. While the review found bonobos to have greater PRH than chimpanzees, this was negated in a subsequent study (Hopkins et al., 2015).

A strengthening of PRH within primates? The roles of task and environment

These results, taken together, seem to lend themselves to the straightforward conclusion that there has been a constant strengthening of PRH since the orangutan divergence. However, detailed analysis shows that whether or not PRH is found in a species partially depends on the task used to assess handedness and the environment in which it is performed. For example, the supposedly unsided orangutans show PLH in the "tube task". In that task, one hand holds a tube from which food is extracted with the other, dominant hand. Orangutans also show PLH in self-touching (Hopkins, 2006; Hopkins et al., 2011). Yet when hand use is assessed in simple reaching or in leading locomotion, orangutans show PRH (Hopkins, 2006). Obviously, such findings complicate any simple conclusion that orangutans show no overall population handedness when tasks are considered as a whole. They indicate that PRH probably has multiple components with differing evolutionary histories.

The testing environment further complicates matters. In the past, Marchant and McGrew criticized the conclusion that chimpanzees show PRH, because the supporting evidence came mostly from captive rather than wild animals (Marchant & McGrew, 2007). If PRH is found only under artificial conditions, it would weaken the argument that it is a derived character with a traceable evolutionary history prior to hominins.

Hopkins (2006) examined the nature of the testing environment in his review and obtained ambiguous results. When handedness was treated as a continuous variable, environment had no effect, suggesting that chimpanzee PRH pertains to both captive and wild apes. However, when handedness was treated as a categorical variable, with animals classified into one of five categories (strongly right-handed, mildly right-handed, ambiguously handed, mildly left-handed, and strongly left-handed), environment did have an effect, with captivity seeming to shift PRH toward a stronger form than that seen in the wild. Nevertheless, PRH was found in both captive and wild animals. Most recently, a study of wild-born orphan bonobos in an African sanctuary tends to confirm the existence of PRH (Neufuss, Humle, Cremaschi, & Kivell, 2017), indicating that it is not just an artifact of close captivity.

That conclusion is fortunate, because a critically important piece of the PRH puzzle comes from a captive chimpanzee study (Wesley et al., 2002). Each of 100 chimpanzees were given six tasks multiple times, allowing their handedness to be assessed for each task. The results were then subjected to a statistical technique known as *factor analysis*, allowing relationships among the tasks to be determined. For example, if individuals who show right-handedness for one task also show it for a second task, while those who show left-handedness for one also show it for the other, the two tasks are related. Factor analysis will assign them to the same group of measures (i.e., the same factor).

Wesley et al. identified two factors in their study. One consisted of the handedness tasks of reaching (picking up food while in a tripedal posture), gesturing (using one hand in a communication gesture), biased feeding (manipulating food placed systematically into the left or right hand), and to a lesser extent, feeding (using one hand to bring food to the mouth while holding other food with the other hand). A second factor consisted of the tube task in its original form, and a modified version in which only one end of the tube was inserted into and affixed to the animal's cage.

Because two different factors were found, they represent two separate components of handedness, with one involving reaching, gesturing, and feeding, and the other involving tube task handedness. Confirmation that these are two separate factors came from a later developmental study, reporting that reaching handedness in young chimpanzees predicted gesturing handedness a decade later, while tube task handedness did not (Hopkins, Russell, & Freeman, 2005).

Components of primate handedness

The significance of the factor analytic results is that they indicate the existence of at least two different components of ape handedness with potentially different evolutionary histories. Examining the components might help account for some of the reported handedness differences between species.

Accordingly, it is important to consider individual tasks, with the expectation that their results may differ from those based on tasks as a group. Considering reaching first – partially comprising one of the factors in Wesley et al. (2002) – both orangutans and bonobos (and in strong trends, common chimpanzees and gorillas) have shown PRH (Hopkins, 2006). Similarly, in studies of chimpanzees and gorillas by Prieur and colleagues, gestures – likewise part of the factor – have shown striking PRH (Prieur, Pika, Barbu, & Blois-Heulin, 2016a, 2016b). Altogether, then, the reaching and gesture component of handedness seems likely to have shown PRH from *before* the divergence of orangutans, not afterward as suggested by tasks-as-a-group analysis.

In contrast, in the tube task – the other factor in Wesley et al. – orangutans have shown PLH (Hopkins, 2006). So have the earlier-diverging lesser apes (Morino, Uchikoshi, Bercovitch, & Hopkins, 2017). Lesser apes have also shown PLH when collecting and drinking water from holes in tree trunks (Morino, 2011), a behavior somewhat paralleling the tube task. Chimpanzees, however, have shown PRH in the tube task, as have gorillas to a seemingly similar though nonsignificant degree (possibly due to a much smaller sample size; Hopkins, 2006). Thus the tube task component of handedness seems to have shifted from PLH to PRH sometime after the orangutan divergence.

Putting these considerations together implicates at least two independent handedness developments in the pre-hominin ancestral line. PRH in the kind of motor activity involved in reaching, gesturing, and feeding is an older character predating the orangutan convergence 17 million years ago. It is likely even more ancient, because communication gestures in baboons and Tonkean macaques, both Old World monkeys, likewise show PRH (Meunier, Fizet, & Vauclair, 2013). These findings suggest that the core of the reaching-gesturing-feeding component of handedness extends back beyond the Old World monkey divergence 32 million years ago. Given the nature of the underlying tasks, this component is perhaps best characterized as unskilled handedness.

In contrast, PRH in the kind of motor activity involved in the tube task apparently postdates the orangutan divergence. However, it probably predates the gorilla divergence, not only because gorillas show a strong tendency toward PRH on the tube task, but also because gorillas are significantly right-handed in the similar task of bimanual feeding. There, one hand (usually the left) provides grip, and the other hand (usually the right) performs delicate food preparation actions (Hopkins, 2006; Tabiow & Forrester, 2013). This seems conceptually similar to the tube task. The cooperative bimanual praxis involved in both seems more highly skilled than reaching-gesturing-feeding, and so this component is perhaps best characterized as skilled handedness.

However, it is important to note that there is evidence of a third component of chimpanzee handedness not captured by the factor analytic study of Wesley et al. *Termite fishing* is a behavior observed in wild chimpanzees, in which a twig is prepared and then inserted into termite mounds. Termites defensively latch on to the twig, and are then pulled out of the mound with it and are eaten. Combining small samples of observations collected in the wild, Hopkins (2006) found a significant PLH, not PRH, for this behavior. However in the largest termite-fishing study to date, there was no significant difference between hands (Sanz, Morgan, & Hopkins, 2016).

Altogether then, termite fishing shows either PLH or no hand difference. Because it is not in accord with the more general PRH that is observed, it presumably represents a third component of chimpanzee handedness. Notably, termite fishing involves tool insertion. It differs not only from reaching-gesturing-feeding, but also from the tube task in that tools, not fingers, perform the required action. This could emphasize spatial perception more than other handedness behaviors and lead to the participation of right hemisphere/left-hand control mechanisms. The importance of this third component of handedness will soon become apparent, when distinctions among unskilled, skilled, and insertion handedness are shown to be relevant to human handedness as well.

Components of human handedness

Human PRH is much more pronounced than that of other primates. Whereas right-handedness is characteristic of about 60% of chimpanzees as assessed in a broad variety of tasks (Hopkins, 2006), it characterizes somewhat over 90% of humans (Cochet & Byrne, 2013; Prieur, Barbu, & Blois-Heulin, 2018). The percentage of human right-handedness is about the same, whether it is assessed through actual task performance, or by questionnaire as more commonly practiced. PRH is universal in all human populations that have been studied, although there is some variation in incidence. Thus, as assessed by throwing, the incidence of left-handedness in Papua New Guinea is over twice that in the United States, with the United Kingdom falling between these extremes (Raymond & Pontier, 2004).

Extrapolating from primate observations, it seems reasonable to expect at least three independent components to human handedness: An unskilled component, a skilled component, and a component involving insertion. This has been confirmed. One of the largest factor analytic studies of handedness, both in terms of participants (1275) and the number of assessed handedness behaviors (47), was conducted in Japan and Canada by Ida and Bryden (1996). Questionnaire data were subjected to factor analysis, and three factors were extracted. One factor involved unskilled handedness and was comprised of 16 items including waving, pointing, and picking up small objects. Another factor reflected skilled handedness and was comprised of 19 items including writing, drawing, hammering, throwing, using a toothbrush, and striking a match. Finally, the third factor involved handedness as assessed by insertion and turning movements. This factor involved 12 items including inserting a key, pulling a key, turning a key, and shuffling cards (which, it should be noted, is an insertion task). The first two factors substantially replicated those of an earlier study involving fewer participants and fewer handedness items. Although the earlier study did not use any of the insertion, turning, or card playing items of the later one, it is intriguing that an item involving winding a stop watch, which involves a turning motion, itself defined a separate factor, apparently prefiguring the emergence of that factor in the later study (Steenhuis & Bryden, 1989).

Thus the same three handedness components found in chimpanzees (skilled, unskilled, and insertion handedness) return strong echoes in human data. This is in spite of the much higher incidence of generalized right-handedness in humans than in chimpanzees.

Clearly, an important question to ask about human PRH is whether it can be attributed to a simple strengthening of the components present in chimpanzees. Some strengthening has certainly occurred, as is evident in the data of Ida and Bryden (1996) and others. For the skilled component, 94% of their participants were right-handed using a

Table 10.1 Three components of PRH, as defined by factor analysis, with inferred hominin trends since the last chimpanzee-human common ancestor. aft = after, bef = before, diff = difference, mya = million years ago.

Component	Appearance	Examples	Hominin trend
unskilled	bef 32 mya	reaching, gesturing, pointing	unchanged
skilled	~ 13 mya	tube task, writing, hammering	PRH → greater PRH
insertion-turning	aft 7.5 mya	termite fishing, key insertion	PLH or no diff → PRH

forced-choice definition of handedness (i.e., splitting those with ambiguous handedness into left- or right-handers). This figure contrasts markedly with the 57% figure found for the skilled tube task in chimpanzees (Hopkins, 2006). Similarly, for the insertion-turning component, 89% of humans were right-handed, compared to between 37% and 54% for chimpanzee termite fishing (Lonsdorf & Hopkins, 2005; Sanz et al., 2016). Thus it is clear that strengthening occurred in these two components of handedness, accompanied by a possible reversal in the insertion-turning component from PLH to PRH. Perhaps an initial involvement of the right hemisphere of the brain in insertion tasks, because of their spatial component, has in us given way to predominant left hemisphere involvement due to motor programming.

Against this background, a strikingly different result was found for the remaining component. For unskilled handedness, 58% of humans were right-handed (Ida & Bryden, 1996). This differs negligibly from the 54% value for chimpanzee reaching, among those individuals showing significant handedness in reaching. It also differs minimally from the 60% value found in all apes for reaching, gesturing, and feeding taken collectively, although ape gesturing taken alone shows markedly more right-handedness (Prieur et al., 2016a, 2016b; Hopkins, 2006).

By the criterion of percent incidence, then, the phylogenetically oldest (unskilled) handedness component appears the least changed in humans, while two newer components related to skilled, precise hand use have undergone the greatest change. Changes in PRH thus seem to parallel increased human capabilities in praxis. Table 10.1 summarizes some of the differences between handedness components.

Genetic foundations

Tracing the evolutionary background of handedness implicitly assumes that it is under genetic influence. Otherwise, it would not be subject to natural selection. But is this a reasonable assumption? In this section, we consider evidence for a genetic foundation of handedness.

Family resemblances

If handedness is under some at least some genetic control, there should be resemblances between family members. Chimpanzee studies generally support this. Thus in wild chimpanzees using tools, or performing the tube task, an individual's handedness is moderately related to that of its mother (Hopkins, 2006; Hopkins, Wesley, Russell, & Schapiro, 2006). Sibling resemblances pertain as well: For handedness assessed by throwing hand, siblings show *concordant* (same-sided) handedness to an extent greater than chance (Hopkins et al., 2005).

These results are certainly consistent with a genetic influence on handedness. Caution should be observed, however, because they may also be consistent with nongenetic factors, i.e., mothers or siblings may introduce environmental pressures toward similar handedness.

In human handedness research, twin studies have a long history, with the first handedness comparison of *monozygotic* (identical) and *dizygotic* (nonidentical or fraternal) twins dating back to research by Siemens in 1924 (Medland, Duffy, Wright, Geffen, & Martin, 2006). Greater concordance in handedness between monozygotic twins is reported, an outcome consistent with a genetic influence although interpretation is clouded by greater similarity in the prenatal and postnatal environments of such twins. Medland et al. (2006) reviewed 34 studies, beginning with Siemens's work, and concluded from concordance rates that about 25% of variation in handedness can be attributed to genetic factors. The remaining 75% is due to unknown environmental factors.

Thus, the scientific literature examining family resemblances is certainly consistent with a genetic influence on handedness, although it cannot be regarded as completely conclusive.

Genetic models

Several specific genetic models of handedness have been proposed. The most influential is the *right shift theory* of Marian Annett, a version of which was first proposed in the 1970s. This states that human handedness is the product of a bell-shaped distribution in combination with a right-shift allele that is present in some individuals and not others. By itself, the bell-shaped distribution would lead to equal numbers of left- and right-handers, as well as a large proportion of persons with mixed handedness. However, according to the model, the distribution is shifted rightward in those with the allele, resulting in a predominance of right-handedness. Furthermore, depending on the task, the subpopulation of individuals with two copies of the allele, one on each of a pair of chromosomes, may be shifted more strongly rightward than those with only one copy (Annett, 1998). More recently, Annett addressed the issue of how the model might account for the reduced PRH in chimpanzees, reporting a right-shift factor similar to that in humans but reduced in strength (Annett, 2006).

The right shift theory makes no distinction between genders, and so the involved gene is presumed to be an *autosomal* one – i.e., a gene located elsewhere than on the sex-determining X and Y chromosomes. McManus and Bryden (1992), however, proposed that there is also a recessive modifier gene located on the X chromosome that can suppress the effect of the autosomal gene. Such a gene, they argued, could account for the existence of a "maternal effect" in which left-handed mothers tend to produce more left-handed offspring than do left-handed fathers (McKeever, 2000; McManus & Bryden, 1992).

A problem with both the autosomal and X-chromosome right shift theories is that testing them largely depends on fitting handedness frequency data to the models. While the fits can be quite good, such data could be viewed as insufficiently weighty to establish the models' validity. This is particularly the case when different studies produce different frequencies. For example, McKeever (2000) reported data showing that in families with one left- and one right-handed parent, left-handed mothers produced left-handed sons (a) more frequently than did left-handed fathers, and (b) more frequently than they did left-handed daughters. Both observations are consistent with a handedness-related gene on the X chromosome, but as McKeever pointed out, earlier studies failed to produce similar findings.

Individual genes

More convincing evidence may emerge from studies relating specific genes to handedness. At least four have been identified so far. Laval and colleagues (1998) measured handedness by the speed of movement of pegs across a pegboard, and found an allele on the X chromosome to be associated with left-handedness. However, the result is not as supportive of the McManus and Bryden version of right shift theory, as initially appears, because right shift theory states that there is no genetic left-handedness, only genetic right-handedness.

A second discovery was that a pegboard measure of handedness correlated to the presence of a specific gene on chromosome 2 (Francks et al., 2003), since designated LRRTM1. So far, though, the relationship has been found only in clinical, and not non-clinical, populations (Beste et al., 2018). Also, the gene acts in a manner quite unlike that predicted by the right shift theories, with greater father-to-child than mother-to-child genetic similarity across multiple samples.

A third gene affecting handedness appears to be located on chromosome 12. It is related to writing and drawing – i.e., the skilled component of handedness (Warren, Stern, Duggirala, Dyer, & Almasy, 2006). That is one of the autosomal chromosomes, which might favor Annett's right shift theory. However, it remains to be seen whether a specific genetic locus can be identified and if so, whether the gene works in right-shift fashion.

Finally, an allele of a gene on chromosome 15, designated PCSK6, has been found to be correlated to handedness as measured by a pegboard task. In an individual, each copy of the allele shifts handedness rightward a small amount. Yet so far, the correlation has been found only in reading-disabled individuals, not in normal readers (Brandler et al., 2013).

It should be noted that if there are several genes influencing handedness, all operating in different fashion, it is unlikely that a simple genetic model of handedness will be found to apply. In that regard, multiple studies of either the entire genome or *exome* (the reduced portion of the genome that codes for proteins) have failed to identify *any* genes with a major influence on handedness (Armour, Davison, & McManus, 2014; Kavaklioglu, Ajmal, Hameed, & Francks, 2016). An effect of PCSK6 is found only when results are combined across studies. The smallness of the effect implies that the genetic component of handedness likely involves many genes, each having a very minor influence (Brandler et al., 2013).

Even so, indirect evidence from family resemblances as well as direct evidence from studies of specific genes does support the existence of genetic influences on human handedness. These discoveries lend weight to the outcome of comparative analysis indicating that human handedness is an elaboration of earlier forms of primate handedness, and that it has been subject to evolutionary forces.

Archaeological evidence

Having considered the evolution and genetics of handedness, and more specifically of PRH, a logical next question is to ask how far back hominins showed evidence of it. As it happens, there is a rich and continuous record of PRH dating back nearly two million years.

The fossils of two hominins from 1.5 to 1.8 million years ago both suggest the presence of right-handedness, though in different ways. An analysis of teeth from a specimen attributed to *Homo habilis*, dating from 1.8 million years ago, indicates an increased number of scratches on the individual's right. Presumably the scratches resulted from holding material in the teeth while cutting with a stone tool held by one hand, usually the right, resulting in occasional accidental contact with the teeth (Frayer et al., 2016). In the other case, a 1.5-million-year-old *Homo erectus* individual, a larger deltoid (shoulder) muscle attachment groove on the right clavicle led to an assessment of right-handedness (Cashmore, Uomini, & Chapelain, 2008). While only two cases are not conclusive evidence of PRH within their time period, they are suggestive.

The oldest *artifactual* evidence of PRH in humans derives from the flakes and cores of stone tools. When modern tool knappers work stone, they tend to hold the *core* (the stone from which the tool is made) in the left hand, striking *flakes* (stone fragments) from it using a striking stone held in the right hand. This distribution of effort undoubtedly traces to the greater skill and precision of the right hand of most people.

The largely left-right distribution of core and striking stone has consequences for the flakes struck from the core. Examination shows that the flakes have a characteristic form involving the *striking platform* (the small flat surface where the blow landed). According to Toth (1985), as the core is rotated clockwise for repeated striking, the striking platform of a flake struck off by the right hand has a scar to the left and a weathered outer surface to the right, when oriented upward. A left-handed knapper would presumably show the opposite orientation of flakes, resulting from holding the core in the right hand, striking flakes off with the left, and rotating the core counter-clockwise. Thus collecting flakes, and analyzing the location of the scar relative to the striking platform, may allow an inference about the handedness of the knapper.

Modern-day counts of flakes produced by right-handed knappers show that the numerical difference between "left-handed" and "right-handed" flakes is not large, perhaps because these tendencies are inherently small in size, but perhaps also because a fatiguing knapper may strike differently or even change hands from time to time. Nicholas Toth, an experienced right-handed knapper, counted his own directional flakes and found that 56% of them were right-handed while 44% were left-handed (Toth, 1985). Similarly, Pobiner (1999) reported that seven right-handed students produced right-left flakes in a 60–40% split. Less strikingly (no pun intended), Bargalló and Mosquera (2014) reported knapping results from eight right-handed and seven left-handed university students in Spain, which can be characterized as a 51–49% right-left split for right-handers and a 50–50% split for left-handers. However, the Spanish splits are based on median values across participants, which may minimize differences compared to the overall frequency counts used by Toth and Pobiner.

The relevance, of course, is that the same analysis can be applied to archaeological collections of flakes. The oldest that have been analyzed were discovered at Koobi Fora, Kenya, and date to 1.5 million years ago. When classified, 57% were found to be right-handed (Toth, 1985), a significant deviation from 50% given the sample size. The figure is not significantly deviant from the 60% proportion produced by the modern right-handed knappers of Pobiner. The Koobi Fora knappers therefore appear to have been predominantly right-handed.

The second-oldest sample is from Ambrona, Spain, dating to about 350,000 years ago, where 61% of the flakes were right-handed (Toth, 1985), again significantly divergent from a 50% value. A sample of Neanderthal flakes, dating to about 180,000 years ago, also showed PRH (Cornford, 1986). An archaeological sample of flakes produced by Australian aborigines, of uncertain age but less than 50,000 years old, showed a right-handed percentage of only 51% (Pobiner, 1999), which did not differ significantly from the modern 60% value. Thus the simplest interpretation of the combined results is that humans have shown consistent PRH from the time of the oldest stone tools that have been assessed, dating from 1.5 million years ago.

Nevertheless, some caution should be observed before accepting these results. Uomini (2006) argued for the rejection of Toth's methodology, in part because she observed no consistency across modern knappers in the direction of rotation of the core. It is not clear, however, how the repeated observations of asymmetry in archaeological assemblies of flakes are otherwise to be explained, except by predominant right-handedness.

Uomini did accept as valid another study of asymmetry in lithic assemblages, that of Phillipson (1997). After analyzing modifications to the edges of 54 handaxes from Kenya, dated at about one million years old, Phillipson concluded that 51 were probably intended to be used by one hand or the other. Of those, 45 (88%) were intended for right-handers, and 11 (12%) for left-handers. Note that besides enjoying a potentially better-grounded methodology than that of Toth, this approach has the additional advantage of specifying specific handedness percentages rather than a general population tendency.

As already indicated, teeth can also indicate handedness through asymmetric scratching, pitting, and chipping by tools. You can simulate this by using your dominant hand to place a pen cap between your teeth, and carefully guiding the pen into it. Then imagine that you missed with an edged stone!

A sample of damaged teeth from *Homo heidelbergensis* in Spain, dating to before 530,000 years ago, revealed that all 20 individuals were right-handed (Lozano, Mosquera, de Castro, Arsuaga, & Carbonell, 2009). Neanderthals, too, show handedness with this methodology, with 29 of 31 individuals (94%) right-handed across two samples (Estalrrich & Rosas, 2013; Fox & Frayer, 1997). It therefore seems that both *Homo heidelbergensis* and *Homo neanderthalensis* showed strong PRH.

Tools also figured in an assessment of handedness by Semenov (1970), reporting that stone scrapers from upper Paleolithic sites (10,000–40,000 years ago) showed traces of greater wear on the right side, indicating right-handed use. Some 80% of scrapers were classified as right-handed, a trend claimed to be visible even in published depictions.

Cave art supports the existence of PRH in *Homo sapiens* 10,000–30,000 years ago in the form of *negative hands*, each of which is the outline of a hand held against the rock and painted by blowing pigment through a tube held by the other, usually dominant hand. Of 507 known negative hands from caves in France and Spain, 77% are outlines of the left hand, implying PRH. Confirming this, the identical 77% figure has been observed in a large sample of university students producing negative hands with a similar technique (who, of course, would be expected to be about 90% right-handed). Furthermore, in these modern would-be troglodytes, throwing and writing handedness was correlated with which hand held the tube (Faurie & Raymond, 2004).

Finally, a study of artistic representation of hand usage over the past 5000 years reveals a stable PRH over that period. Instances of one-handed use of a tool or weapon were tallied on the assumption that models held these artifacts in their preferred hands. Across the years, with only minor and seemingly random variation, 93% were right-handed (Coren & Porac, 1977).

In summary, the archaeological evidence pertaining to human handedness is consistent: PRH has existed over at least the past 1.5–1.8 million years, probably on the order of 90% of individuals. In their general direction, these conclusions conform to expectations from the observed PRH in apes, and they suggest that the quantitative shift, from the approximately 60% PRH found in apes to 90% in humans, dates back at least to the beginning of the genus *Homo*. It may date even further back, as archaeology is silent on the matter before 1.8 million years ago.

Why handedness?

Perhaps the most important question about handedness, specifically PRH, is why it exists at all. The answer is likely to imply something important about the evolution of human cognition as expressed through hand movement.

It is proposed here that PRH reflects the evolution of mechanisms of praxis that brought the timing and sequencing of movement under control of the left hemisphere of the brain. Having begun as a simple mechanism underlying the relatively gross movements of reaching, gesturing, and pointing, it became progressively differentiated into a more skilled component supporting precision movement about 13 million years ago, and then, sometime after 7.5 million years ago, one showing sidedness in the complex insertion and turning manipulations of tools. Throughout this evolutionary history, operation of left hemisphere mechanisms of praxis presumably drove a preferential use of the right hand simply because of the predominance of contralateral motor connections. The advantage was a system that operated more smoothly and with less delay than one requiring cross-hemisphere coordination.

At present it is difficult to reach definite conclusions about the evolution of brain changes that supported increasing manual skill in general and PRH in particular. However, three brain areas have been identified that will very likely prove relevant in future research.

First, monkey and human unskilled movements have in common an activation of primary motor cortex opposite the hand (Rizzolatti et al., 2002). It *may* be, therefore, that the basis of the most primitive form of PRH is asymmetry in the hand area of the primary motor cortex, possibly dating to before 32 million years ago. This asymmetric anatomy appears to exert influence even over skilled handedness, although that likely involves other anatomical regions as well. Thus in chimpanzees, not only reaching and feeding but also tube task handedness, all appear to be related to asymmetry in the volume of the hand area of the primary motor cortex (Dadda, Cantalupo, & Hopkins, 2006; Taglialatela, Cantalupo, & Hopkins, 2006).

A second highly relevant brain area is a portion of the ventral premotor area in Old World monkeys known as F5. This area appears to be anatomically homologous to human Brodmann area 44 (Figure 10.4), located in the inferior frontal gyrus (Rizzolatti et al., 2002). While area 44 exists on both sides of the brain, in the left hemisphere it is part of Broca's area and plays a major role in the production of speech. Notably, gesture,

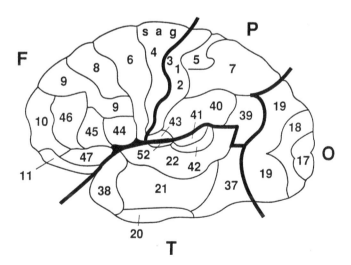

Figure 10.4 Brodmann areas of the human brain. Lobes: F = frontal, P = parietal, T = temporal, O = occipital.

Source: Brodmann areas and lobes of the left hemisphere, licensed from Elsevier under STM Permissions Guidelines.

hammering, and termite-fishing handedness – representing all three of the handedness components in Table 10.1 – are related to asymmetry in the chimpanzee inferior frontal gyrus, with it tending to be larger on the left (Hopkins, Russell, & Cantalupo, 2007; Taglialatela et al., 2006).

F5 is of note because it is the location of mirror cells, which in the monkey are active during the production of specific actions such as grasping, holding, and tearing. They are also active when the same actions are observed in others. In humans, area 44 becomes active not only during speech production, but also when manipulating objects and especially when executing precision grips, as we often do when using tools. Simply observing tool use, or even tools themselves, also results in area 44 activation (Rizzolatti et al., 2002). Thus it appears to be one location of mirror cells in humans, with neighboring area 6 likely representing another (Cerri et al., 2015; Cook, Bird, Catmure, Press, & Heyes, 2014; de la Rosa, Schillinger, Bülthoff, Schulz, & Uludag, 2016).

Significantly, in chimpanzees, use of a precision index finger–thumb grip is more frequently performed by the right hand than by the left (Hopkins et al., 2005). This seems similar to human right-handedness in object manipulation and tool use. Therefore, it may be that the evolution of asymmetry in area F5/44, favoring the left hemisphere, is what supported the fuller emergence of the more skilled form of PRH that is observed in primate communicative gestures, and that in humans supports the use of simple tools. As indicated earlier in this section, comparative analysis suggests this development occurred about 13 million years ago.

However, the area itself is obviously older than its asymmetry. Macaques have an area F5/44 (Petrides & Pandya, 2009), suggesting that it emerged no later than the time of the Old World monkey divergence 32 million years ago. That put a critical anatomical substrate into place long before the development of skilled handedness. Its seeds may date back even further than that, because mirror neurons have been found in the marmoset homolog of area F5 (Suzuki et al., 2015), and the New World monkey divergence was 45 million years ago.

A final brain area that has been implicated in primate handedness is the "*pli-de-passage fronto-parietal moyen parietale*" (PPFM). This is a gyrus buried along the central sulcus, connecting the precentral and postcentral gyri; i.e., under the dark line between Brodmann areas 4 and 1–2–3 in Figure 10.4. In chimpanzees, differences in speed of performance of a simulated termite-fishing task correlate to anatomical asymmetry of the PPFM, such that individuals who are faster with the right hand show a larger PPFM as well as a larger inferior frontal gyrus in the left hemisphere. Individuals faster with the left hand show the opposite, a larger PPFM and inferior frontal gyrus in the right hemisphere (Hopkins et al., 2017). Unfortunately, it is not yet known whether this result is specific to the insertion-turning component of handedness, or generalizes to the unskilled and skilled components as well.

While asymmetries of the primary motor cortex, F5, and PPFM are correlated to primate handedness, it is not yet clear how they can account for mutual independence of the three components of handedness (unskilled, skilled, and insertion-turning). That issue must remain to future research.

Why left-handedness?

In spite of the overwhelming right-handedness of human populations, nearly 10% of people remain left-handed. That has probably been the case for well over a million years. What accounts for the maintenance of this minority?

One popular view is that creativity is more often associated with left-handedness than right-handedness. Could it be that a "creative edge" is what maintains left-handers in the population? Some studies report increased percentages of left-handers in music or the visual arts, or higher scores on tests of creativity or divergent thinking in left-handers relative to right-handers. However, a literature review shows that, on balance, there is no evidence for increased creative excellence in left-handers in music, the visual arts, or literature (Heilman, 2005). It therefore seems unlikely that the maintenance of left-handedness can be accounted for by increased creativity, if there has been no increased production of creative products by left-handers.

In contrast, one hypothesis receiving empirical support is that in violent societies, left-handers enjoy a survival advantage because of their unorthodox style in confrontational fighting. The offensive and defensive hands, which in most right-handers are the right and left, respectively, are more often left and right in left-handers. Unfamiliarity with this reversal puts right-handers at a disadvantage compared to their left-handed opponents, who are more likely to have seen instances of their adversaries' opposite style. Consistent with the unorthodox style hypothesis, sizeable increases in the incidence of left-handers are observed in interactive sports like martial arts, table tennis, and ice hockey, relative to noninteractive sports like swimming, skiing, and archery (Raymond, Pontier, Dufour, & Møller, 1996).

More recently, the analysis of homicide rates and the incidence of left-handedness in eight traditional societies has lent further support to the hypothesis by finding a strong correlation between the two, as shown in Figure 10.5 (Faurie & Raymond, 2005). Left-handedness increases as homicide rates increase, as predicted by the hypothesis that left-handers have an advantage in confrontational conflict.

By any reasonable interpretation of the daily news, even in developed societies, old-fashioned arm's-length murder remains a popular means of dispensing with adversaries. Of the 63,661 murders committed in the United States in 2011–2015, 23% were by way of knives, blunt objects, beatings, drownings, stranglings, or asphyxiations – and

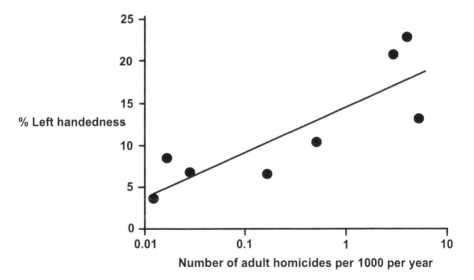

Figure 10.5 The correlation (r = +.83) between homicide and left-handedness, across eight traditional societies.

Source: Original figure.

that does not even include the 48% committed with handguns, many of which were surely arm's-length confrontations (FBI, 2018). Thus it seems possible that confrontational conflict continues to play a role in maintaining left-handedness even in modern Western society.

Conclusion

The view developed in this chapter is that handedness has multiple components whose causes ultimately trace to the same roots, namely a preponderance of left hemisphere involvement in praxis. Taking an anatomical point of view, lesions to the primary motor cortex in the left hemisphere can cause a paralysis of the right hand. Damage to the supplementary motor cortex, particularly in the left hemisphere, can lead to disruption of movement rhythms. Although the premotor cortex of the right hemisphere integrates spatial cues into movement, later in movement learning, left hemisphere influence predominates. Notably, the ventral portion of the premotor cortex is one important site of mirror cells that respond to observation of the actions of others, and that fire when the observer makes the same motions. The parietal lobe plays an indirect role in movement through spatial computation and sensory feedback evaluation, but also a more direct role, with the intraparietal sulcus integrating spatial cues with motion control. Finally, the cerebellum corrects movement errors.

The aforementioned areas emerged over many millions of years of evolution in the human ancestral line. Comparative approaches suggest the beginnings of the cerebellum date back 400–500 million years. The primary motor cortex dates from over 160 million years ago, and interconnections between supplementary, premotor, and parietal motor areas date from 83 million years ago. By 45 million years ago, the premotor cortex exerted control over distal musculature. Finally, enlargement of the cerebellum relative to the rest of the brain likely started about 32 million years ago.

A striking aspect of human praxis is the existence of population right-handedness (PRH), with about a 90% incidence. Its roots extend at least as far back as the ape–Old World monkey divergence 32 million years ago, and possibly further. PRH likely began with a simple anatomical imbalance, perhaps in the hand area of the primary motor cortex, that produced a slight handedness effect in reaching, gesturing, and feeding.

Later developments, it is proposed, built upon the initial imbalance, exploiting areas increasingly working in tandem with the primary cortex. At some point in the human-ape line, subsequent to the divergence of orangutans 17 million years ago but prior to that of gorillas 10 million years ago, a more skilled component of praxis emerged in the left hemisphere. It likely exploited a neurological substrate, namely ventral premotor cortex, in the production of communication gestures as well as in tool-related learning and manipulation. Further developments in manual praxis involved the kinds of spatial manipulations involved in insertion and turning motions. Their control may have started in the right hemisphere in the common ancestor of humans and chimpanzees prior to 7.5 million years ago, but transferred to the left in humans as part of the motor programming functions of the human left hemisphere.

Investigations of genetic influences on human handedness have a long history, but they have most recently focused on specific genes. Four have already been implicated in handedness, lending weight to the susceptibility of handedness to evolutionary influences.

Archaeological evidence, while sometimes controversial, consistently shows hominin PRH as far back as inferences can be made, about 1.5–1.8 million years ago. It consists of lateralized tooth scratches, asymmetric muscle attachments to bone, directional flaking of stone tools, and depictions of hands in caves and historical art.

PRH probably exists because praxis mechanisms evolved to bring the timing and sequencing of movement under control of the left hemisphere, eliminating the need for their cross-hemisphere coordination. Both monkey and human simple reaching involves the activation of the primary motor cortex opposite the hand, and in chimpanzees, reaching and feeding handedness is related to primary cortex asymmetry. The premotor cortex plays a critical role, with all components of PRH in chimpanzees proving to be related to inferior frontal gyrus asymmetry. Notably, the gyrus is also a site of mirror neurons.

Left-handedness, however, has been consistently maintained at roughly 10% of the human population, begging the question of what advantage it confers. The preliminary view that it affords a "creative edge" appears not to be true. Its maintenance may depend instead on the existence of confrontational fighting. By using an unfamiliar, unorthodox style, left-handers enjoy an advantage over right-handers. Thus the higher the level of homicide in traditional societies, the higher the incidence of left-handedness. Left-handers are also overrepresented in confrontational sports like martial arts and ice hockey.

11 Tools and planning

Tools were once considered uniquely human artifacts. However, scientists have known since 1917 that chimpanzees both use and manufacture tools. In that year Wolfgang Köhler reported that captive chimpanzees at Tenerife in the Canary Islands used sticks to retrieve food. They sometimes inserted a thinner tube inside a thicker one to extend their reach, or even chewed the sides off a plank of wood so it could be used as the insert. The animals also stacked boxes to reach suspended food, and developed a "climbing jump" using a long stick, holding it vertically and climbing quickly before it tipped over (Köhler, 1927).

Such behaviors imply the use of a number of cognitive faculties: Spatial processes and memory to be sure, but possibly also planning and the understanding of causality. In this chapter we consider the use of tools by great apes and their development of tool cultures through social learning. We examine the relationship between brain size and tool use, as well as the evolution of grip. Hominin tools and tool cultures are described, and ape and human tool behaviors are compared. Last considered are the role of planning in tool manufacturing and the understanding of causality, and changes in the brain that evolved to support them.

Great ape tool use

Modern research has revealed extensive tool use by chimpanzees in natural circumstances. Beginning in the 1960s, Jane Goodall reported from the Gombe Forest Preserve in Tanzania that chimpanzees use twigs for "termite fishing", sticks to retrieve honey from underground bees' nests, and leaves as sponges (Van Lawick-Goodall, 1971). Subsequently, a wide range of chimpanzee tools have been identified, including leaves used as scoops, stones as hammers and anvils, and sticks as digging tools, levers, hooks, bone marrow picks, and pounding clubs (McGrew, 1992; Sanz & Morgan, 2007). Crude spears are also used to hunt galagos, as described in Chapter 9.

Furthermore, not all tool uses are practical in the sense of aiding feeding or drinking. Thus assemblages of stone have been found inside tree hollows, thrown there by chimpanzees during ritual-like display behaviors that include pant hoots and drumming on the tree by the feet or hands (Kühl et al., 2016).

Based on these observations, it seems reasonable to suppose that the chimpanzee-human common ancestor was proficient in a limited vegetation- and stone-based tool culture 7.5 million years ago. Sticks trimmed and sharpened for insertion seem within the capabilities of such a creature. So do sticks employed to hook otherwise inaccessible objects, leaves wadded as sponges, and stones used as hammers and anvils.

Might such capabilities have existed even earlier? If so they should be observed in other apes. In this regard gorillas seem enigmatic, in captivity using and sometimes modifying sticks into probes, weapons, and rakes but almost never employing tools in the wild (McGrew, 1992; Parker, Kerr, Markowitz, & Gould, 1999). A notable exception is a female observed using a stick as a depth-tester when wading a pool, and as a prop while collecting aquatic plants (Breuer, Ndoundou-Hockemba, & Fishlock, 2005).

Tool use by orangutans is rare in the wild (Byrne, 2004), but common and even creative in captivity (McGrew, 1992) where it is perhaps unleashed by boredom. Nevertheless, wild orangutans do use sticks to probe bees' nests for honey, and to extract insects from nests and seeds from fruit (Fox, Sitompul, & van Schaik, 1999).

The use of tools by the other great apes, even if limited, might lead us to conclude that tool use is phylogenetic. However, before we make that leap we might consider a warning by W.C. McGrew (1992, p. 196):

> Termite fishing may just as well have been invented in 1959, the year before Jane Goodall arrived, or a million years ago. It may well have been invented only once by some chimpanzee Edison . . . and then diffused across Africa, or it may have been re-invented 100 or 1000 times.

McGrew's point seems well taken. To what extent might primate tool use be cultural and recent, and not phylogenetic and ancient?

Primate tool cultures

One reason to believe primate tool use may be cultural is that different chimpanzee groups have different tool practices. For example, after comparing sites in Africa, McGrew (1992) noted substantial variation in chimpanzee use of the oil palm. On one hand, in Guinea, stone anvils and hammers were employed to crack palm nuts to remove the nutritious kernels. But in Gabon, hammers were not used and the palm fruit was consumed whole, with the uncracked nuts passing through the digestive system. Both contrast with the Ivory Coast, where palm fruit was available but ignored, even though hammers were used to crack open other nuts.

Explaining such differences is difficult based on phylogeny alone. Instead, such variations could represent tool cultures, with some societies passing practices from one individual to another, and with other societies failing to develop them at all.

Defining culture

Whether or not chimpanzees show evidence of culture may depend on how culture is defined. A full definition might involve assessing a behavior's innovation, dissemination, standardization, durability, diffusion, and tradition, among other attributes (Davidson & McGrew, 2005). Andrew Whiten and colleagues, however, have focused on just two characteristics: *Cultural variation*, the extent to which a behavior differs between ape communities, and *diffusion*, the extent to which behavioral similarities across communities are attributable to social transmission (Whiten et al., 2001).

With respect to the first of these, Whiten et al. identified 39 behaviors qualifying as cultural variants in chimpanzee communities stretching across Africa. Each variant was habitual in at least one location and absent in at least one other. While not all involved tool use (e.g., the "rain dance", a display at the start of heavy rain), many did (e.g., using a pestle to pound palm crowns, or leaves to clean the body).

Diffusion, the second characteristic, is difficult to assess. As Davidson and McGrew (2005) stated, "Processes of social transmission are extraordinarily difficult to document for animals that cannot speak to us" (p. 806). Nevertheless, the availability of chimpanzee communities stretching from West to East Africa allowed Whiten et al. the opportunity to identify possible instances by examining whether contiguous communities showed similar behaviors. Of several tool use behaviors, nut hammering was found to be consistent with a single origin followed by diffusion to contiguous communities. Other tool behaviors were found in noncontiguous communities and thus were more consistent with multiple origins. For example, the use of sticks in ant dipping, clubbing, levering, and termite fishing were judged likely to have been invented in more than one location (Whiten et al., 2001). It is notable that these behaviors are simpler than nut hammering, which involves assembling multiple parts and sequencing actions. Perhaps simple behaviors are reinvented, while complex ones tend to be invented once and socially transmitted.

In orangutans, cultural diffusion has been reported by van Schaik and colleagues across locations in Sumatra and Borneo. They identified 24 likely cultural variants, a number of them involving tools (e.g., wiping the face with squashed leaves and using a stick to scratch the body). There proved to be a pronounced correlation with distance: The closer two locations were, the more they resembled one another with respect to cultural variants. That is exactly the result expected of diffusion (van Schaik et al., 2003).

Social learning

Diffusion presumably occurs through *social learning*, the acquisition of new behaviors by observing others. A detailed example was provided by a study of wild chimpanzees at Bossou in Guinea (Biro et al., 2003). The community was already familiar with palm nuts, and cracking with hammer and anvil was customary. When unfamiliar coula and panda nuts were provided, it was primarily juveniles who initially undertook their cracking, although a number of adults eventually did as well. Notably, nut-cracking attracted interest among others, with juveniles frequently observing their peers.

Juvenile-led innovation has been observed in several other contexts, including potato washing by Japanese macaques (Choleris & Kavaliers, 1999). While not a general primate characteristic, it notably exists in the social and investigative activities of chimpanzees, and in investigative uses of objects by capuchins (Perry, Barrett, & Godoy, 2017). Yet in the Bossou study, learning how to crack the new nuts mostly involved individuals observing their peers or older animals, not younger ones (Biro et al., 2003). Together, these observations suggest that juveniles are more likely than adults to introduce object-related cultural innovations, but are also more often ignored. Perhaps the combination encourages "prudent innovation" – change is allowed, but only slowly in case of pitfalls. In any case, the parallel with the introduction of new human fads and customs seems irresistible.

The acquisition of tool use in an orangutan was described by Fox et al. (1999). Andai, a juvenile, often observed her mother use sticks to probe tree holes. Two years later, she made her own tools, which she used to probe holes that her mother had just left. Two years after that, she was a fully competent toolmaker and user who found her own holes to probe. Significantly, during the learning phase, the mother was once observed to make her probes in an inefficient way, by shortening them on both ends rather than one end, and Andai proceeded to make a tool in the same manner.

The use of an inefficient technique seems a sure tip-off to its acquisition through observational learning (Fox et al., 1999). A similar example is the imitation by younger chimpanzees of a manually disabled adult's inefficient technique for self-grooming, involving rubbing against a vine held fast by the feet (Hobaiter & Byrne, 2010).

Recently, a chimpanzee community in Uganda afforded a unique opportunity to witness the spread of a novel tool behavior. A flooded waterhole was apparently the impetus for the dominant male to craft a moss-sponge, which he then used to drink water. Over six days, seven more chimpanzees were observed doing the same. Six had been explicitly observed watching the male doing so, and the seventh acquired the crafting behavior immediately after using a discarded moss-sponge (Hobaiter, Poisot, Zuberbühler, Hoppitt, & Gruber, 2014).

Most of these examples represent relatively passive social learning: One individual engaged in a behavior, and another learned by watching. Active teaching, involving one individual intervening to further the learning of another, is much rarer in primates. However, it has been seen in chimpanzees in the form of "tool transfers". Skilled females have been observed preparing termite-fishing probes before handing them to their child or sibling. Subsequently, the recipient "fishes" for a longer period, with more nest insertions, than in periods before they were handed a tool. The presumption is that at minimum, providing an expertly prepared tool affords an enhanced opportunity to learn termite fishing. But it may also transfer knowledge about the appropriate raw materials from which to make a probe (Musgrave, Morgan, Lonsdorf, Mundry, & Sanz, 2016).

In conclusion, there can be little doubt that cultural transmission through social learning helps spread ape tool culture. However, it seems equally clear that phylogeny must play a role as well. First, while social learning may be widespread among animal species, apes excel at it and may be the only animals to understand requests to "do this" (Zentall, 2006). The implication is that the biologically based intelligence of apes has much to do with cultural transmission. Second, as the Whiten et al. study indicates, many tool behaviors are continually reinvented in different ape communities. That, too, grounds tool behaviors in the basic intelligence of apes. When these considerations are paired with the evolutionary changes in praxis and handedness outlined in the previous chapter, the conclusion appears unavoidable that primate tool use is ultimately phylogenetic – *and* cultural.

Brain size and tool use

The idea that intelligence engenders tool use is supported by a relationship between tool use and brain size in birds (Lefebvre, Nicolakakis, & Boire, 2002). Across bird species, the use of tools is associated with a larger *residual brain size*, which is the amount of brain over and above that predicted by body weight. This concept is similar to that of the encephalization quotient (EQ). In birds, the relationship is accounted for by the relative size of the neostriatum, one of the brain areas believed to be equivalent to the neocortex in mammals (Lefebvre et al., 2002). However, tool use is widespread among animals, and it may not be related to brain size in all animal groups (see Box 11.1).

Intriguingly, brain size also appears informative in the case of capuchins. As New World monkeys, they seem unlikely candidates for high intelligence, but they nevertheless show EQs equivalent to those of macaques (Table 11.1). Wild capuchins use stone hammers and anvils to crack provided palm nuts. This is accomplished not with a free-fall drop of the hammer, but with a full follow-through of the arms and shoulders, sometimes recruiting muscles of the lower body. The behavior is known to predate provisioning by humans, using the nuts of wild palms (Fragaszy, Izar, Visalberghi, Ottoni, & de Oliviera, 2004). Capuchins also hammer hard-shelled fruit (De Moraes, Souto, & Schiel, 2018).

Termite fishing affords a second behavioral resemblance between capuchins and chimpanzees. Wild capuchins have repeatedly been observed tapping a termite nest, inserting

Box 11.1   Tool use in birds and other nonprimates

Although apes provide the most applicable comparisons to human tool use, we now know that many nonprimates use tools as well. Among birds, corvids (crows, ravens, jays, magpies, etc.) comprise a family with relatively large brains and that show diverse tool behaviors. These include holding food with a skewer, dropping prey on hard surfaces, using sticks to extract food from holes, and crafting and using hooks to retrieve food (Lefebvre et al., 2002; Rutz, Klump, et al., 2016; Rutz, Sugasawa, van der Wal, Klump, & St. Clair, 2016).

Tool use has also been found in about 30 aquatic species, including cephalopods, crustaceans, and fish. The use of coconut shells for protection by octopuses has been observed, much like hermit crabs use gastropod shells as temporary housing. Hermit crabs also sometimes carry sea anemones on the shells, using their stinging cells to repel predators and their weight to help maintain balance. The squirting of water jets can be viewed as tool use and is employed by many fish and squid to attack prey, to burrow, or to attach eggs. Some fish are also known to use stone or coral anvils to crack shellfish or to remove spines from sea urchins. Others lay their eggs on leaves, then protect them by subsequently moving the leaves from danger. Sea otters use stone hammers and anvils to extract the meat of shellfish, and bottlenose dolphins sometimes "wear" sponges as protection while foraging (Brown, 2012; Mann & Patterson, 2013).

Among tool-using terrestrial mammals are elephants, which sometimes throw objects at others as part of aggressive displays (Bentley-Condit & Smith, 2010), and which modify and use branches to ward off flies (Deecke, 2012). A brown bear has been observed picking up a barnacle-encrusted rock in its paws, and using it to rub its face, perhaps to relieve itching or to remove food debris (Deecke, 2012).

Even insects have been observed using tools. Some ants use leaves as carriers for food (Brown, 2012). Sawfly larvae defend themselves by applying tree-derived acids to their attackers (Bentley-Condit & Smith, 2010).

Nonprimate animals may offer fertile ground for examining the relationship between brain size and tool use. Corvids, which we have seen show diverse tool behaviors, are notably large-brained among birds (Emery, 2006). On the other hand, preliminary analysis suggests that in fish, there is little relationship between tool use and "residual" brain size after correcting for body weight (Brown, 2012). In the future, determining the conditions under which brain size predicts, or does not predict, tool use may provide further insight into the evolutionary origins of our own tool usage.

a stick while rotating it, and removing the stick to eat the attached termites. Impressively, the tapping appears to rouse the termites, increasing the yield, while the rotation reduces stick breakage. Neither of these elaborations of technique have been reported in chimpanzee termite fishing (Souto et al., 2011).

Generally speaking, brain size fairly well predicts tool use in primates. Macaques, the encephalization equivalent of capuchins, have been observed using hammerstones to

Table 11.1 Comparative encephalization quotients (EQs) among living primates, two extinct Eocene primates, and a living, related tree shrew. mya = millions of years ago.

Group	Species	Common name	EQ
Homo	*Homo sapiens*	Modern human	64
Great apes	*Pan troglodytes*	Common chimpanzee	18
	Pan paniscus	Bonobo	18
	Gorilla gorilla	Gorilla	18
	Pongo pygmaeus	Orangutan	18
Lesser apes	*Hylobates lar*	White-handed gibbon	9.2
	Symphalangus syndactylus	Siamang	9.3
Old World	*Cercopithecus aethiops*	Vervet monkey	6.4
monkeys	*Macaca arctoides*	Stump-tailed macaque	8.3
	Macaca fascicularis	Crab-eating macaque	6.5
	Macaca mulatta	Rhesus macaque	8.1
	Papio anubis	Olive baboon	11
New World	*Aotus trivirgatus*	Northern owl monkey	2.7
monkeys	*Ateles geoffroyi*	Geoffroy's spider monkey	10
	Ateles panicus	Red-faced spider monkey	9.5
	Callithrix jacchus	Common marmoset	1.9
	Cebus albifrons	White-fronted capuchin	7.9
	Cebus apella	Black-capped capuchin	7.7
	Saguinus oedipus	Cottontop tamarin	1.8
	Saimiri sciureus	Common squirrel monkey	3.9
Strepsirrhini	*Eulemur fulvus*	Common brown lemur	2.4
	Indri indri	Indri	3.3
Plesiadapiforms	*Microsyops annectens*	(extinct species, 47 mya)	0.7
	Ignacius graybullianus	(extinct species, 55 mya)	0.4
Tree shrews	*Tupaia glis*	Common treeshrew	0.8

Note: Based on a 0.28 exponent. *Homo sapiens* value from Table 5.1. *Microsyops* brain and body masses from Silcox, Benham, and Bloch (2010). *Ignacius* brain and body masses from Silcox, Dalmyn, and Bloch (2009). All other values derived from Williams (2002), using means of male and female brain and body masses where available.

open oysters and, more recently, to crack palm nuts (Proffitt et al., 2018). Spider monkeys, with slightly higher EQs, have been reported using sticks as body scratchers (Lindshield & Rodrigues, 2009). Baboons, still higher in EQ, throw sticks at rivals during aggressive displays (Cheney & Seyfarth, 2007, p. 187). Perhaps we should be impressed that a gross measure of encephalization predicts as well as it does, in that tool-relevant brain mechanisms will almost certainly be found to also vary across species.

In summary, a case can be made that tool use is influenced by a species' general intelligence and that this is related to encephalization. As we saw in Chapter 10, an additional role is played by the evolution of specific brain mechanisms relative to praxis. But as the next section outlines, the physical hand itself is a third factor affecting the evolution of tool use and manufacture.

The evolution of grip

Precision grip refers to the delicate holding of a small object between the index finger and thumb. Noting that anthropologists in the 1950s and 1960s thought precision grip was a human characteristic, Pouydebat, Coppens, and Gorce (2006) set out to determine the extent to which it is possessed by other primate species. What they found was startling. Given the opportunity to pick up small objects, not only humans but apes, Old World monkeys, and even New World monkeys used precision grips a significant proportion of the time. Furthermore, Old World and New World monkeys used it more often than apes, with the result that nonhuman species noted for frequent tool use (i.e., chimpanzees, orangutans, and capuchins) did not adopt precision grips more often than others.

Nevertheless, our precision grip differs from that of many other primates. We pinch an object between the pads of our thumb and index finger, a fully opposable grip. Many other primates have a partially opposable grip, using a side or tip of a digit as part of it (Figure 11.1). At least in part, this is due to the animals having long fingers and short thumbs (Drapeau, 2015).

There can be little doubt that a fully opposable thumb played an important role in our acquisition of complex tool use and manufacture. Strictly speaking, however, it is not unique to humans. Baboons and geladas have been observed making pad-to-pad grips, and it is said to be an ability of most terrestrial cercopithecines, a classification that also includes macaques (Marzke, 1997). What may make it *seem* unique is that it is not a characteristic of other apes (Almécija, Shrewsbury, Rook, & Moyà-Solà, 2014; Marzke, 1997; Pouydebat et al., 2006). If a relatively long thumb is required, that condition at least may have first appeared in our direct ancestral line 10 million years ago in our common ancestor with gorillas (who, as apes go, are long-thumbed). In that case, the short

Figure 11.1 Orangutan using a precision grip involving the forefinger and thumb tip.

Source: Orangutan using precision grip.jpg, CC BY-SA 4.0, William H. Calvin.

thumb of the other great apes is a derived character reflecting below-branch suspension (Almécija, Smaers, & Jungers, 2015; Drapeau, 2015).

Human hand dexterity

When thinking about our ability to use tools, it is useful to consider the overall dexterity of the hand in adopting a range of grips and postures. Marzke and Marzke (2000) noted that the ability to cup the hand around objects of different shapes seems to be a specifically human capability. Such dexterity is thought to be in part the product of distinctive hand joints and bone surfaces. Because we have the highest ratio of thumb length to index finger length of any primate, we also better accommodate different shapes and can rotate them between thumb and fingers. In addition, the pads at the ends of our fingers are uniquely stabilized at their tips but mobile elsewhere, allowing conformation to different shapes.

A final distinction is that humans have a horseshoe-shaped crest on the distal segment of the thumb as well as a small round bone embedded in the tendon attached to it. Both features greatly increase the leverage the tendon can exert, improving the thumb's gripping power (Marzke & Marzke, 2000). Indeed, humans can exert considerable force in "precision pinch" grips, while other primates cannot. Their grips might instead be called "precision holding" (Tocheri et al., 2003).

Homo habilis had the aforementioned tendon bone, as well as broad fingertips and a wrist suggesting a fully opposable thumb. As a result, stones would have been held in a strong, stable grip (Marzke & Marzke, 2000). In fact, the changes probably began earlier than that. Analyzing the distal thumb segment of *Orrorin tugenensis*, Gommery and Senut (2006) concluded that it possessed the horseshoe-shaped crest identified by Marzke and Marzke (2000) as the attachment point for a tendon capable of high leverage. If so, the fossil extends the evolutionary history of a powerful thumb back to six million years ago.

The earliest hominin tools

Vegetation must have accounted for much of the toolmaking material employed by the earliest hominins. Primate tool cultures commonly show use of twigs and sticks as probes, levers, hooks, spears, and digging tools, and of leaves as sponges and scoops. These were almost certainly within our ancestors' capabilities after they split from the chimpanzee line. Fossil evidence may never be found, however, because such tools would not have been cached in quantity, and vegetation usually decays rapidly.

Chimpanzee observations suggest that the earliest hominin tools made of permanent materials were probably natural stones used as hammers and anvils. In fact, possible stones and hammers dating to 3.3 million years ago have been identified from a site near Lake Turkana in Kenya (Harmand et al., 2015). Hominin hammers and anvils have also been discovered at Olduvai, dating to about 1.8 million years ago (Diez-Martin et al., 2010).

Traditional tool use among living Australian hunter-gatherers provides potentially valuable insights into other early hominin tools. When the Australian hunter-gatherers use stone tools, it is primarily to make wooden objects such as spears and digging sticks. Sharp edges are taken as they come: As natural edges if available, but as flakes broken off stone cores if not. Flakes with a right-angle edge are used to smooth surfaces on shaft tools like spears. Although worn flakes are sometimes resharpened, more often new flakes are selected. All of these kinds of tools would be very difficult to identify in archaeological contexts. They have unworked edges for the most part, and the casual nature of their acquisition scatters them across sites (Hayden, 2015).

Worked stone tools

The earliest known hominin tools made of worked stone, like the earliest possible hammers and anvils, come from near Lake Turkana. Consisting of cores and flakes, their working is so crude that a naïve eye might not recognize them as tools. Nevertheless, there is evidence for transportation from a raw material site 100 meters away, and for multiple fractures of stone from cores. At 3.3 million years old, they substantially predate the genus *Homo*, so that either *Australopithecus afarensis* or *Kenyanthropus platyops* is considered their likely maker (Lewis & Harmand, 2016). They have been termed Lomekwian tools after the site at which they were found, in recognition of distinctive fracturing and their status as an apparent waypoint between pounding and flaking technologies (Harmand et al., 2015).

Gona and Oldowan tools

The next oldest tools are 2.6 million years old. They were found at Gona, Ethiopia, and were possibly crafted by *Australopithecus garhi*. They have been described as "cores, whole and broken flakes, angular and core fragments, a small number of retouched pieces and in some instances unmodified stones transported to sites" (Semaw, 2000, p. 1198). Flakes predominate, not only for the obvious reason that multiple flakes must be struck from a core to make it into a tool, but because obtaining the sharp-edged flakes themselves would often have been the maker's object. The flakes show swellings just below the point of impact that clearly indicate human agency. Some cores show pitting consistent with use as hammerstones, although that can also occur from weathering. To the less-trained eye, however, the flaked tools are the most identifiable as produced by hominins (Figure 11.2, especially the right and bottom views showing multiple removals).

The Gona tools resemble the Oldowan toolkit, named for Olduvai Gorge in Tanzania, where 1.9-million-year-old tools were found and attributed either to australopithecines or to *Homo habilis*. In recent years, however, Oldowan-like tools have also been found in the Republic of Georgia in Eurasia. They date to about 1.8 million years ago (Mgeladze et al., 2011). They are thought to have been produced by *Homo erectus* following what was possibly the earliest hominin migration out of Africa.

Roughly contemporaneous with these tools are cut marks on bone, from Bouri, Ethiopia, dating to 2.5 million years ago. Hammerstone impact marks on a shattered bovine tibia also indicate the extraction of bone marrow. These discoveries are from layers associated with fossil remains of *Australopithecus garhi*, suggesting that it was the toolmaker (de Heinzelin et al., 1999). A claim of even earlier cut marks dating to 3.4 million years ago (McPherron et al., 2010), was initially criticized on the basis that they may have resulted from trampling (Domínguez-Rodrigo, Pickering, & Bunn, 2010). However, the claim has been made plausible by the subsequent reporting of actual stone tools from 3.3 million years ago, and studies indicating that trampled bones do not show similar cut marks (Thompson et al., 2015).

Figure 11.2 Examples of tools from Gona, Ethiopia.

Source: File:Pierre taillée Melka Kunture Éthiopie fond.jpg, CC BY-SA 4.0, Didier Descouens.

The role of cognition

Did worked stone tools originate 3.3–3.4 million years ago in a sudden flash of insight that spread through cultural means? Or did they emerge more gradually? It is too early to tell, but either way, it seems safe to assume that vegetation-based tools long preceded worked stone. Apparently a lengthy period of evolution was necessary to bridge the conceptual gap between stick tools and worked stone tools. Yet if strong grips emerged early, possibly in *Orrorin* six million years ago, and hunting even earlier, then the gap was apparently not related to shifts in grip or diet. The conclusion seems inescapable that something cognitive occurred at the time of the australopithecines that allowed a transition from vegetation-based and natural stone tools to worked stone tools.

A clue to that "something" comes from an attempt to teach *knapping* (shaping stone through systematic chipping) to a captive pygmy chimpanzee named Kanzi. When striking off flakes using a bimanual technique, Kanzi appeared not to calculate striking angles or to have an idea of what should be accomplished (de Beaune, 2004). In other words, Kanzi may not have been able to generate and sustain a mental model of a stone tool and how it is formed.

Such a model would include the idea that overlapping chip removals, struck at the proper angle off a prepared core, leave a nearly continuous, extremely sharp edge. Forming that idea requires a high degree of three-dimensional spatial processing, and probably spatial imagery. It seems qualitatively different, in part because of its complexity but also because the end result is less immediate, from the spatial processing required to prepare any chimpanzee tool. Perhaps it is a shift in spatial ability and spatial memory that we should look to for causes of the transition to worked stone, a theme visited in the next chapter.

Alternatively, it may be that the ability to mentally order the necessary chip removals – i.e., planning – was more important. Even the intention to make worked tools involves planning, because the necessary materials must first be gathered. In the case of Oldowan tools, matching the type of selected rock to available outcrops indicates that raw materials were sometimes transported over distances in excess of 10 km. Furthermore, they were probably selected for the durability of tool edges (Braun, Plummer, Ferraro, Ditchfield, & Bishop, 2009). That is surely a testament to the planfulness of the toolmakers.

Later tools

Tools evolved much as species did, with more successful ones replacing less successful ones. In general, later stone tools are characterized by increasing complexity and diversity.

Acheulean tools

After Lomekwian and Oldowan tools, the next tool culture to arise was the *Acheulean*, beginning about 1.8 million years ago (Table 11.2) and named for the archaeological site of St. Acheul in France. Handaxes first arose in this culture (Kooyman, 2000).

Caution should be observed when identifying particular tool industries with particular periods of time, because the period during which an industry existed depends on the geographic location. Nevertheless, it is useful to give approximate dates, as in Table 11.2, in order to interpret other cultural remains (e.g., artwork, needles, and hearths) associated with the cultures.

Table 11.2 Some prominent tool cultures, with very approximate ages and geological periods (kya = thousands of years ago). Ages largely derived from Gowlett (1992b) and Jones, Martin, and Pilbeam (1992).

Industry	Approx. age	Period
Lomekwian	3300 kya	Lower Paleolithic
Oldowan	1000–2600 kya	Lower Paleolithic
Acheulean	150–1800 kya	Lower Paleolithic
Mousterian	15–150 kya	Middle Paleolithic
Aurignacian	27–40 kya	Upper Paleolithic
Gravettian/Périgordian	12–27 kya	Upper Paleolithic
Solutrean	18–21 kya	Upper Paleolithic
Magdalenian	12–17 kya	Upper Paleolithic
Hamburgian/Ahrensburgian	8–13 kya	Upper Paleolithic
Clovis	8–12 kya	Upper Paleolithic

The names of the tool cultures generally reflect the geographic locations from which they were first or best known. Thus the Oldowan industry is named for the Olduvai Gorge in Tanzania. Nevertheless, it should not be assumed that an industry is found only in the location for which it was named, because most spread widely. As for why the Lower, Middle, and Upper Paleolithic periods are so named, "think dig". When digging down, the upper layers are the youngest, the middle layers are the next youngest, and the lower layers are the oldest.

Acheulean tools originated in Africa, where they are first known from 1.8 million years ago (Diez-Martin et al., 2015; Lepre et al., 2011). However, they spread widely throughout Africa, Europe, and Asia as far as India and Korea (Gowlett, 1992b). There is evidence that the spread was literal, because variability in the tools is highest in Africa and lowest in areas furthest from Africa, exactly as one would expect from transmission through repeated migrations (Lycett & von Cramon-Taubadel, 2008).

These tools were substantially more complex than their predecessors, incorporating symmetry around one or two axes, and were often heavily worked so that little or no raw stone surface was left exposed (Gowlett, 1992a; Kooyman, 2000). The high degree of symmetry of some Acheulean handaxes strikes many modern viewers as quite beautiful, appealing to an aesthetic that their makers may have felt as well (Figure 11.3). Yet they were also efficient tools, probably notably more efficient at butchery than the previously used stone flakes (Galán & Domínguez-Rodrigo, 2014).

Figure 11.3 Acheulean handaxe from Spain.

Source: Bifaz en mano.jpg, CC BY-SA 2.5, José-Manuel Benito.

One late development in the Acheulean tradition was the *Levallois technique*, involving removing many flakes from a core in preparation for striking off one desired flake that with minimal or no *retouching* (secondary flaking along the edge) became the sought-after tool (Gowlett, 1992b). The advantage to this technique was that through careful preparation of the core, a tool exactly the desired size and shape could be obtained, with a sharper edge than was possible using earlier techniques (Kooyman, 2000).

What Acheulean tools were used for, besides butchery, has been a point of controversy. However, analysis of *phytoliths* (microscopic mineral particles formed by plants) found on the working surfaces of Acheulean handaxes from about 1.5 million years ago indicate that they were used for chopping wood (Dominguez-Rodrigo, Serrallonga, Juan-Tresserras, Alcala, & Luque, 2001). The finding is important in revealing that woodworking occurred at a remote time from which wooden artifacts themselves have not survived. However, there is also evidence that points of Acheulean age, in this case 500,000 years ago, were hafted to shafts to make spears, becoming the first known instances of *composite tools* (tools made of more than one material). Presumably they were used for hunting (Wilkins, Schoville, Brown, & Chazan, 2012). The varied uses of the Acheulean handaxe have led to its being described as the "Swiss army knife" of its age (Calvin, 2002).

Mousterian tools

The *Mousterian* toolkit, named for Le Moustier, France, is largely though not exclusively associated with Neanderthals. The emphasis was on discoids and on Levallois flakes used as scrapers. Small stone points with true tangs, suggesting hafting onto a shaft, developed out of this tradition and are found in northwest Africa, dating to about 22,000–45,000 years ago (Gowlett, 1992b; Straus, 2001). Also, arrow points made of bone, more than 60,000 years old, have been reported in South Africa. They are thought to have been inserted in a shaft and secured with a wrapped collar (Backwell, d'Errico, & Wadley, 2008).

Aurignacian tools

Aurignacian culture, which extended from Spain to Russia (Gowlett, 1992b), was marked by an increase in *microblades* (small blades less than 5 cm in length) and *microliths* (small blades with at least three distinct sides). The latter were mounted in series to form a continuous edge. By thus combining stone with a mounting material, a composite tool was created (Kooyman, 2000). Aurignacian culture also created bone points. Bone and antler were common toolmaking materials, although wood was undoubtedly used as well (Gowlett, 1992b).

Gravettian and Solutrean tools

Gravettian culture (called *Périgordian* in France) was characterized by *backed blades* (blunted on one edge, probably for ease of use) and the common utilization of scrapers (Gowlett, 1992b). The co-occurring *Solutrean* culture achieved the most sophisticated of all Paleolithic stone tools, the laurel-leaf point, named for its characteristic shape (Figure 11.4). It was made by pressure-flaking, in which flakes are detached by pressure rather than with blows, representing another technological advance (Gowlett, 1992b; Kooyman, 2000).

Figure 11.4 A Solutrean laurel-leaf point.

Source: Biface feuille de laurier.JPG, public domain, Calame.

Magdalenian tools

Magdalenian culture introduced harpoons, lance-heads, and spearthrowers, made from bone and antler. The spearthrowers were often decorated with carvings of animals, birds, and fish. Bone needles, complete with eyes, attest to the wearing of clothing (Clark, 1967; Gowlett, 1992b), although the invention of clothing must have substantially pre-dated this culture (see Box 11.2). Microlithic-backed stone blades were also common (Kooyman, 2000).

Hamburgian and Ahrensburgian tools

The *Hamburgian* and *Ahrensburgian* cultures are known for their many organic remains, preserved in both caves and open-air sites. Analysis of animal remains and projectile points yields the conclusion that Hamburgian hunters shot animals (usually reindeer) from the side, often while a herd passed through a valley entrance. Some debate remains whether spearthrowers or bows and arrows were used, but evidence leans toward the

Box 11.2   The great denudation

Important tool uses have included the preparation of hides and textiles for clothing and shelter, and the creation of fire. To some extent, these thermoregulatory uses of tools must be linked to the loss of body hair in the hominin line, or in other words to what Markus Rantala (2007) has picturesquely called "the great denudation".

In the past there has been little consensus on whether the use of clothing, fire, and portable shelters accompanied the loss of hair, or followed it much later as hominins moved into colder climates. To begin addressing this question, Rogers, Iltis, and Wooding (2004) examined changes in the MC1R gene. The gene affects skin color and was selected on the assumption that the loss of body hair must have been accompanied by darkening skin as protection against the sun. Using a molecular-clock approach, the researchers concluded that the great denudation occurred about 1.2 million years ago. Interestingly, this date corresponds to a split in the human louse line, with different subspecies of lice beginning to diverge at about that time (Light & Reed, 2009). Certainly the loss of human body hair would have had implications for body parasites.

But at what point was clothing adopted? Based on a later genetic divergence between human head lice and body lice, it has been argued that clothing first appeared 83,000–170,000 years ago. Thus the much earlier denudation is assumed to have confined lice to the head, which then moved to the body and genetically diverged once clothing was adopted. The 83,000- to 170,000-year-old date is believed to be prior to the passage of modern human beings out of Africa, suggesting that clothing was invented before the removal of *Homo sapiens* to cold climates (Toups, Kitchen, Light, & Reed, 2010). However, this leaves unclear how earlier hominins were able to survive in the cool parts of Europe before 500,000 years ago (Hosfield, 2016).

Direct evidence of clothing is very limited beyond a few thousand years ago. The oldest known clothing dates back 20,000 years, and dyed flax fibers about 33,000 years. Less direct evidence includes eyed needles at 30,000–35,000 years, and bone awls at 76,000 years or more (Hoffecker, 2005; Kvavadze et al., 2009; Rantala, 2007).

Although scraped hides likely extend back about 300,000 years (Kvavadze et al., 2009; Rantala, 2007), hides can be used for either clothing or shelter (Toups et al., 2010). If they were then used for clothing, it may have been in a simple form involving draping a single layer loosely over the body, providing only a small amount of warmth. Complex clothing, consisting of fitted pieces sewn together to more tightly fit the arms, legs, and torso, might then be inferred from the much later awls and needles. They would have been warmer, and they would have allowed layering for even greater warmth (Gilligan, 2010). Perhaps it was this form of clothing that appeared 83,000–170,000 years ago.

The material evidence relating to the use of fire is also limited. In Africa, it may date as far back as *Homo erectus* 1.5–1.6 million years ago. Yet a heating purpose cannot necessarily be assumed because fire can also be used for light, protection against predators, cooking, and socializing (Beaumont, 2011; Gowlett, 2016). In colder European climates there is no evidence of the intentional use of fire until after 450,000 years ago (Roebroeks & Villa, 2011). Thus the story of the great denudation, and its implications for clothing, shelter, and heating, is likely to prove a complicated one to unravel.

former, at least until the later Hamburgian period. The somewhat later Ahrensburgian culture definitely had arrows as is known from the remains of their notched shafts (Terberger, 2006).

Clovis tools

Finally, the Clovis culture is specific to central North America and is named for a site near the town of Clovis, New Mexico. It is known for elaborately worked stone points with a fluted base that allowed firmer attachment to a shaft. Clovis hunters took down large game including mammoths, mastodons, camels, and bison (Gowlett, 1992b; Waguespack & Surovell, 2003).

Comparison of ape and human tool behaviors

With respect to tool use and manufacture, what humans and other apes may have in common is the ability to create, remember, and execute a *script*. This can be defined as a contextually dependent series of actions leading to a goal. For instance, a nut-cracking script might involve searching for a hammer, transporting it to an appropriate site, locating an anvil, bringing nuts to the anvil, cracking the nuts between the hammer and anvil, and consuming the nuts. Observations of ape tool use clearly show an ability to follow such scripts, not just in nut-cracking but in the preparation of "fishing" tools, hunting spears, leaf sponges, and honey probes (Parker, 2004).

Where humans and other apes differ is in script complexity. We can incorporate more steps in the sequence of actions needed to reach the goal, and in the number of alternative actions accommodated during the sequence (Parker, 2004). This notion of a human-ape difference in script complexity is related to the qualitative differences, already highlighted, in the planning and spatial processing humans and apes bring to toolmaking. Stone tools require more steps than simple vegetation-based tools, increasing both the complexity of the script and its memory demands.

Similar differences in planning and execution over time are seen in two other aspects of human tool manufacture (Byrne, 2004). First, composite tools, made from multiple component parts, are found in all known human cultures and indicate planning for assembly. Second, humans manufacture many tools whose sole use is to make or assist other tools, the ultimate in long-term planning. Add to these elaborations the wide diversity of materials used by humans, and animal tools seem "poor things in comparison" (Byrne, 2004, p. 32). A likely reason for that is that apes lack the degree of memory-intensive spatial planning needed to seek diverse materials, transform them by way of protracted processes, and if necessary combine them into composite tools.

Increased hominin planning

An increase in hominin planning appears to have emerged by at least 2.6 million years ago. Although Oldowan tools seem simple to us, possibly even within the grasp of a nonplanning mind, meticulous reconstruction of Oldowan cores from their constituent flakes implicates a manufacturing process in which up to 30 flakes were struck off in sequence. Furthermore, strike angles were maintained throughout the sequence, and the technique was repeated across cores (Byrne, 2004).

Other data, derived from later Acheulean examples, show that stone tools were rarely completed in one location. Instead, a core might be produced at one place through initial flaking, then removed to a second place for further flaking (Hallos, 2005). This shows a striking resemblance to modern adze-making on the island of New Guinea. Adze-making involves knapping rough forms at a quarrying site, followed by transportation to a village and then more detailed stonework (Stout, 2002). Such operations extend the time frame over which the overall goal must be remembered and over which subgoals are attained (Hallos, 2005).

There seems nothing of similar complexity in ape behavior, even if it is true that chimpanzees engage in a limited degree of planning when they prepare a "fishing" tool at one site for transport to another (Byrne, 2004). In fact, even lemurs, removed from humans by some 73 million years of evolution, show evidence of planning when they grasp objects with an uncomfortable grip, in anticipation of rotating them to a more comfortable end position (Chapman, Weiss, & Rosenbaum, 2010). But both of these behaviors represent much simpler planning than is involved in stone tool manufacture.

Is the evolutionary emergence of complex tools yet another manifestation of increasing brain size? The makers of the first known stone tools are thought to have been *Australopithecus afarensis* or *Kenyanthropus platyops*. The encephalization quotient of the latter is unknown due to an unknown body size, but afarensis had an EQ of 23 (see Table 5.1). Interestingly, *Homo floresiensis*, the diminutive hominin popularly known as "the Hobbit", had a very similar EQ of around 24, and it too manufactured stone tools (Morwood & Jungers, 2009). These EQ values are distinctly larger than those for the great apes (Table 11.1).

However, raw encephalization tells only part of the story. As it turns out, the brain of *floresiensis* may offer a unique insight into human advances in planning.

Brain mechanisms of planning

In hominins, traces of brain structure are often left behind on the inside surfaces of fossilized skulls, having been impressed by pressure into developing bone. They can be recovered by creating an *endocast*, either a physical latex rubber molding of the inside of the skull or a virtual representation constructed by computational means using imaging data.

A remarkable virtual endocast has been recovered in the case of the Hobbit. One of its most noticeable features is a large bulge on each side of the midline at the *frontal pole* (foremost part) of the frontal lobe (Figure 11.5), much more pronounced than in the modern human brain. The location corresponds to Brodmann area 10 in our prefrontal cortex (see Figure 10.4), which has both a lateral extent (i.e., to the side of the midline), and a medial extent (i.e., on the inner surface of each hemisphere). It appears safe to conclude that *floresiensis* had a relatively enlarged area 10 (Falk et al., 2005).

Why is area 10 particularly prominent in this species? We saw in Chapter 5 that *floresiensis* may represent an instance of island dwarfism, the body size reduction that sometimes occurs due to resource limitations on isolated islands. A possibility is that area 10 is enlarged in the Hobbit because it was relatively preserved when brain size shrunk in company with shrinking body size. Such preservation might occur if area 10 was so critical to the species' survival that it underwent positive selection during the dwarfing process.

The rise of area 10

That area 10 played an important role in hominin evolution is strongly suggested by comparative analyses undertaken by Katerina Semendeferi and colleagues (Semendeferi, Armstrong, Schleicher, Zilles, & Van Hoesen, 2001). They found that humans have a greatly enlarged area 10, comprising 1.2% of brain volume compared to 0.5–0.7% in the great apes and only 0.2% in gibbons. Combining these findings with other data,

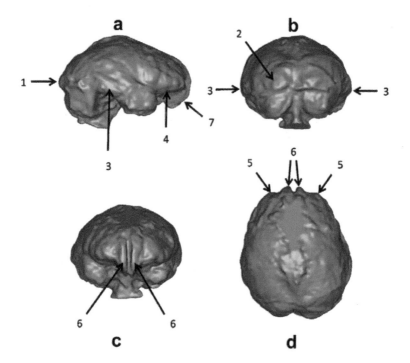

Figure 11.5 View of the *floresiensis* endocast from the (a) side, (b) back, (c) front, and (d) top. Note the substantial bulges, labeled as feature 6, at the frontal pole.

Source: Virtual endocast of LB1, licensed from Elsevier under STM Permissions Guidelines.

Semendeferi et al. proposed that in an initial ancestral state now approximated by gibbons, area 10 occupied a limited area nearest the eyes. This was then followed by two expansions: One in a common ancestor of great apes and humans, and the other more specifically in humans when area 10 greatly expanded in size and connectivity with other brain areas (Semendeferi et al., 2001). In concert with this, area 10 matures more slowly than other brain areas in both chimpanzees and humans, and more slowly in humans than in chimpanzees (Teffer et al., 2013).

The expansion's nature is clarified by the discovery that while area 10 comprises 1.2% of our brain volume, it has only 0.3% of the brain's neurons (Barton & Venditti, 2013). That means there must be unusually profuse connections between area 10 neurons, taking up volume that might otherwise be filled by cell bodies. Indeed, pyramidal cells in area 10 are found to have larger *dendritic arbors* (the treelike branchings of dendrites) relative to several comparison areas (3, 18, and 4; see Figure 10.4 for these areas). The same is also true of chimpanzees (Bianchi et al., 2013). It appears, therefore, that area 10 has long been under selective pressure for increased cellular interconnectivity, going back at least to the chimpanzee-human common ancestor.

The cognitive role of area 10

If area 10 expanded in connectivity during ape evolution and in size during hominin evolution, we might expect that it serves a critically important role in cognition. In fact, that appears to be true. For one thing, it is involved in human *working memory*, a form of memory that holds and manipulates information over the short term. Thus a review of brain-imaging studies found that working memory activates the lateral portions of area 10 (Gilbert et al., 2006). While other cortical areas may play an overall larger role in working memory (Cabeza & Nyberg, 2000), area 10 more specifically evaluates its contents (Leung, Gore, & Goldman-Rakic, 2005). In addition, the anterior portion of the area is involved in rapidly switching between tasks (Gilbert et al., 2006).

The role of area 10 in evaluating the content of working memory and in task switching has clear applicability to the manufacturing of stone tools. Knapping involves keeping in mind a goal state, evaluating the emerging tool in terms of that state, and repeated switching between evaluative and action-oriented phases. The area 10 processing load would have been particularly demanding in a small-brained species like *floresiensis* (and perhaps not in modern large-brained knappers, as suggested by the silence of area 10 in the brain-imaging results of Stout, Toth, Schick, & Chaminade, 2008). Together, the evaluation and task switching functions of area 10 seem almost synonymous with planning, if planning is understood to include not just the initial plan but its updating over time.

Other brain-imaging research confirms a role of area 10 in planning. In human experiments, the Tower of London task is one of the most commonly used planning tasks, involving moving beads one at a time to transform a beginning state to a desired goal state (Figure 11.6). Problems vary in complexity; for example, from one to seven moves. While area 10 activation is not always apparent overall when the task is performed (Boghi et al., 2006), research usually finds that its activation increases with increasing difficulty (Schall et al., 2003; van den Heuvel et al., 2003). Also, better performance is associated with increased activation in the area (Unterrainer et al., 2004). Both findings indicate area involvement in the task.

Further supporting the role of area 10 in planning, Roca and colleagues found that damage to it impaired the ability to multitask while operating a virtual hotel; specifically, to perform several subgoals while keeping in mind an overall goal. Patients with damage elsewhere in the brain had less difficulty (Roca et al., 2011).

It has been pointed out that area 10 is probably involved in some other functions as well; for example, emotion (Gilbert et al., 2006), alertness (Coull, Frackowiak, & Frith, 1998), evaluating the implications of pain for behavior (Peng, Steele, Becerra, &

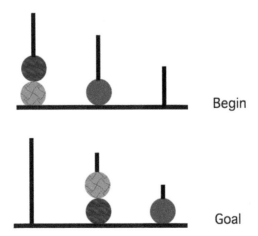

Figure 11.6 A three-move Tower of London problem.

Source: Original figure.

Borsook, 2018) and "mentalizing", or *theory of mind* (i.e., ascribing mental and emotional states to others; Gilbert et al., 2006). Nevertheless, its involvement in planning, shown not just in working memory, task switching, and multitasking research but in a task of a sequential nature, argues strongly for its role in the manufacturing of tools. That the area was apparently preserved during the reduction of the Hobbit's brain testifies to its importance to hominins.

Understanding causality

Much of the planning using area 10 may only involve *heuristics* (rules of thumb that tend toward the solution of a problem). The similarity heuristic, for example, entails making the current state more similar to the desired end state. Followed to its conclusion, this often results in a correctly crafted tool or a solved Tower of London problem. However, more complex problem solving can be achieved if an understanding of causality is also brought to bear. For example, in some of the more complex Tower of London problems, a move must be made that decreases the similarity between the current and end states, in the interest of allowing future moves that will again increase similarity. This requires understanding causal relationships across multiple moves: "If I do this, then I can do that and that".

Wolpert (2007) has argued that compared to us, apes have a very limited notion of causality. To be sure, they can learn to associate an action with an outcome, but they often lack insight into why one leads to the other. Thus Wolpert cited work by Povinelli, Reaux, Theall, and Giambrone (2000), indicating that a chimpanzee may understand that tool contact is necessary to reach food, but not understand that a hook at the end of the tool can be used to pull the food within reach. Also, when raking food toward them with a pit obstructing the path, few chimpanzees reliably select a pathway that avoids the food falling into the pit. These failures suggest a very limited understanding of causality. Yet Wolpert (2007) did allow that there is occasionally some understanding, "at the edge, even just over the edge" as he put it (p. 59), as in the assembly of hammers and anvils and the leveling of the anvils for nut-cracking.

The same "just over the edge" conclusion may apply to the solving of mazes, which younger chimpanzees can plan out to two steps (Völter & Call, 2014). Some notion of causality must be present, as the animal must understand that the solution of one step affects the next. However, this does not detract from the general conclusion that their causal understanding is limited.

It is not clear that area 10 supports causal understanding, so much as it supports the comparison of the current and end states of a problem in working memory, an evaluation of the discrepancy, and any task switching needed to reach a solution. There is, however, another brain area that seems more clearly related to the understanding of causality. In humans, but not in macaques, a portion of the inferior parietal lobe becomes active in the left hemisphere during the observation of tool use. More importantly, its activity is contingent on understanding the goal of using the tool. For example, the area is activated when participants view a small rake dragging an object, but not when they view the rake moving alone without an object. Thus the area, more specifically the anterior portion of the supramarginal gyrus (Brodmann area 40, see Figure 10.4), appears to grasp causal relations between the intended use of the tool and the results of using it (Peeters et al., 2009).

Furthermore, tasks requiring the understanding of spatial causality without the use of tools, such as deciding whether a ball striking another caused it to move, activate the supramarginal gyrus on the *right* side (Straube & Chatterjee, 2010; Wende et al., 2013).

Thus the gyrus seems generally involved in understanding physical causality, with its left-versus-right manifestation depending on specific aspects of the task.

An understanding of causality is important to planning because it produces solutions when simple heuristics fail. Thus we have no difficulty solving the rake-and-pit problem failed by chimpanzees, because we understand that dragging the rake straight toward us will cause the food to fall into the pit, and that moving the rake sideways at the right place will avoid that. Understanding causal relations is one of the things allowing us to innovate when solving problems.

The fact that humans, but not macaques, have the localized parietal lobe mechanism for understanding physical causality dovetails with the observation that beginning with the genus *Homo* about 2.4 million years ago, the inferior parietal cortex visibly acquired new gyri. They were specifically the supramarginal gyrus and the angular gyrus, Brodmann areas 40 and 39 (Glover, 2004; Tobias, 1995). It appears, therefore, that humans acquired a new brain area to allow a deeper appreciation of physical causal relationships. Together with the planning functions of area 10, these changes helped launch us on the pathway to ever more sophisticated technologies.

Conclusion

Apes and some monkeys show remarkably diverse tool manufacture and use, which can take on a cultural dimension and show evidence of diffusion across neighboring communities. Social learning is the likely mechanism of diffusion, with innovations often introduced by younger animals. However, most copying of behavior proceeds either peer-to-peer or from older to younger individuals. In apes, social learning is aided by an ability to understand requests to "do this".

Across primate species, there is some indication of a positive relationship between encephalization and the ability to use tools. Thus primate tool use appears to be both phylogenetic *and* cultural in nature.

While precision grip is more widespread among species than was previously appreciated, humans, unlike other apes, have a truly opposable thumb that can make pad-to-pad contact with the index finger. The relatively long thumbs that are possibly a precondition for this, may date back as far as our common ancestor with gorillas 10 million years ago.

Compared to other primates, humans have evolved both greater hand dexterity and more forceful grips. Specific evolutionary acquisitions include a hand more capable of cupping various shapes of objects, a longer thumb relative to the fingers, tip-stabilized finger pads that are otherwise mobile, and a horseshoe-shaped crest in the distal thumb that increases its gripping power.

The result, in part attributable to such influences, was the emergence of worked stone tools by 3.3 million years ago, developing out of what was presumably an earlier vegetation- and hammer-and-anvil-based tool tradition. Their maker was probably either *Australopithecus afarensis* or *Kenyanthropus platyops*. In contrast *Australopithecus garhi* may have manufactured the tools found at Gona, Ethiopia, dating from 2.6 million years ago.

It seems likely that the biggest factors in the emergence of worked stone tools were cognitive in nature. Stone tool design progressed from the relatively crude Lomekwian and Oldowan styles to the highly symmetrical and heavily flaked Acheulean style about 1.8 million years ago. Composite tools adopting other materials are first known to have appeared about 500,000 years ago but, by about 45,000 years ago, incorporated a variety of forms and materials.

The mental model of a stone tool requires a high degree of three-dimensional spatial processing, and achieving a satisfactory product requires a mental script including planning and plan revision. Although apes follow scripts in their tool manufacture and use, human scripts are more complex, involve more spatial planning, and are more likely to defer success or failure into the future. Here there appears to be a strong influence of encephalization, with nearly the same EQ achieved by the likely makers of Lomekwian tools and by the *floresiensis* toolmakers on the island of Flores. The relative preservation of Brodmann area 10 in the dwarfed brain of the latter likely reflects its importance in planning, through its evaluation of the contents of working memory and its role in task switching.

In summary, then, it appears that tool use and manufacture on the one hand, and cognitive evolution on the other hand, have long had intertwined histories. Primate tool use may depend in part on reaching macaque or capuchin levels of encephalization. However, changes of brain function and associated structure, and not just size, are likely to have been equally important. In humans, the development of brain areas such as the prefrontal cortex and the supramarginal gyrus was probably key. The latter in particular seems to be involved in comprehending physical causality. It and neighboring areas in the inferior parietal lobe became visible in the human brain beginning with the genus *Homo*, presumably aiding in the manufacture of increasingly sophisticated tools.

12 Spatial perception

When an early *Euglena*-like organism first chased the sun by altering movement of its flagellum in response to the shading of a photoreceptor, a long evolution of spatial behavior was begun. Today, our capacities to identify the location, depth, motion, and quantity of objects represent some of our most important perceptual abilities. The result is that we can successfully navigate an environment much more complex than anything encountered by *Euglena*.

In this chapter we examine several kinds of spatial information, consider how it is processed by complex organisms with emphasis on humans and other primates, and trace evolutionary changes as far back as data permit us.

Location

The first kind of spatial information is location. Early on, it was sensed using the simple optical properties of the eye. Thus a lensless pit eye was capable of determining the general direction of light through the alignment of photoreceptors with both the light source and the eye's aperture (see Figure 7.1). As the pit deepened and the aperture shrunk, approximating a pinhole eye, localization became more precise because of a substantially improved image on the retina. Finally, in the lensed eye there was a sharply imaged, one-to-one mapping of source points to retinal points. Thus in a straightforward evolutionary sequence, successive improvements in the structure and physics of the eye led to a more finely tuned ability to locate a light source, all from changes in the periphery of the nervous system.

Of course, the periphery tells only part of the story, and we must look to the eye's projections to the brain and to the brain itself to complete it. In the human eye, the output of some 130 million receptors are funneled to the brain over 1.2 million optic nerve axons, an average compression ratio of over 100:1. As we have seen, much of the compression is accomplished by compromising peripheral vision. Thus massive numbers of receptors in the periphery funnel down to relatively few ganglion cells, producing substantial information compression but poor acuity. In comparison, there is little funneling from receptors in the fovea to a relatively large number of ganglion cells, producing little compression and high acuity.

These patterns are repeated at V1, the primary visual cortex, where a *cortical magnification factor* is apparent. This is a mathematical relationship between the acuteness of an area of vision and the amount of cortex serving it, so that more cortex is devoted to sharper areas of vision. As a result, much larger numbers of cortical neurons are devoted to an area of the fovea than to a same-size area of peripheral vision (Born, Trott, & Hartmann, 2015).

In turn, acuity determines the precision of spatial localization. This can be measured by *Vernier acuity*, the ability to detect whether two line segments are continuations of the same line or rather are offset from one another. Vernier acuity is much worse in the periphery than in central vision. An impression of this can be gained by averting central vision away from the lines in Figure 12.1, which makes detection of the offset much more difficult. Thus the ability to spatially localize the lines is positively related to the degree of cortical magnification for their location on the retina (Greenwood, Szinte, Sayim, & Cavanagh, 2017).

Relative acuity across primate species is fairly well understood, using a version of Vernier acuity in which two lines are perceived if the gap is discriminable but one line if not. Measured in that way, acuity in apes (including humans) and macaques is nearly twice as good as it is in strepsirrhines and New World monkeys (Ross, 2000).

Evolutionarily speaking, what happened is that in diurnal anthropoids – including apes, humans, and macaques – the cornea and lens became reduced in size relative to the rest of the eye. That allowed their positioning further from the retina, increasing the *focal length* (the distance required for an image to come into focus). In turn that magnified the image on the retina (Ross & Kirk, 2007). The greater the magnification, the greater the acuity, because any given part of the image was spread over more receptors. Night vision degraded, however, because smaller corneas and lenses meant less light-gathering power.

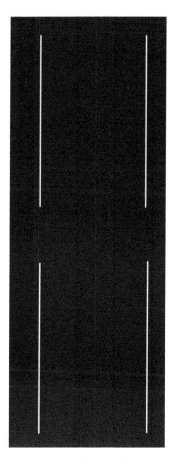

Figure 12.1 Demonstration of Vernier acuity. The left lines are aligned, while the right ones are slightly offset.

Source: Original figure.

But in diurnal anthropoids that did not pose a significant problem, because life's critical activities were performed in daylight (Ross & Kirk, 2007). These changes to the cornea and lens must have taken place close to the origin of anthropoids 72 million years ago.

Higher brain mechanisms for recognizing location

So far we've really only considered how the appropriate conditions were created for the brain to recognize location. Although V1 maintains a direct correspondence between a source's position on the retina and the location of cortical activation, further processing is needed to locate the source relative to the self or to the environment.

The brain region most involved in referencing locations to the self is the parietal lobe, the upper rear portion of the brain. It receives input from V1 partly by way of the immediately adjacent area V2 (Brodmann area 18, see Figure 10.4), along what is called the *dorsal visual stream*. The dorsal stream is sometimes referred to as the "where" pathway to emphasize its role in spatial processing (Figure 12.2).

Judgment of location relative to the body uses what is called the *egocentric frame of reference*. Thus deciding whether an object is to the left or right of our position, and to what degree, refers to ourselves and is therefore ego-oriented. For example, Neggers and colleagues asked participants to judge the position of a vertical bar superimposed on a horizontal bar, in relation to the middle of the body. Activation was found in the superior parietal lobe, Brodmann areas 5 and 7. This activation was not found when the positions were judged in relation to the center of the horizontal bar, an environmentally referenced judgment using what is called the *allocentric frame of reference*. That uses the *ventral visual stream* of processing, located in the temporal lobe (Figure 12.2; Neggers, Van der Lubbe, Ramsey, & Postma, 2006), including the hippocampus (Box 12.1).

Other studies indicate that the area of activation involved in egocentric location judgments may be narrowed down to area 7 and the bordering posterior intraparietal sulcus (Zaehle et al., 2007). Furthermore, when hemispheric differences are found, the right hemisphere is more involved than the left (Iachini, Ruggiero, Conson, & Trojano, 2009). Right hemisphere lateralization is also evident in visual field studies, in which a dot is

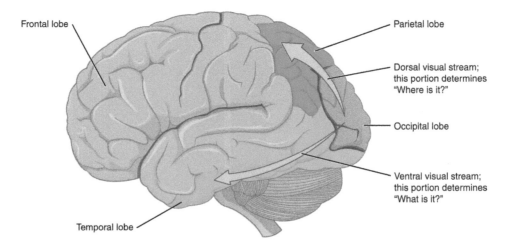

Figure 12.2 The dorsal and ventral visual streams.

Source: 1424 Visual Streams.jpg, CC BY-SA 3.0, OpenStax College.

Box 12.1   Place cells, grid cells, and the hippocampus

Located in the medial temporal lobe, the hippocampus is a ridge of gray matter roughly in the shape of a sea horse; thus its name, which is Latin for sea horse. It makes up part of the ventral stream of processing and is the site of *place cells*, neurons that respond to particular locations in the environment. Place cells in turn connect to *grid cells*, neurons that respond to locations in an overall grid representation of the environment. Because the representation is environmental – the grid is in effect a map – it is allocentric. Grid cells can be found both in the hippocampus and in the neighboring cortex (Jacobs et al., 2013).

Both types of cells have been found in humans undergoing electrode placements in their brains to locate the source of epilepsy prior to possible surgical treatment (Ekstrom et al., 2003; Jacobs et al., 2013). They have also been found in rhesus macaques (Killian, Jutras, & Buffalo, 2012), rats and mice (Fyhn, Hafting, Witter, Moser, & Moser, 2008), and bats (Yartsev, Witter, & Ulanovsky, 2011).

Furthermore the grid cells of bats are very similar to those of rats, for example in grid layout, cell layering, and the correlation of cell firing with movement velocity (Yartsev et al., 2011). That suggests that they, and presumably place cells as well, have a very long evolutionary history dating back beyond the last common ancestor of bats, rats, and humans – which lived about 97 million years ago (Graphodatsky, Trifonov, & Stanyon, 2011).

There is evidence that the size of the hippocampus is related to how much it is used in spatial processing of the environment. Birds that cache food for later retrieval, such as the marsh tit and black-capped chickadee, have larger hippocampi than related species that do not cache food (for the examples given, the bridled titmouse and the Mexican chickadee). Similar results apply in rodents (Roth, Brodin, Smulders, LaDage, & Pravosudov, 2010). It has even been seen in humans: University students who better learned a mental map of campus, as evidenced in a task requiring them to point to relative locations while blindfolded, had larger posterior hippocampi than those who performed more poorly (Schinazi, Nardi, Newcombe, Shipley, & Epstein, 2013). This confirmed other studies examining navigation in taxi drivers (Woollett & Maguire, 2011).

---

presented in the left or right visual field and the task requires subsequent report of the location. Accuracy is generally higher in the left visual field, projecting to the right hemisphere (Boles, Barth, & Merrill, 2008).

It is no accident that egocentric location processing emphasizes the dorsal pathway and the parietal lobe, while allocentric processing emphasizes the ventral pathway and the temporal lobe. Recall that the parietal lobe is also the principal location of somesthetic processing. In representing the body's surface, SI is highly egocentric. Thus evolution appears to have unified egocentric representation by placing the tactile and visual frames of reference in proximity within the same lobe. Indeed, in the macaque, many neurons in the superior parietal lobe respond to both visual and somesthetic stimulation (Gamberini et al., 2018; Gamberini, Galletti, Bosco, Breveglieri, & Fattori, 2011). In contrast, allocentric location processing depends heavily on referencing objects in the environment. As we shall see in the next chapter, object recognition is largely performed in the temporal lobe as part of the "what" pathway (Figure 12.2).

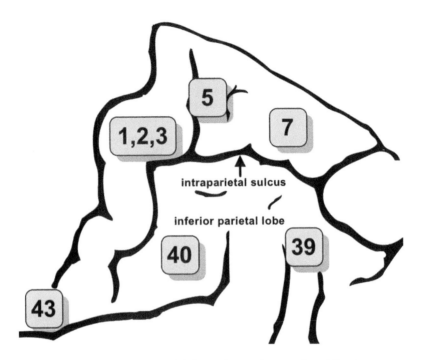

Figure 12.3 Major features of the parietal lobe, referenced to numbered Brodmann areas.

Source: Adaptation of Gray726-Brodman.svg, public domain, Henry Vandyke Carter.

In the macaque, the intraparietal sulcus (Figure 12.3) has neurons that code the coordinates of visual stimuli in a head-centered, egocentric frame of reference. There is substantial sensitivity of the sulcus to motion of both the stimulus and the body, providing spatial updating that prepares the animal for movement (Grefkes & Fink, 2005).

The use of area 7 and the intraparietal sulcus for egocentric processing may trace to early primates. Galagos, which are strepsirrhines, have a large posterior parietal area, while our close nonprimate relatives the tree shrews do not (Kaas, 2013). However, egocentric processing is performed by species as diverse as mice, birds, and insects (Benhamou & Poucet, 1996; Bruck, Allen, Brass, Horn, & Campbell, 2017), presumably using other neural mechanisms.

Depth

Other processes are devoted to localization in a dimension not yet considered, the one involving positions nearer to or farther from the observer. Depth can be computed from many cues, but four are particularly powerful. Three are *monocular cues* because they can be seen by a single eye, as opposed to the fourth, which is a *binocular cue* (requiring two eyes to see).

The monocular cues are *occlusion*, in which an object whose image partially blocks another is judged closer; *relative size*, in which the images of two similar objects are compared and the larger is judged closer; and *height in the visual field*, in which the images of objects lower in the visual field are usually judged closer than those higher in the visual field (Glueck, Crane, Anderson, Rutnik, & Khan, 2009). All emerge from location processing, as well as more general pattern recognition processes covered in the next chapter.

Here, attention focuses on the binocular cue specifically. It is called *binocular disparity*, the discrepancy between images from the two eyes. In order to compare images across eyes for the purpose of detecting disparities, it is necessary for them to come together in the primary visual cortex (V1). As we have seen, after the optic nerves leave the eyes, they split with the inner halves crossing over and the outer ones passing straight back (Figure 7.6). As a result, the inner half of the one eye and the outer half of the other eye, both surveilling the same visual field, project to the same hemisphere of the brain. The outputs from their images come together and are compared in V1, which has cells that directly detect binocular disparity.

Across species, the use of binocular disparity as a depth cue depends heavily on the spatial arrangement of the eyes. A minimal requirement is that the eyes be capable of imaging the same object at the same time, achieving *binocular convergence*. This is not as trivial as it seems, since many species (e.g., horses) have eyes located on the sides of the head nearly opposite one another. Because such species have only a small overlap between what the two eyes see, binocular disparity cannot be used as a depth cue in most of the visual field. Other species, however, including primates, have their eyes shifted forward so they face the same direction. Images with much greater overlap are produced, providing disparity cues throughout the visual field.

Binocular vision has a deep history within primate ancestry. Even the earliest primates showed substantial binocular convergence. Convergence can be measured as the *binocular field of view* (BFOV), i.e., the opening angle of the joint field of view. If you close your left eye and look straight ahead, you can move a fingertip left until it disappears behind your nose about 70° left of center. If you repeat the process with the right eye, the fingertip disappears about 70° right of center. Both eyes can see everything within those limits, a joint field of view of about 140°.

Rats at 40–60° and tree shrews (close relatives of primates) at 60° have BFOVs less than half that amount. In contrast, the early Eocene primate species, *Teilhardina asiatica*, dating to 55 million years ago, had an estimated BFOV of about 110–120°. That value is not much less than the 127–128° value of tarsiers, the 135° value of galagos, the 138–146° values of New World owl and squirrel monkeys, and even the 140–160° values of macaques and modern humans (Heesy, 2004; Ross, 2000).

Thus binocular convergence was largely in place during an early phase of primate evolution, increasing only modestly during our descent. It testifies to an origin in nocturnal predation, as it is believed to have been used by early primates to counter camouflage by detecting depth disparities between an insect and its surroundings (Heesy, 2008; Ni, Wang, Hu, & Li, 2004). In fact, predators in general, not just primates, tend to have front-facing vision, because the overlap between the eyes provides good depth perception to discern and attack prey. In contrast, prey animals tend to have side-facing vision with little binocular overlap, for maximal surveillance of the environment.

Another factor is that larger bodies have greater separation between the eyes, increasing disparity cues and improving depth perception. In fact, in our ancestral line, eye separation increased dramatically as body sizes increased. Thus *Teilhardina asiatica* weighed only about 1 ounce and had an inner *interorbital distance* (the space between the eyes, as measured between their nearest edges inside the skull) measuring only 4 mm (Ni et al., 2004). In contrast, the typical interorbital distance in modern humans is on the order of 20–30 mm (Mafee et al., 1986).

Depth perception, however, requires further cortical processing beyond mere convergence between the eyes in V1. Tsao and colleagues (2003) used random-dot stereograms

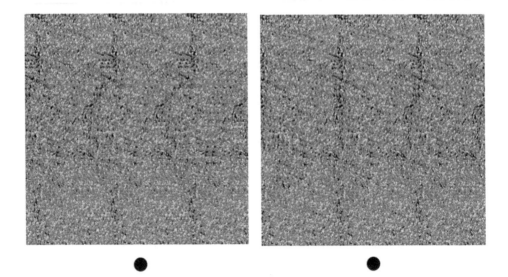

Figure 12.4 A random-dot stereogram. Slightly crossing the eyes so that the squares lay atop one another and fuse, produces a perception of abstract shapes in depth.

Source: Adaptation of Stereogram Tut Random Dot Shark.png, CC BY-SA 3.0, Fred Hsu.

(simpler but otherwise similar in structure to Figure 12.4) to examine brain mechanisms of depth perception in both humans and macaques. Binocular fusion of two stereogram views resulted either in a perception of depth or of no depth. When depth was present, there were widespread increases in cortical activation that were strongest in both species in area V3a and the *caudal* (rear) intraparietal sulcus (Tsao et al., 2003). Area V3a is in the occipital lobe bordering on the parietal lobe, and like the intraparietal sulcus is thus part of the dorsal "where" stream of processing.

However, to some extent depth is also processed along the ventral "what" pathway (see Figure 12.2; Georgieva, Peeters, Kolster, Todd, & Orban, 2009). There it is used to recognize objects in depth rather than their location as such (Verhoef, Vogels, & Janssen, 2016). Perceiving 3D shapes uses this pathway.

Motion

Other cues for perceiving depth come from motion. If one object moves to occlude another, partially or completely hiding it, then the occluding object is presumed to be closer to the observer. More subtly, as we move toward objects, ones that are closer show substantial moment-to-moment change in the size and position of their image at the retina, while those farther away show relatively little change. Vehicle occupants notice this in the form of *optic flow*, the experience of objects rushing past in the periphery while the view of the road far ahead seems to change much more slowly. However, optic flow is more than just a depth cue, because it also provides essential information as to *heading*, i.e., the direction of movement. Given stationary objects, eyes, and head, the point in space from which flow seems to emanate is the direction in which the observer is moving.

Both monkeys and humans use optic flow, and show at least somewhat similar patterns of cerebral activation. In humans, studies using two different brain-imaging technologies, *PET* (*positron emission tomography*) and *fMRI* (*functional magnetic resonance imaging*; for both, see Box 12.2) found that human observers passively viewing optic flow showed activation in area MT/V5 (see Figure 13.1), located in part of the inferior temporal sulcus (Sunaert, Van Hecke, Marchal, & Orban, 1999). The same was found during the active determination of heading, although parietal activation was also observed (Peuskens, Sunaert, Dupont, Van Hecke, & Orban, 2001). These results indicate that MT/V5 is sensitive to the movement present in optic flow, while the parietal lobe performs the spatial computations needed to determine heading.

For their part, macaques have an equivalent of human area MT/V5 that is sensitive to motion. It responds most to coherent motion in one direction, although incoherent motion provokes a response of lesser degree (Zaksas & Pasternak, 2005). Comparative studies suggest that this area exists in all primates including galagos, but not in species related to primates such as tree shrews. It therefore appears to have been a primate innovation (Kaas, 2004b). However, the ventral parietal cortex of monkeys is also involved in perceiving optic flow, with most neurons responding to expansion but some to contraction (Bremmer, 2005). This area may differ somewhat from that used by humans to perceive heading, which is more dorsal than ventral in location (Peuskens et al., 2001).

Box 12.2 PET and fMRI imaging

Positron emission tomography, or PET, involves injecting a radioactive substance into the bloodstream, either intravenously (e.g., radioactive glucose) or via respiration (e.g., radioactive oxygen). For safety reasons, substances with short half-lives are used to limit the amount of radiation emitted.

When brain areas are active due to the performance of a task, blood flow increases locally, perfusing the tissue with the radioactive substance. As radioactive decay occurs, positrons are emitted. Each positron is annihilated on contact with an electron, emitting a pair of photons with each pair member at a 180° angle to the other. An apparatus outside the head detects these pairs, and computer calculations work out their point of origin. As a result, over a short period of time enough pairs are detected that an image can be constructed of activity in the brain. However, this method does not produce a physical image of the brain on which the activation image can be superimposed. Instead, activation is usually specified in *Talairach space*, a set of X, Y, and Z coordinates that relate it to structures in an "average" brain.

Functional magnetic resonance imaging, or fMRI, provides more precise localization of function by overlaying an image of brain activity on an image of the physical brain. It works by first using an intense magnetic field to induce spinning protons in some of the brain's hydrogen atoms to align with lines of force. This is followed by a radio frequency pulse that deflects the protons, causing a wobbling of their spins. Detectors of electrical currents outside the head time the realignment of protons with the magnetic field once the radio pulse is turned off. As it happens, this timing is affected by whether the protons are in gray matter or white matter, allowing construction of a physical image of the brain. It is also affected by the level of blood oxygenation, indicating which brain areas have been more active. The resulting functional image can then be overlaid on the physical image, allowing activation to be localized precisely in relation to brain structures.

Quantity

The ability to judge quantities has clear adaptive value, for it can help determine which tree has more fruit, or which way to run when escaping groups of predators. In the absence of verbal counting, which is presumably a recent addition to hominin capabilities, quantity perception must rely on some type of spatial representation.

Studies of quantification in modern humans indicate the existence of three separate processes. When varying numbers of stimuli such as dots are presented briefly, and observers are asked to specify their number as quickly as possible, a three-limbed function is produced relating reaction time (RT) to numerosity (Figure 12.5). The first limb is from 1 to 3 or 4 dots, showing a modest increase in RT with each additional dot. The second is from 3 or 4 to about 7 dots, with a steep increase per dot. Finally, the third limb is marked by negligible increases in RT from about 7 dots on.

There has been little controversy over the nature of the second and third limbs. The second has been held to be due to an explicit counting process that involves verbalization and working memory, two time-intensive components that account for the substantial increase in RT as the number of dots increases. The third limb is due to an estimation process that kicks in when the number of dots is too large to be counted within the brief memory of the display. In fact, when the display is left on until response, RTs continue to rise with an increasing number of dots – i.e., estimation does not occur – because counting continues to completion. Because brief display of a large number of dots involves a single numerical estimation, RT is not much affected by the number of dots (Mandler & Shebo, 1982).

Subitizing in humans

Greater controversy has surrounded the first limb of the numerosity curve. Although held by consensus to reflect a *subitizing* process (an early perceptual process that rapidly enumerates small numbers of items), attempts to further characterize it have ranged widely.

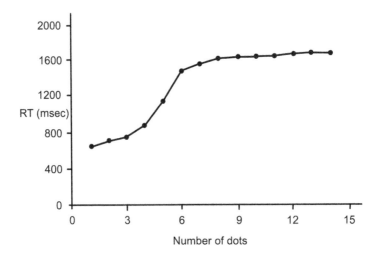

Figure 12.5 The numerosity curve, relating the number of dot stimuli to the time required to enumerate them.

Source: Original figure.

They have included explanations based on the large proportional differences that exist among small numbers, making them more discriminable than differences among large numbers (Averbach, 1963); the ability of small numbers of items to form patterns such as triangles and lines (Mandler & Shebo, 1982); a "fast counting" process specific to small numbers (Gallistel & Gelman, 1992); and a preattentive tagging of item locations that is limited in capacity and therefore specific to small numbers (Trick & Pylyshyn, 1993).

However, more recent research strongly supports the "fast counting" interpretation of subitizing (Boles, Phillips, & Givens, 2007). The research was motivated by the observation that extracting number information from a specific type of bargraph involves the same perceptual process as extracting number from dot clusters (Figure 12.6). Thus both are better recognized in the left visual field (LVF) as opposed to the right visual field (RVF), indicating a right hemisphere processing advantage. Significantly, the visual field differences are correlated so that observers with a large LVF advantage on one also tend to show a large LVF advantage on the other (Boles, 1991). Furthermore, trainees who practice bargraphs show substantial transfer of training to dot clusters but not to other numerical tasks (Boles, 1997), and simultaneous performance of the two tasks produces greater interference than pairing either with a different task (Boles & Law, 1998). All of these observations indicate that the two tasks share a common process.

The experiments of Boles et al. (2007) examined the relationship between the tasks in more detail, beginning with the plotting of numerosity curves. The dot clusters showed the classic multi-limbed function (Figure 12.7). The bargraphs, in contrast, showed an intriguing scallop-shaped function in which small increases in RT occurred at values starting at 0, 4, and 8 – i.e., the values represented by the horizontal reference lines (Figures 12.6 and 12.7). This pattern of small RT changes with increasing number suggested that the same subitizing process underlies the enumeration of small numbers of dots and the enumeration of bargraphs. The suggestion was strongly confirmed by other relationships in the results between the bargraphs and the subitizing range (1–4) of dot

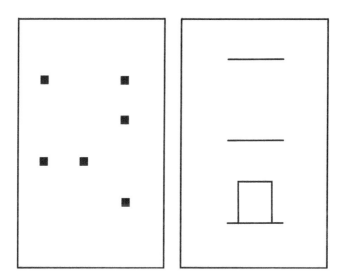

Figure 12.6 Examples of dot clusters and bargraphs that use the same perceptual process. The horizontal reference lines for the bargraph are at the 0, 4, and 8 levels, with the example reflecting the value "2".

Source: Original figure.

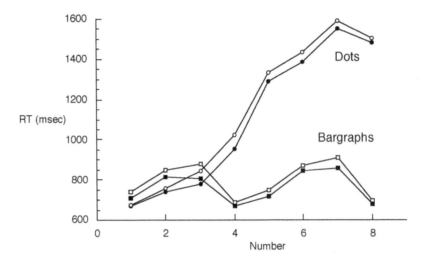

Figure 12.7 RTs to dot clusters and bargraphs in Boles et al. (2007). Closed symbols LVF, open symbols RVF.

Source: Response times from Study 1, reprinted by permission from Springer Nature Customer Service Centre GmbH, David B. Boles, Jeffrey B. Phillips, Somer M. Givens.

clusters that were not found between bargraphs and the higher range (5–8) of dot clusters (Boles et al., 2007).

The importance of these observations is that by establishing that subitizing underlies bargraph recognition, the nature of subitizing is revealed by the pattern of bargraph RTs. The scallop-shaped RT function is a telltale indication that subitizing is a rapid, nonverbal counting process operating on analog quantities. In the case of bargraphs, the analog quantities are imaginary intervals or steps above or below the reference lines, while in the case of dot clusters they are the dots themselves. Because subitizing uses analog quantities, it seems reasonable to classify it as a spatial process. Furthermore, consistent with subitizing's spatial nature, brain-imaging studies implicate right parietal lobe involvement, involving either the intraparietal sulcus (Cutini, Scatturin, Moro, & Zorzi, 2014) or area 39 at its junction with the temporal lobe (He, Zuo, Chen, & Humphrey, 2013).

Subitizing in other primates

Do other primates subitize? The most comparable study is of a chimpanzee named Ai, who was trained to associate the Arabic numerals 1–7 with the appropriate numbers of dots in a display, and to press the appropriate numeral (Murofushi, 1997). She was subsequently timed in her responses to varying numbers of dots. Her results were strikingly similar to those from human participants. Tested on 1 to 7 dots, a subitizing range of 1 to 3 or 4 dots was clearly differentiated from a "counting" range of 3 or 4 to 6 dots (Figure 12.8).

The results indicate that chimpanzees likely engage in the same analog-based subitizing process as humans. What they do in the "counting" range is less clear since they presumably do not have the same verbal processes as humans. Murofushi observed that with the larger numbers Ai's eyes tended to move back and forth between the display and the response numerals, as if taking additional decision time – a possible cautionary note

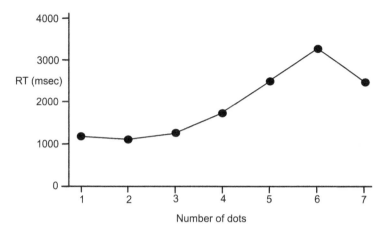

Figure 12.8 Numerosity curves for a chimpanzee responding to 1–7 dots. RT = Reaction time.
Source: Original figure.

for research on human enumeration in suggesting that large increases in RT can in fact be due to processes other than verbal counting.

There are indications that macaques show much the same behavioral pattern of results when enumerating items, as well as at least some parietal lobe involvement. Nieder, Freedman, and Miller (2002) used a somewhat different experimental design requiring rhesus monkeys to compare quantities of successive patterns of 1–5 dots. The resulting numerosity curve nevertheless showed a strong resemblance to that found for humans and chimpanzees, except that the subitizing range was limited to 1 to 2 dots rather than 1 to 3 or 4 dots (Nieder et al., 2002).

Nieder et al. also recorded from single neurons in monkeys performing the task, and they found that 7% of inferior parietal cells, and about a third of randomly selected prefrontal cells, showed activation that varied with the number of dots. Furthermore, individual neurons were "tuned" to a particular number, so that some responded with peak activity to one dot, others to two dots, and so on, indicating an individual-neuron-based sensitivity to numerosity. The involvement of inferior parietal cells is generally consistent with the human literature cited earlier.

Because there is at least some inferior parietal involvement in subitizing in macaques, it appears that use of the area for that purpose traces back at least to the origin of Old World monkeys about 32 million years ago. How much further it extends beyond that is presently unclear.

Orienting

Spatial perception is enhanced by our ability to shift eyes and attention toward a source of stimulation; in other words, *orienting* to objects. This allows more precise estimates of location and depth. The parietal lobe plays a major role in orienting. In humans, both *saccades* (eye movements toward a source of stimulation) and *pursuit eye movements* (eye movements following a moving object) involve an area in the middle of the intraparietal sulcus including a portion of Brodmann area 7 (Culham, Cavina-Pratesi, & Singhal, 2006; Olson, Gatenby, Leung, Skudlarski, & Gore, 2003).

It is sometimes called the *parietal eye field*, and one of its roles is to update spatial information in response to movement (Culham et al., 2006). In addition, it may actually initiate eye movements (Pierrot-Deseilligny, Ploner, Müri, Gaymard, & Rivaud-Péchoux, 2002), although others take the position that it maintains eye movement accuracy (Rafal, 2006). As we have seen, galagos have a large posterior parietal lobe, and so the involvement of the intraparietal sulcus and area 7 may trace back to the early history of primates.

Eye movements also involve a second important area, the *frontal eye field*, located in the frontal lobe. There is increasing consensus that it is located near the rear of the middle frontal gyrus (see Figure 10.4), more specifically at or close to the intersection of the superior frontal sulcus with the precentral sulcus (Rafal, 2006). Research suggests it plays a larger role in triggering voluntary saccades (i.e., to a stimulus long present) than in triggering reflexive saccades (i.e., to the sudden appearance of a stimulus; Rafal, 2006). This may be a matter of degree, since bilateral lesions of both the frontal and parietal eye fields appear necessary to cause a severe, long-lasting effect on both voluntary and reflexive saccades (Pierrot-Deseilligny et al., 2002).

Spatial attention, the ability to attend to objects or locations independent of eye position, involves parietal areas that overlap with those controlling eye movements. For humans, the posterior parietal cortex (Brodmann area 7) was highlighted as critical to spatial attention by Posner and colleagues (Posner, Walker, Friedrich, & Rafal, 1984). They presented lesion evidence indicating that the area is involved in the engagement and disengagement of attention to positions in space. The bordering intraparietal sulcus is also activated by the preparation and movement of attention to the periphery, even if the eyes are kept fixated in a central location (Astafiev et al., 2003).

A second parietal area involved in spatial attention is the inferior parietal lobe. When it is damaged, a syndrome called *visuospatial neglect* (or visual neglect, or spatial neglect) can result. The patient fails to orient to objects opposite the side of the lesion. The syndrome most often (but not exclusively) follows right hemisphere damage, thus producing a left-sided neglect.

The symptoms can be quite striking, as when patients draw objects with their left sides missing, or bisect lines far to the right of their actual midpoints. A companion phenomenon, *pseudoneglect*, can be observed in normal individuals by requiring the bisection of lines staggered down a page. Bisections are typically made slightly to the left of true center, opposite to the effect seen in most neglect patients. This result is attributable to an intact right hemisphere attentional mechanism that overattends to its half of space, resulting in overestimation of line length on that side (Boles, Adair, & Joubert, 2009; McCourt, Garlinghouse, & Reuter-Lorenz, 2005).

Several studies have mapped the lesions of patients with visuospatial neglect. Taken together, they find that the critical damage is either in the posterior portion of area 40 or the adjoining area 39 (e.g., Mort et al., 2003; Ptak & Schnider, 2011; Verdon, Schwartz, Lovblad, Hauert, & Vuilleumier, 2010). Since both area 40 (the supramarginal gyrus) and area 39 (the angular gyrus) border on and are inferior to the intraparietal sulcus (see Figure 12.4), it is an interesting question whether they represent the same or a different area than that identified in attention shifting. The available evidence suggests that it is in fact different. Astafiev et al. (2003) found that after accounting for areas involved in saccades, a small portion of Brodmann area 7 could be attributed to spatial attention specifically. In contrast, the lesions involved in neglect appear to lie in the ventral portions of areas 39 and 40 and thus at some remove from area 7 and the intraparietal sulcus. Therefore, areas 40/39 and 7 may well support separate components of spatial attention.

The evolutionary background of orienting

Evidence from primates suggests that the orienting mechanisms found in humans have a long evolutionary history. In macaques, eye movements involve a lateral intraparietal location believed to correspond to the human parietal eye field. However, there may be a small difference in that the area is located in the lateral intraparietal sulcus in monkeys but at its medial wall in humans. Monkeys also have a frontal eye field, but at a location somewhat anterior to ours (Astafiev et al., 2003; Culham et al., 2006; Grefkes & Fink, 2005).

Further parallels are found in attention. Just as human attention and eye movements involve similar parts of the intraparietal sulcus, macaque attention and eye movements both involve the lateral intraparietal sulcus (Astafiev et al., 2003; Grefkes & Fink, 2005). There is even an indication of a similar mechanism involved in spatial neglect, and in a species with an older evolutionary divergence from humans. Specifically, in marmosets, neglect is produced by lesions of the parietal lobe, extending down to include part of the temporal lobe. The affected animals have difficulty locating rewards opposite their lesions (Marshall, Baker, & Ridley, 2002). While the critical part of the lesioned area producing this behavior is not known, the result raises the strong possibility that monkeys have equivalents to both of the attention-related parietal areas found in humans.

Reaching and grasping

Orienting allows us to "home in" on locations and objects, not just as an aid to judging location and depth, but also for the purposes of reaching and grasping. In both macaques and humans, the most posterior part of the superior parietal lobe is involved, on its medial surface. In humans that is the medial part of area 7, known as the *precuneus* (Gamberini et al., 2018; Vesia et al., 2017).

Grasping *independent* of reaching, however, appears to involve the anterior intraparietal sulcus, and immediately below it, the superior *rostral* (forward) portion of area 40 (Frey, Vinton, Norlund, & Grafton, 2005; Vingerhoets, 2014). This introduces an inferior parietal component to the processing of spatial location, perhaps because grasping is more object-oriented than the more purely egocentric tasks eliciting superior parietal activation.

The two parietal areas principally involved in recognizing location, areas 7 and 40, both showed evolutionary change in our immediate ancestry. A bulge of the parietal region, observed in *Homo sapiens*, has been attributed to an expansion of the precuneus of area 7. The bulge, and the expansion, are not observed in chimpanzees. Nor are they evident in Neanderthal endocasts (Bruner, 2010; Bruner, Preuss, Chen, & Rilling, 2017). For its part, the supramarginal gyrus, the location of area 40, is evolutionarily recent and has no analog in macaques. Nevertheless, the anterior infraparietal area is involved in grasping in both humans and monkeys (Vingerhoets, 2014).

Navigation

As perceptual information is progressively processed in the "where" pathway in the parietal lobe and then in the frontal lobe, its coding becomes more action-oriented, matching percept to motion. All of this activity comes together in our ability to navigate our environment.

As we learn to navigate, memory becomes increasingly important. It allows us to recall the locations of landmarks, and to associate with them actions such as turning left or right. Generally speaking, when we find ourselves in a new environment, whether it is a new building or a new city, most of us initially build *route knowledge*, assembling knowledge of landmarks and the associated turns to learn pathways from one point to another. As learning progresses, we gradually develop more sophisticated *survey knowledge*, which is the knowledge of how landmarks are configured relative to one another. At this point a true cognitive map develops, allowing us to deviate from established pathways and find new routes (Wolbers, Weiller, & Büchel, 2004). However, there are individual differences in the formation of cognitive maps. Some individuals begin to acquire information about spatial relationships and start building a cognitive map on first exposure to a new environment, essentially bypassing the route knowledge phase. Others never proceed beyond route knowledge to build a cognitive map (Ishikawa & Montello, 2006).

There is good evidence that many chimpanzees develop true cognitive maps. In African forest habitats, they follow nearly linear paths to food trees scattered among many thousands of other trees, with the vector varying depending on the relative locations of the tree and themselves. They also decelerate when nearing the goal. The implication is that the chimpanzees "know" where they are going on a mental map (Normand & Boesch, 2009). Similar results were found by Janmaat, Ban, and Boesch (2013), who followed the foraging behavior of five female chimpanzees for several weeks. They reported that 13% of close inspections of food trees were preceded by vector-like goal-directed approaches.

Brain mechanisms in navigation

Not surprisingly, the parietal lobe plays a major role in the acquisition of navigation-relevant knowledge. Shelton and Gabrieli (2002) created a computer-based virtual convention center, a market place, and a park, with individual human participants assigned one venue for route learning and another for survey learning. Route learning used a virtual walk-through of the venue, while survey learning used an aerial perspective to show multiple landmarks simultaneously. Scans of brain activity were made during learning, using fMRI. Greater activity was found for route learning than for survey learning in portions of bilateral Brodmann areas 5 and 7, and right area 40, all in the parietal lobe. Survey learning produced greater activation than route learning in another portion of area 7 bilaterally. As might be expected, other areas were also implicated and were probably related to memory, visual object perception, and movement components of the complex task.

Wolbers et al. (2004) created a virtual town consisting of buildings and roads, and imaged brain activity while participants both learned the environment by navigating it, and retrieved information concerning it. The development of route as opposed to survey knowledge was encouraged by emphasizing landmarks. It was confirmed by checking maps drawn after the scans, which showed accurate learning of landmark pairs but poor survey learning. Activity in inferior locations in both parietal lobes, Brodmann areas 39 and 40, was associated with better learning. Learning-related activation was also found in the medial frontal gyrus and medial *retrosplenial cortex* (cortex immediately posterior to the corpus callosum), perhaps related to movement and memory, respectively (Wolbers et al., 2004).

Another approach to uncovering the brain areas critical to navigation uses correlational methods to relate activation to navigation performance. Moffat, Elkins, and Resnick (2006) examined route learning in a virtual building, and found a correlation between navigation accuracy and activation in parietal area 7. Correlations were also reported for widespread areas including the cerebellum, inferior and middle frontal gyri, and the medial temporal and cingulate cortex, among others.

Together these studies indicate that much of the spatial information used in navigation is processed by the parietal lobe, involving both posterior and inferior locations. The complex nature of navigation tasks, however, makes it impossible to ascribe them as a whole to a particular brain area due to the involvement of other areas in memory, object perception, and movement, all critical to navigation.

Comparable primate data are limited. However, Sato, Sakata, Tanaka, and Taira (2006) made recordings from individual neurons in the medial parietal cortex of Japanese macaques, while the animals navigated a virtual building with a joystick, using previously trained route knowledge. About a third of the neurons were found to be sensitive to movement in the navigation task, most of them in a highly selective manner (e.g., responding only to a left turn at a particular point). To confirm these results, an *agonist* (a drug that temporarily binds to neural receptors and prevents their normal function) was next injected at the sites, and navigation was subsequently found to be impaired. Thus, like humans, monkeys appear to make use of parietal mechanisms during navigation. To the extent that those mechanisms involve the posterior parietal lobe, especially area 7, they may have originated early in the primate line.

The independence of spatial processes

Given that the parietal lobe is involved in so many spatial functions, it is tempting to deemphasize the differences between them and to view the lobe as a general-purpose spatial processor. By this view, it might be regarded as relatively undifferentiated in function, with perception of location, depth, motion, and quantity of objects, and our ability to orient to them, merely representing minor variations on an overall theme.

There are good reasons to think otherwise, however. First, as already reviewed, different abilities draw on different parts of the parietal lobe. That indicates that at least to some extent, spatial processes are independent of one another. Thus for example, area 7 in the superior parietal lobe processes location in an egocentric frame of reference and is involved in reaching, but does not seem to be involved in grasping or in subitizing. Inferior parietal cortex, areas 39 and 40, is involved in grasping and subitizing but not in egocentric location or reaching. As another example, the intraparietal sulcus dividing superior and inferior parietal cortex is involved in a number of functions, but different portions are used for different things: e.g., the posterior portion for egocentric location and depth, the middle portion for eye movements, and the anterior portion for grasping. This involvement of different parietal areas in different abilities argues for a differentiated parietal lobe housing a number of discrete processes.

The independence of spatial processes is also supported by a series of experiments described by Boles (2002). These were conducted with the purpose of examining what relationships, if any, exist between a number of spatial tasks. The tasks included recognizing the number represented by a bargraph; judging the intensity of facial emotions; recognizing the location of a flashed dot; bisecting lines; judging whether a line was above or below a fixated point; estimating large numbers of dots; and judging which bar of a cross lay on top of the other. The tasks were all administered to a large number of participants, and visual field differences were assessed. All but one showed an advantage of the right hemisphere over the left, as evidenced by faster or more accurate responses to the LVF compared to the RVF.

Nevertheless, even though the same hemisphere was involved in nearly all of the tasks, the correlations among the task measures were all essentially zero! In other words, knowing the size of a person's hemispheric advantage for one task did not at all predict its size for any of the other tasks, constituting strong evidence of the independence of spatial processes (Boles, 2002). Taken together with the evidence of differing localization of spatial processes, this indicates that the parietal lobe is differentiated, housing numerous discrete spatial processes. It will no doubt prove challenging to identify all of them and to work out their evolutionary backgrounds.

Evolution of the parietal lobe

The involvement of the parietal lobe in both humans and monkeys is the common thread running through most of the spatial perception literature. This section addresses the extent to which this region of the brain has been conserved, and to what extent it has changed during hominin evolution.

Gross size comparisons between monkey and human parietal cortex are fraught with difficulty, because not everyone agrees on cross-species anatomical markers that separate parietal from occipital and temporal cortex (Semendeferi & Damasio, 2000). Nevertheless, there are some indications of a general expansion. Thus, while occipital cortex appears to be only about half as large in humans as would be extrapolated from primate data for a brain our size (Deacon, 1992b; de Sousa et al., 2010), parietal and occipital cortex taken together are very nearly in line with such extrapolations (Semendeferi & Damasio, 2000). The implication is that while the entire sector has maintained a constant size relative to the rest of the cortex, the parietal lobe has expanded relative to the occipital lobe (Grefkes & Fink, 2005). A quantitative analysis of primate and human brains, based on 29 anatomical landmarks, has pointed to a specific widening of the parietal lobe in the medial-to-lateral direction, in humans as compared to chimpanzees. Unfortunately, the anterior-to-posterior direction could not be adequately assessed, again because the border between the parietal and occipital lobes was unclear (Aldridge, 2011).

Indications of the parietal expansion are seen in hominin remains. Endocasts from *Australopithecus afarensis* show a more developed parietal lobe relative to apes as of 3.2–3.5 million years ago. Subsequently, there was a parietal widening, particularly in the inferior parietal lobe, which is visible in *Homo habilis* endocasts dating to about 1.8 million years ago (Tobias, 1995).

Bruner (2004) compared three species within the genus *Homo*: *Erectus*, *neanderthalensis*, and modern *sapiens*. Endocast measurements indicated that relative to *erectus*, *neanderthalensis* experienced a general upward expansion of the brain that was not limited to the parietal area. *Sapiens*, in contrast, specifically experienced "extreme enlargement of the parietal contour", accompanied by an occipital decrease (Bruner, 2004, p. 290). As a result, the occipital lobe shifted to a position further under the parietal lobe, in effect rotated underneath by the parietal expansion. Rotating with it was the cerebellum, moving from a position under the rearmost portion of the occipital lobe to its present position nearly under the temporal lobe (Bruner, 2004).

Taking all of these observations together, there appears to have been a continuous expansion of the parietal lobe relative to the rest of the brain, beginning sometime prior to the earliest australopithecines for which we have endocasts (3.5 million years old), through to modern *sapiens*. In the process, some areas of the lobe appear to have changed more than others. In modern humans the inferior parietal area has about twice the cortical surface area, relative to the rest of the cortex, as it does in macaques (Orban, Claeys, et al., 2006). Also, as previously noted, beginning with the genus *Homo* the inferior parietal cortex acquired new gyri, specifically the supramarginal and angular gyri, Brodmann areas 40 and 39 (Tobias, 1995).

Accompanying the expansion, there has been a "medialization" of some parietal areas in humans, relative to more lateral positions in macaques (Grefkes & Zink, 2005). Enlargement of the precuneus (medial area 7) was part of the expansion, and it was exclusive to *sapiens* and not *neanderthalensis* (Bruner et al., 2017). Finally, although it is not known to be directly related to the expansion, there is some indication that modern humans have a larger number of functionally distinct areas in the intraparietal sulcus than do macaques (Orban et al., 2006).

Conclusion

Our capacities to identify the location, depth, motion, and quantity of objects, and to orient to them, are among our most important spatial perception abilities.

Although the optics of the eye itself allow a one-to-one mapping of source points to the retina, the greater compression of information from the periphery limits acuity and in turn our ability to locate objects. This is repeated at the cortex, where a given area at the periphery engages less cortex than the same area in central vision.

Information passing along the dorsal visual stream to the parietal lobe allows the location of objects in the egocentric frame of reference. In the parietal lobe, Brodmann area 7 and the bordering posterior intraparietal sulcus, especially in the right hemisphere, are most involved, bringing the tactile and visual egocentric frames of reference into close proximity. Macaques likewise use the intraparietal sulcus for egocentric location processing. Because a large posterior parietal lobe exists in galagos but not tree shrews, the use of the region for egocentric processing may have originated in early primates.

The allocentric frame of processing, in contrast, references locations to the environment. It uses the ventral stream of processing, especially the hippocampus and neighboring cortex. Place cells respond to particular locations in the environment, and grid cells to locations in a grid representation of the environment. These mechanisms date back at least as far as the common ancestor of bats, rats, and humans about 97 million years ago.

Depth processing involves a number of cues, but binocular disparity is one of the most powerful. Outputs from the two eyes converge in the primary visual cortex (area V1), which has cells that detect binocular disparity. Forward-facing eyes, found even in early primates, allow a great deal of overlap in the two images and are a feature of predators that use depth perception to discern and attack prey. Body size also has an important effect, with large bodies having increased eye separation, which helps better resolve depth. In the brain beyond area V1, depth is processed by the superior occipital lobe and the caudal intraparietal sulcus as part of the dorsal stream.

Motion provides cues about both depth and heading, the latter through optic flow. Area MT/V5 is sensitive to the movement present in optic flow, but parietal cortex appears to perform the spatial computations needed to determine heading. Macaques seem to use similar areas. Comparative research suggests that MT/V5 is a primate innovation, as it does not exist in tree shrews.

In humans, reaction time reveals that the perception of quantity in brief visual displays involves three processes: Subitizing, used for 1 to 3 or 4 items; counting, used for 3 or 4 to 7 items; and estimation, used for more than 7 items. Although several different explanations have been proposed for the fast processing of the subitizing range, research using both dot and bargraph displays show that both use the same subitizing process. Because quantities are only implied in the bargraphs in the form of imaginary quantities above or below reference lines, subitizing must involve the fast counting of analog quantities.

A chimpanzee named Ai has shown a very similar pattern, as have macaques. These outcomes imply that chimpanzees and macaques share our subitizing and estimation processes, and that they have a kind of slow "counting" process probably based on eye movements. Monkeys use the inferior parietal lobe for these processes, as do humans at least for subitizing.

The parietal lobe is also involved in integrating spatial analysis into action. The control of reaching and grasping involves the precuneus (medial area 7), while grasping without reaching involves the anterior intraparietal sulcus and the superior rostral part of area 40. Macaques seem to use much the same general areas even though, strictly speaking, they do not have an area 40. Relative to monkeys, we have shifted some areas to more medial positions and have evolved a bulge of the parietal region that has been attributed to an expansion of the precuneus.

We also use the parietal lobe in orienting to objects. The parietal eye field, located in the middle of the intraparietal sulcus and part of area 7, updates spatial information in response to movement, allowing the accurate targeting of eye movements. Spatial attention, which is the ability to attend to objects or locations independent of eye position, is controlled by the posterior parietal cortex (area 7) and by the inferior parietal lobe (areas 40 and 39). When the latter is damaged, spatial neglect can occur, involving the overlooking of objects on one side (usually the left) of a display.

Macaques similarly use an intraparietal location for controlling eye movements, but more laterally than the medial location used by humans. Spatial neglect has been observed in marmosets following lesions that include the inferior parietal lobe, similar to humans.

In most cases, humans learning to navigate a new environment first build route knowledge from landmarks, and then gradually develop more sophisticated survey knowledge that allows formation of a true cognitive map. Chimpanzees also use cognitive maps when they follow linear paths to food trees. During route learning in virtual environments, activation is found in areas 5 and 7 bilaterally, and 40 in the right hemisphere, with one study finding bilateral activation of both 40 and 39. During survey learning, a different portion of area 7 bilaterally is used. One study of the use of route knowledge in macaques found involvement of neurons in the medial parietal cortex.

The tempting view that the parietal lobe is an undifferentiated spatial processor is refuted by two lines of evidence. First, different areas of the lobe serve different processes. Second, experiments using a number of spatial tasks show that the extent to which each draw on the right hemisphere is independent of the others. Thus the parietal lobe is differentiated and houses a number of spatial processes, each of which may have a separate evolutionary history.

Finally, there is evidence of a parietal lobe expansion during human evolution, relative to the remaining cortex. The lobe at least widened, its lengthening being hard to address due to the lack of agreed-upon landmarks. A well-developed parietal lobe is apparent in *Australopithecus afarensis* endocasts from 3.2 to 3.5 million years ago. Widening of the inferior parietal lobe is visible in *Homo habilis* about 1.8 million years ago. Thus there appears to have been a continuous expansion of the parietal lobe relative to the rest of the brain, beginning at least 3.5 million years ago and extending down to modern *sapiens*. It included the appearance of new gyri in the inferior parietal lobe beginning with the genus *Homo*. The number of functionally distinct areas probably increased relative to macaques, and some areas shifted to more medial positions in the parietal lobe.

13 Pattern recognition

Throughout our evolutionary history, survival has depended on distinguishing among objects in the environment, and within the class of objects that are living things, among predators, prey, and *conspecifics* (other members of our species). We depend particularly heavily on visual recognition, which is achieved by a series of processing stages along the ventral "what" pathways of the brain (Figure 13.1).

In this chapter, we trace the progress of information processing along these pathways, achieving the recognition of features and contours on the way to developing a representation of general form, and ultimately of objects and their classification. Similarities and differences between ourselves and other species are highlighted, allowing inferences about the evolutionary origins of our pattern recognition processes.

Localized processes

Once visual information enters the cerebral cortex in area V1, it proceeds in "feed forward" fashion through a series of locations in the occipital and temporal lobes. However, anatomical studies indicate that every such location is directly linked to most others (Kolb & Whishaw, 2009), resulting in connections that are partially serial (e.g., V1 to V2 to VP to V4, as suggested by Figure 13.1) but are also often capable of bypassing intermediate levels (e.g., V1 directly to V4).

Nevertheless, the system has a "stages of processing" aspect to it. Single-cell recordings reveal that information arriving at V1 in the macaque produces activity serially in V2, V4, the posterior inferotemporal cortex (in humans, located just anterior to LO in Figure 13.1), and the anterior inferotemporal cortex (just forward of that), in phases each about 10 msec apart. About 20 msec after that, the prefrontal cortex activates (Thorpe & Fabre-Thorpe, 2001). Humans show similar "feed forward" processing, although the timing differs (Isik, Meyers, Leibo, & Poggio, 2014).

All eutherians have multiple visual areas in the cortex, but the number and arrangement differ dramatically across species. Hedgehogs, with low reliance on vision, have only two or three (Rosa & Krubitzer, 1999). Galagos have six (Wong & Kaas, 2010), while cats and apes, both highly visual, have 20 to 30 or more (Rosa & Krubitzer, 1999). In the case of humans, the locations shown in Figure 13.1 are only the initial ones in a larger network of visual areas.

There is strong evidence that in the early stages of cortical processing, primate neurons respond to features of a visual stimulus. These could be a particular line slant or color at a specific location. At later stages, neurons respond to specific configurations such as a particular face or object.

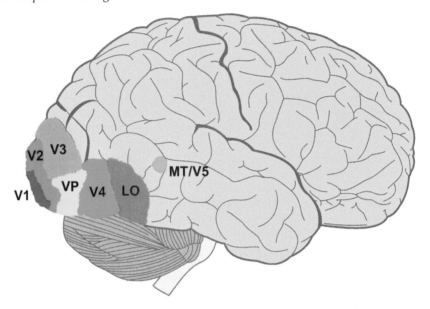

Figure 13.1 Major visual areas in the "what" pathway of the posterior lateral surface of the brain. Some extend to the medial surface as well.

Source: Adaptation of Human-brain.SVG, CC BY-SA 3.0, Hugh Guiney.

Features and contours

The initial cortical site of feature processing, V1, is also known as Brodmann area 17, striate cortex, and primary visual cortex. Its neurons are sensitive to line orientations located at specific retinal positions. They therefore perform a feature analysis by representing a scene in terms of small segments of contours (Pasupathy & Connor, 2001).

These *simple cells* are the first in a cascade of feature detectors that starts in V1 (but see Box 13.1). *Complex cells*, mostly found in V2, combine the outputs of multiple simple cells, and thus respond to orientation over a larger area and to movement in a particular direction. Through further integration of outputs, many *hypercomplex cells* respond to edges that have both an orientation and an end. Some cells in V1 and V2 also respond to curvature, color, and luminance (Nunez, Shapley, & Gordon, 2018; Spillmann, 2014).

Thus a wide range of information is coded by V1 and V2 cells: Orientation, movement, curvature, color, luminance, and spatial frequency (Box 13.1) at a minimum. Recent evidence suggests it may be even wider just at the V1 level alone. Specifically, neurons in the *superficial layer* of V1 (the layer of cells closest to the surface of the brain, also known as layer 1) appear to combine outputs from deeper V1 cells, to produce responses to particular angles and junction types. Indeed, such cells make up about half of the neurons in the superficial layer (Tang et al., 2018). Perhaps it should not be surprising that it codes such complexity, because the superficial layer is mostly composed of dense connections with other cells (Schmolesky, 2018) allowing information to be pooled.

Two problems must be solved early on in pattern recognition. These are (a) how to extract an object from its background, known as the *figure-ground segregation* problem, and (b) how to specify the object's contour. These are closely related problems, because segregating figure from ground requires identifying an object's contour among visual clutter.

In monkeys, and presumably in humans, figure-ground segregation begins in V1 with cells that combine the outputs of edge-sensitive feature detectors. This allows the

Box 13.1   Feature analysis or spatial frequency analysis?

David Hubel and Torsten Wiesel won a Nobel Prize in 1981 for their investigations of visual information processing. Recording from the visual cortex of cats and monkeys, they reported feature detection by simple cells. Specifically, the cells were found to respond to an edge or bar stimulus at a particular orientation at a particular point in space. They also discovered complex and hypercomplex cells (Spillmann, 2014).

However, in subsequent years a competing view has strongly emerged. This view maintains that the cells involved in early stages of visual pattern recognition can better be characterized as spatial frequency detectors than as feature detectors. The optimal stimulus for such detectors is a repeating sine wave pattern at a particular orientation – a grating, not a bar (Figure 13.2).

The spatial frequency model of visual information processing has enjoyed tremendous success, with many experimentally validated implications that go well beyond what is possible to describe here (De Valois & De Valois, 1990). One basic confirmation is that many visual cortex cells that respond strongly to bars respond even more strongly to gratings (Albrecht, De Valois, & Thorell, 1980).

Nevertheless, there is good evidence that people really do detect features. For example, the time to determine whether two patterns are the same or not is affected by the number of features they share (Bagnara, Boles, Simion, & Umiltà, 1982). If letter patterns are artificially stabilized on the retina, their features – mostly lines and curves – perceptually disappear and reappear over time (Topolski & Inhoff, 1995). Features also sometimes "migrate" from one object to another. Thus a red "O" and green "B", flashed either at the same place with a brief delay or in different places simultaneously, might on occasion be perceived as a green "B" and a red "O" (Botella, Privado, de Liaño, & Suero, 2011; Prinzmetal, 1981).

The apparent answer to the feature-versus-spatial frequency puzzle has two parts. First, about 5% of cells in V1 show larger bar than grating responses (Albrecht et al., 1980). This raises the possibility if not probability that there are V1 edge detectors that exist outside the spatial frequency domain. Other features to which

Figure 13.2 A spatial frequency stimulus (grating), representing a sine wave pattern at a particular orientation. The frequency is specified by the number of repetitions per degree of visual angle.

Source: Sine pattern.png, CC BY-SA 3.0, Setreset.

V1 neurons are sensitive include orientation, color, and bar repetition (von der Heydt, Peterhans, & Dürsteler, 1992).

The second part of the answer is that by responding to a bar, even if the response is less than that to a grating, spatial-frequency-sensitive cells are nevertheless sensitive to an edge feature. From their output, and from the output of the other cells responding more specifically to features, more complex representations can be developed. As information passes through the ventral "what" system, there is increasing sensitivity to contours and colored abstract shapes (Pasupathy & Connor, 2001, 2002). Thus both spatial frequency analysis and feature analysis have roles in play in any full account of the early stages of visual pattern recognition.

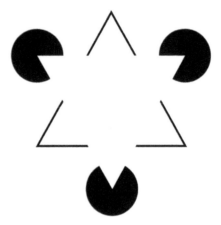

Figure 13.3 Cells in V2 extract the contour of the implied (but not physically present) borderless triangle.

Source: Kanizsa triangle.svg, CC BY-SA 3.0, Fibonacci.

detection of boundaries. This is followed by a "filling in" of the figure relative to the ground, i.e., an enhancement of its boundary accompanied by coloring of the interior. That process increases in intensity as visual input passes from V1 to V2 and V3/VP (Lamme, Rodriguez-Rodriguez, & Spekreijse, 1999; Seymour, Williams, & Rich, 2016). It appears to finish when area LO feeds input back to V1/V2. It is so powerful that it can extract a contour when parts are missing or even when a contour is merely implied and not physically present (see Figure 13.3; Wokke, Vandenbroucke, Scholte, & Lamme, 2013). In humans, figure-ground segregation is better performed by the right as opposed to the left hemisphere (Boles, 2002; Glezer, 1995).

Similarities across species in the anatomy of V1 and V2 lead to the conclusion that both derived from an early mammalian ancestor (Rosa & Manger, 2005). The distribution of V1 is very wide among mammals including monotremes and marsupials, indicating an origin at least 170 million years ago. V2 may have had later eutherian origins, about 160 million years ago (Rosa & Krubitzer, 1999). Unfortunately, for the most part we do not know whether the two areas play the same functional roles across distantly related mammalian species as they do in primates. An exception is that since cats were used in the feature analysis experiments of Hubel and Wiesel, it can probably be assumed that the sensitivity of V1 neurons to line orientations has largely been preserved over the last 95 million years. However, an initially disordered layout of neurons in V1 became orderly only about 84 million years ago in the common ancestor of tree shrews and primates (Kaas, 2012).

General forms

From V2 a further processing step brings visual information to VP and V3 (also known as V3v and V3d to emphasize their locations in the ventral and dorsal streams), as well as to V4. All are involved in the processing of form, with evidence linking VP and V3 to dynamic form, and V4 to color form (Kolb & Whishaw, 2009). Neurons in areas VP and V3 respond to specific shapes at depths corresponding to the eyes' fixation point and beyond, and also to small objects embedded in clutter (Rosa & Manger, 2005).

The VP/V3 complex is found in both primates and nonprimates, with at least traces of it found in tree shrews, ferrets, and cats, but perhaps not in marsupials like the opossum. At present, the best interpretation of the evidence is that like V2, VP/V3 originated with eutherian mammals about 160 million years ago (Rosa & Manger, 2005).

Studies of form recognition are hampered by the virtually infinite variety of forms that could be used as stimuli. Finding a shape that evokes a strong neuronal response doesn't preclude the possibility that another shape might evoke an even stronger response. Nevertheless, using a large variety of shapes often enables the identification of feature combinations to which a cell is sensitive. In macaque area V4, some cells are sensitive to particular curvatures at particular orientations relative to an object's center. An example might be a concave curve to the left of center.

However, other cells in V4 are sensitive to certain *combinations* of curvatures, enclosing forms of a particular color. These cells respond to fairly specific objects, for example a red "squashed raindrop" (Pasupathy & Connor, 2002). Even so, it is probably the combination of features that evokes the response, not the object itself, so that other objects with similar features would evoke a similar response. Thus there is nothing special about a red squashed raindrop beyond its particular assemblage of features. Other objects with different features would evoke a response in other V4 cells. If nothing else, this illustrates the subtlety of information coding in the ventral stream: At any particular point in processing, representations may not match up closely with what we end up perceiving, i.e., specific objects, but rather with abstract combinations of features.

V4 builds upon earlier contour processing by combining parts of an object (Pasupathy & Connor, 2001, 2002), thus achieving a generalized, abstract representation of it. The area also resolves depth information passed along from areas V1 and V2. Presumably this is in the service of developing a 3D representation so that the abstract object can be perceived in depth (Verhoef et al., 2016).

Tracing the evolutionary origins of V4 is complicated by uncertainty in identifying corresponding locations in monkeys and humans. While a color-sensitive V4 exists in both humans (Hong & Tong, 2017) and macaques (Pasupathy & Connor, 2001, 2002), the evidence for homology is ambiguous at a neuroanatomical level (Orban, Van Essen, & Vanduffel, 2004). In any case, little if any effort has been devoted to uncovering the origins of V4 prior to the divergence of Old World monkeys.

A stronger case for homology can be made for the next cortical stage of pattern recognition beyond V4. This involves a correspondence between inferotemporal (IT) cortex in macaques and the "lateral occipital complex" (LOC) in humans. Human lateral occipital (LO) cortex, illustrated in Figure 13.1, is usually considered part of the LOC (Denys et al., 2004). However, LOC also includes posterior portions of the inferior temporal gyrus (see Figure 10.3) as well as the middle fusiform gyrus on the underneath side of the cortex. That allows it to be characterized as a large area straddling the border between the occipital and temporal lobes (Denys, et al., 2004).

The human LOC appears to be homologous with the monkey IT, both anatomically (Denys et al., 2004; Orban et al., 2004) and functionally. Functionally, both areas respond similarly to two-dimensional, three-dimensional, and moving patterns. Also, both show a tendency to respond less in their anterior portions than in their posterior portions to scrambled patterns, indicating increasing sensitivity to complex combinations of features as processing moves forward in the brain (Denys et al., 2004). In this respect, IT and LOC both resemble the function of area V4, differing from it in that larger proportions of their cells show sensitivity to complex feature combinations (Tanaka, 1997).

Assuming that IT and LOC really are homologous, what is their evolutionary history? Human LOC appears to have a greater monopoly over form processing than macaque IT, in that the macaque parietal lobe also participates in recognizing visual form (Denys et al., 2004). While this could indicate decreased specialization of brain areas in macaques relative to humans – thereby "spreading out" processing to a larger degree – contrary observations apply to the processing of motion and form. There, monkeys show less overlap between motion and form processing areas than do humans, suggesting *increased* specialization of brain areas.

The answer to this apparent contradiction may lie with tools. Specifically, habitual tool use requires the integration of motion with two- and three-dimensional representations of tools, potentially accounting for the greater overlap of motion and form processing in humans (Denys et al., 2004). Viewed in this way, the greater overlap is actually an indication of increased specialization for tool use. Of course, even if this explanation is correct, it is difficult to say to what extent the species difference is due to human evolution, or to dissimilar tool traditions – i.e., differences in learning.

Objects

We have seen that information accumulates about objects as visual input passes in a forward direction from V1 on through the monkey IT and human LOC areas. The next step is to achieve specific neural responses to specific concrete objects.

One possibility is that a given neuron in the monkey IT cortex uniquely responds to a specific object. To early researchers this seemed quite plausible, because one of the most striking characteristics of the ventral system was the specificity that many cells seemed to exhibit. In the early years of research on the ventral "what" pathway, such neurons were called *gnostic cells*. Subsequently, they became popularly known as *grandmother cells* because a single cell might specifically respond to one's grandmother (Bowers, 2017; Gross, 2002).

An alternative to grandmother cells is a distributed network of neurons that produces a specific response to a particular object only as a population. According to this formulation, each cell within the network responds to a wide variety of objects, but every object produces a unique pattern of activation of some cells and not others.

In fact, both of these possibilities are problematic. For grandmother cells, one problem is that there would need to be a specific cell for every possible object we can recognize, itself a mind-boggling prospect. In addition, the cell would somehow need to respond to that object regardless of orientation, size, and color (Gross, 2002). *Perceptual invariance* is the ability to recognize an object regardless of those and other physical transformations. While it is an important achievement of the ventral "what" system operating as a whole, it seems unlikely that a single cell recognizes a single object in all of its possible transformations.

For distributed networks, the major problem is that a network by itself does not seem well suited to assess its own state. For example, how can the network decide which specific object corresponds to a particular pattern of activation across cells, unless there is a cell to recognize each pattern of activation – in other words, a grandmother cell?

No current solution provides neat answers to these problems, but there are suggestions that the key to object recognition lies somewhere between the extremes of grandmother cells and fully distributed networks. Hung, Kreiman, Poggio, and DiCarlo (2005) combined an artificial intelligence application and single-cell recordings to estimate the number of IT neurons needed to recognize an object. They recorded responses to 77 different objects from 300 neurons and fed them in varying numbers into a mathematics-based computer program that in turn "learned" to recognize the objects from the neural responses. It was found that reasonably high performance (72% correct) could be achieved by feeding 256 responses into the program. This result suggests that assemblies of IT neurons numbering in the hundreds are sufficient to achieve object recognition (Hung et al., 2005).

Even in this approach, however, the "grandmother" problem was alive and well, because in the artificial intelligence implementation, each of the 77 objects had a "classifier" – in effect, a grandmother cell – to determine whether an identification could be made. On the other hand, an important feature of the study was that good performance was obtained even when object position or size changed (Hung et al., 2005). Together, these observations indicate the potential feasibility of a hybrid solution that combines a distributed network to achieve perceptual invariance, thereby greatly reducing the number of object instances that must be recognized, together with grandmother cells to identify that reduced number.

It seems intuitively obvious that perceptual invariance must be achieved at a relatively "high" level of the system, following substantial integration of neural activity. Studies of brain activity support this intuition. Within both macaque IT and human LOC, the more anterior sectors, at the furthest remove from earlier stages of processing, show the greatest perceptual invariance. However, only a subset of cells in these areas achieve invariance, indicating that the object recognition system maintains both invariant and variant representations of objects (Kovács et al., 2003; Sawamura, Georgieva, Vogels, Vanduffel, and Orban, 2005). Again, this makes intuitive sense, because we can clearly differentiate among views while simultaneously recognizing that they are of the same object.

The increased integration of neural activity as processing proceeds forward within IT/LOC cortex leads to an expectation that information should become increasingly integrated across sensory modalities as well. In fact cells in the LOC area have been found to respond to objects presented not just visually but also tactilely (Hernández-Pérez et al., 2017). However, they are in the minority, and it seems more appropriate to characterize LOC as making available a wide range of representations of objects. These include: (a) specific objects of specific position, size, viewing angle, contrast, and/or texture; (b) specific objects independent of such factors or even of modality; (c) generic objects of a general shape and color; and even (d) generic objects of no particular shape and color (Desimone, Albright, Gross, & Bruce, 1984). Presumably the development of a wide variety of representations is in the service of ultimately achieving, by way of their integration, unambiguous recognition of the object.

Processes within IT/LOC appear to be evolutionarily well conserved, with an origin sometime prior to 32 million years ago. However, the usefulness of this conclusion is limited by the absence of other comparative data that could put a more definitive limit on the emergence of the area.

Object classes

Recognizing an object as a unique instance of a larger class, for example your grand-mother as a member of the class of people, is aided by specializations of the ventral stream that are class-specific. Particular subareas within the occipito-temporal cortex appear to be sensitive to particular object classes such as faces or letters. Thus the human fusiform gyrus, particularly in the right hemisphere, is sensitive to faces. In part this is known because of single-cell recordings from patients in whom electrodes were implanted to localize the source of epileptic seizures. A face recognition "patch" was found to be located largely, though not exclusively, in the lateral and posterior portion of the fusiform gyrus. Other patches in the fusiform gyrus and its vicinity proved responsive to objects such as cars, butterflies, flowers, and letters (Allison, Puce, Spencer, & McCarthy, 1999).

The important question, of course, is whether particular patches are sensitive to specific object classes and *not* others. This seemed to be the case in the embedded electrode study. Across participants, face patches were generally located anterior to object patches, and letter patches were located near the occipito-temporal sulcus as compared to the fusiform location of face patches. Additional evidence that face patches are face-specific was reported in an fMRI imaging study by Rhodes, Byatt, Michie, and Puce (2004), using lepidoptera (butterfly and moth) experts. They found little overlap in the specific locations activated during lepidoptera recognition and face recognition tasks, even though both tasks used the fusiform gyrus.

McCarthy, Puce, Belger, and Allison (1999) examined the effects of altering faces on the response of face-specific patches in epileptic patients. Patch responses didn't vary across color versus black-and-white renditions of faces, or to larger versus smaller sizes, indicating that they responded to faces irrespective of color or size. The patches also responded to cat and dog faces, with reduced sensitivity. Inverting human faces (turning them upside down) likewise reduced their response, more so in the right hemisphere than the left. Because inversion presumably impacts the configuration of a face more than its features, the result was viewed by the authors as implying configural processing of faces in the right hemisphere, but piecemeal processing of faces in the left hemisphere. Finally, there were strong indications that internal parts of faces like the eyes, nose, and mouth were separately processed from the full face, at sites close to the face patches.

Visual agnosia

Varieties of *visual agnosia*, the inability to recognize objects presented visually, likewise support the existence of cortical patches that are sensitive to particular object classes. Cases have been reported in which the recognition of living things is deficient even though nonliving things are identified normally. The opposite has also been reported (Humphreys & Forde, 2001). Other object classes that sometimes show deficient recognition, while other classes are preserved, include faces, tools, animals, and fruits and vegetables (Capitani, Laiacona, Mahon, & Caramazza, 2003; Tranel, Damasio, & Damasio, 1997).

Prosopagnosia is the name given to face-specific visual agnosia. It is typically due to lesions of the fusiform gyrus, usually in the right hemisphere (Kanwisher & Yovel, 2006). Tranel and colleagues have reported lesion data indicating that deficits in recognizing tools, faces, and animals involve damage to differing portions of the temporal cortex, although in some cases occipital and parietal areas are also involved (Tranel et al., 1997).

Comparisons between species

The localization of object class patches appears to be similar in humans and monkeys. Macaques have face-specific patches of inferior temporal cortex (Lafer-Sousa, Conway, & Kanwisher, 2016), and direct stimulation of their IT cortex results in a biasing of decisions toward seeing faces and away from seeing nonfaces (Afraz, Kiani, & Esteky, 2006). However, their face patches are less widespread than those in humans, perhaps because of cortical expansion during human evolution that pushed some of them in the ventral direction (Lafer-Sousa et al., 2016). Face patches are also seen in marmosets, which are New World monkeys (Hung et al., 2015).

Furthermore, just as humans have difficulty recognizing upside-down faces, presumably due to the disruption of configural information (see Figure 13.4), so do chimpanzees,

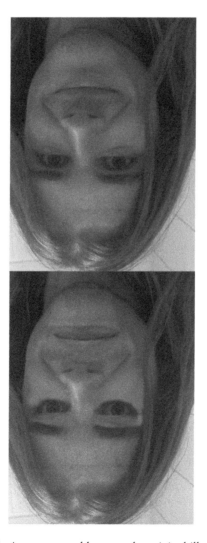

Figure 13.4 The Thatcher illusion, so named because the original illusion used pictures of British Prime Minister Margaret Thatcher. Examine the pictures, then turn the page upside down and reexamine them.

Source: Thatcher.PNG, public domain, Albert Kok.

macaques, and New World monkeys (Calcutt, Rubin, Pokorny, & de Waal, 2017; Taubert, Van Belle, Vanduffel, Rossion, &Vogels, 2015; Wilson & Tomonaga, 2018). However, the effect of inversion may be weaker and more methodology-dependent in monkeys than in apes (Parr, 2011).

Crows and pigeons, in contrast, fail to show a face inversion effect (Brecht, Wagener, Ostojic, Clayton, & Nieder, 2017). However, sheep recognizing other sheep *do* show the inversion effect. Strikingly, like humans, they also show a right hemisphere advantage in recognizing faces, and as face familiarity increases, they become less sensitive to features and seem to use more configural information (Peirce, Leigh, daCosta, & Kendrick, 2001; Peirce, Leigh, & Kendrick, 2000).

These results suggest that primate taxa diverging from one another as far back as 45 million years ago inherited a configuration-based and orientation-specific face recognition system from a common ancestor. The sheep data suggest that this system dates back to at least 95 million years ago, but the bird data suggest it postdates 300 million years ago. However, orientation specificity may have strengthened in apes, including humans.

Global versus local bias

Viewing the Thatcher illusion, we only sense that something is catastrophically wrong when the faces are in their familiar upright configurations, not when they are inverted even though the same feature distortions are present. The illusion suggests that we are more sensitive to configural distortions than we are to featural ones.

Could it be, then, that we are generally more sensitive to global configurations than we are to features? Experiments using hierarchical stimuli (Figure 13.5) suggest that we are. In the typical experiment, participants are asked to recognize either a large global letter or small local ones, with either conflicting or neutral ("+") information at the other level. Thus, in two conditions of the typical experiment, participants are asked to recognize a global letter and are given either conflicting or neutral stimuli (left and center examples). In the other two conditions, they are asked to recognize the local letters, and again are given either conflicting or neutral stimuli (left and right examples).

Two major findings emerge from such studies (Figure 13.6). First, the global level is responded to faster than the local level. Second, having conflicting information at

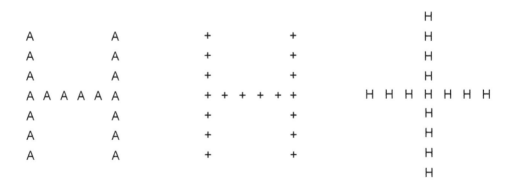

Figure 13.5 Illustration of hierarchical stimuli. Left: Conflicting, Center: Global neutral, Right: Local neutral. As shown, in the neutral conditions a nonletter plus sign substitutes at one level.

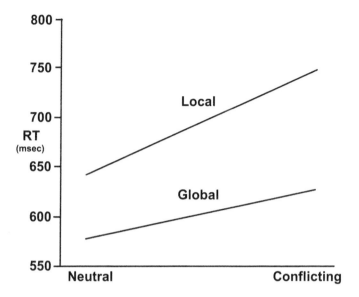

Figure 13.6 Mean RTs to the hierarchical stimuli.

Source: Licensed from Elsevier under STM Permissions Guidelines.

the other level slows local-level responses more than it does global responses. Together, these two phenomena are sometimes called the *global precedence effect* because both presumably result from the faster processing of global information. Thus, if the global level provides conflicting information, it interferes more because of its earlier availability. These basic results have been replicated many times (e.g., Blanca & López-Montiel, 2009; Boles, 1984; Boles & Karner, 1996; Dale & Arnell, 2013).

Global and local processing in other primates

Similar hierarchical stimuli have been used to study pattern recognition in several primate species. Hopkins and Washburn (2002) presented a match-to-sample task to common chimpanzees, first showing as the sample a hierarchical letter stimulus, followed by the same hierarchical stimulus as the target and a different hierarchical stimulus as the foil. The task was to select the matching target while rejecting the foil, which could differ from the sample at the global level, the local level, or both. Essentially, reaction time reflected the time to reject the mismatching level or levels. It was found that the chimpanzees rejected the global level more quickly than they did the local level. However, when the same task was given to rhesus macaques, no difference was found between global- and local-level rejections.

Fagot and colleagues used a somewhat different stimulus set and task for humans, chimpanzees, and baboons (Fagot & Deruelle, 1997; Fagot & Tomonaga, 1999). Although the results depended somewhat on other experimental manipulations such as the size and sparsity of the display, the general result was that humans rejected the global patterns faster than the local patterns, chimpanzees showed little difference, and baboons rejected local patterns faster than global patterns. However, a follow-up study found that chimpanzees rejected global patterns faster than local ones when the global pattern was made more salient (Fagot & Tomonaga, 1999).

Finally, capuchins and tamarins have been subjected to similar testing and found to make faster or more accurate local rejections than global rejections (De Lillo, Palumbo, Spinozzi, & Giustino, 2012; Neiworth et al., 2014; Truppa, De Simone, & De Lillo, 2016). One study of capuchins even addressed the interference of one level on another. Thus Spinozzi, De Lillo, and Salvi (2006) used hierarchical stimuli in a matching-to-sample study, with global matching and local matching performed on separate days. Percent errors rather than reaction time was the measure, and the results are shown in Figure 13.7.

Although little difference was found between global and local matches when stimuli were consistent, a big advantage of local matches was revealed when stimuli were conflicting. This indicates interference of conflicting information at the local level on global matches, but not vice versa, opposite that of the human results of Figure 13.6. Capuchins appear to show not a global but a local precedence effect.

Looking across all of these comparative studies, it appears that the modern human bias toward global visual patterns is reflected at least partially in chimpanzees, but that Old World monkeys (baboons and macaques) and New World monkeys (capuchins and tamarins) have a reversed, local bias. The data indicate that a quite striking evolutionary shift has been occurring since we diverged from Old World monkeys 32 million years ago, extending past the chimpanzee-human common ancestor 7.5 million years ago.

What exactly was it that shifted? Fagot and Tomonaga (1999) attribute the species differences to the greater ability of chimpanzees and humans, relative to baboons, to perceptually group stimuli into global forms. Recognizing global form in hierarchical stimuli like those in Figure 13.5 (or, for that matter, in Figure 13.3) exploits the proximity and continuity of small elements to extract contours that don't actually exist. This may be something that bigger, more interconnected brains are better at, starting with those of chimpanzee size.

Certainly an enhanced ability to extract larger forms from loosely associated features improves pattern recognition, especially when the larger contours are obscured. In our evolutionary past, global bias may have improved the recognition of partially obscured

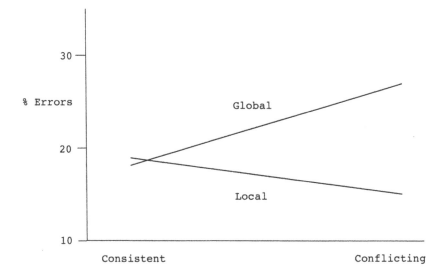

Figure 13.7 Capuchin percent errors to hierarchical stimuli.

Source: Original figure.

prey or predators. It may have aided navigation over global terrain covered by seasonally changing local vegetation. It may even have allowed better global imagery of a planned tool being assembled from local parts. Seeing the forest before the trees (Navon, 1977) had definite advantages in many situations.

Other pattern recognition processes and areas

Cortical areas besides those along the V1-inferotemporal route also participate in visual pattern recognition. For example, the superior temporal sulcus (see Figure 10.3) contains other face-sensitive patches, and responds to visual cues communicating face and body movement. It also has a high proportion of "polysensory" cells that respond not just to vision but to other senses as well (Allison et al., 1999; Desimone et al., 1984).

In both monkeys and humans, the intraparietal sulcus (Figure 10.3) may play a role in cross-modal matching, for example in matching a visual shape to a tactile one (Grefkes, Weiss, Zilles, & Fink, 2002). The LOC is also active during visual-tactile integration (Lacey, Tal, Amedi, & Sathian, 2009). *Perirhinal cortex* (cortex located in the medial temporal lobe) appears to play a role in extracting the meaning of polymodal object representations (Holdstock, Hocking, Notley, Devlin, & Price, 2009).

Then, too, the auditory and somesthetic modalities have their own pattern recognition systems. These are mostly localized, respectively, in the superior temporal and parietal lobes. The auditory system plays a primary role in language and will be further described in that context in Chapter 15.

Conclusion

The "what" pathway supports much of our visual pattern recognition. Located in the ventral portions of the occipital and temporal lobes, it initially uses both spatial frequency and feature analyses. Cells in the superficial layer of V1 appear to combine features so that they are sensitive to complex angles and junction types. By the time information has "fed forward" and been processed in V2, figure has separated from ground, and contours have been extracted. Areas VP and V3 (also known as V3v and V3d) enhance boundaries and color them in.

Comparative analyses indicate that V1 originated at least 170 million years ago with the emergence of mammals, while V2 as well as VP/V3 originated about 160 million years ago with eutherian mammals.

Neurons in V4 are sensitive to combinations of curvatures organized relative to an object's center, thus achieving a representation of an object's combined parts. It also incorporates depth so that a 3D representation is achieved. The next stage, involving the homologous areas of the LOC in humans and IT cortex in macaques, seems to build on these representations through sensitivity to complex feature combinations.

The LOC/IT areas, and possibly V4, trace at least as far back as the emergence of Old World monkeys 32 million years ago, but little is known about their earlier origins. There seems to have been some divergence between macaques and humans in that humans show an increased specialization of LOC relative to monkey IT, yet greater integration of motion and form information. This may reflect increased specialization for tool use.

Recent research has estimated the number of inferotemporal neurons necessary to recognize an object in the hundreds rather than the much larger numbers of a fully distributed network. However, they don't completely escape a problem associated with "grandmother" cells, namely the need for a dedicated unit to recognize each object. On the other hand, a strength is that networks can achieve a substantial degree of invariance, representing the same object regardless of transformation in such physical characteristics as position or size.

Neurons at this level are organized into "patches" that are specific to object classes such as faces, cars, butterflies, flowers, and letters. Face patches are largely though not exclusively located in the fusiform gyrus, especially in the right hemisphere. Damage to this area can cause prosopagnosia, a visual agnosia for faces. New World and Old World monkeys, chimpanzees, humans, and even sheep all appear to have the same configuration-based, orientation-specific mechanism for recognizing faces, something not shared by pigeons. This suggests an origin of at least 95 million years ago, but how much more recently than our 300-million-year-old common ancestor with birds is unclear.

Experiments using hierarchical stimuli indicate that humans are more sensitive to global configurations than to features. We respond faster to the global level, and conflicting information at that level interferes more with feature processing than the reverse, two phenomena that together are called the global precedence effect. Chimpanzees appear to show a weaker form of the effect, while Old World and New World monkeys show an opposite, local precedence effect. Thus an evolutionary shift from local to global precedence may have started about 32 million years ago, down to later than 7.5 million years ago. It may be that bigger, more interconnected brains are better able to use the proximity of small elements and their apparent continuity to extract global form.

14 Memory

Skilled handedness, tool use and manufacture, spatial perception, and object recognition have something in common: They require a highly evolved ability to remember. In our day-to-day lives we constantly access information in memory that represents the meaning of environmental stimuli. We use memory to select action, and to execute it after recalling learned motor patterns. In a real sense, memory can be viewed as a prerequisite for all perceptual and cognitive activity.

In general, memory is defined as the ability to retain information from past events. It is phylogenetically ancient in the form of *habituation*, a decline in response to repeated stimulation. Even single-cell organisms show declining responses on repetition, for example by ceasing contraction in response to repeated pressure changes in the surrounding water (van Duijn, 2017). Because habituation involves retaining information from past events – i.e., an event has occurred repeatedly in the past and so response is reduced – it qualifies as a memory phenomenon, although a very primitive one (Moore, 2004).

A more restrictive definition of memory requires *association*, the arbitrary pairing of a stimulus with a response. In classical conditioning, associations are learned when a conditioned stimulus (e.g., a ringing bell) shortly precedes an unconditioned stimulus (e.g., meat), which evokes a reflexive response (e.g., salivation in Pavlov's dogs). Soon the conditioned stimulus itself produces the response without any intervening unconditioned stimulus, showing that an arbitrary association has been learned between the stimulus and the response. In theory, the same stimulus could arbitrarily be paired with any other response and conditioned in similar fashion.

Association is what most people mean by memory. For example, when we learn a new face, we learn arbitrary facts associated to the face: A name, that this is our new boss, that this face's office is down the hall. There is nothing "built in" or reflexive about any of those associations; they have to be learned as arbitrary bits of information attached to a stimulus.

Like habituation, conditioning in various forms is phylogenetically ancient. It has been observed in snails, wasps, and birds as well as in kittens, lambs, rats, monkeys, and a host of other mammals (Moore, 2004). Significantly, conditioning has been observed in planaria (Hutchinson, Prados, & Davidson, 2015), a flatworm whose common ancestry with our own traces back almost to the origin of bilateria during the Cambrian explosion. Thus the origins of association likely extend back at least to 520 million years ago. However, it may not extend much before that, as jellyfish are said to be incapable of conditioning using arbitrary stimuli (Alexander & Challef, 2000), even though habituation occurs (Johnson & Wuensch, 1994).

These observations probably indicate the importance of relatively well-developed nervous systems in conditioning. While cnidarians including jellyfish have a diffuse neural net, planaria and some other flatworms have greater centralization in the form of

central ganglia. Flatworms also show a differentiation of neurons into sensory, association, and motor types (Hickman et al., 1997), unlike cnidarians whose neuron subtypes
resist clear functional categorization (Havrilak et al., 2017). A specialization of neurons
into functional types, and their aggregation into larger units such as ganglia, may be
prerequisites for conditioning.

One problem in uncovering the evolutionary origins of memory is the sheer variety
in forms of learning that have been investigated. Moore (2004) suggested that there are
dozens of biologically distinct kinds of learning, including 10 different forms of conditioning. However, his analysis proceeded largely on logical grounds, and it is unclear to
what extent so many different kinds of learning actually involve different abilities.

In any case, it seems indirect to trace memory's evolutionary origins by examining
learning abilities rather than types of memory. Whether or not an animal learns using
a particular method is likely to depend on a number of factors. They include sensitivity to the stimuli used, the ability to make the required response, and even the species'
predisposition toward associating certain stimuli to certain responses and not others
(Seligman, 1970).

Cognitive psychology offers a more direct approach by positing the existence of several empirically dissociable *memory stores*, or kinds of memory. This structural approach
eliminates the problems of a learning-based approach by focusing on the nature and fate
of the information that is remembered, not the method by which it became represented
in memory.

For the most part, the approach taken here will be to describe each component of
memory from the human perspective, accompanied by a comparative analysis to identify
evolutionary trends.

Sensory memory

Sensory memory is a very short-lived trace that preserves much of the physical form of
stimuli and lasts at most several seconds. The visual form of sensory memory, known
as the *icon*, may be the best understood. It is a copy of visual stimulation that lasts half
a second or less in brightly lit conditions but up to four seconds in the dark (Sperling,
1963). It becomes apparent when a flashlight is waved in a circle in the dark. The resulting smear makes the circle appear complete when in actuality, at any instant the flashlight
itself only emits a point of light on the circle. The icon is thus a very brief memory that
forms a bridge from one instant of time to the next, creating a continuous visual experience. It is responsible for our sense of continuity when watching a 24-frame-per-second
Hollywood movie.

The peripheral component

The icon is due to the aftereffects of activity at various levels of visual processing, including the periphery of the nervous system. Sakitt and Long (1979) established the retina
itself as the location of one component of the icon. In one of their experiments, the task
was to judge the location of a red dot on a gray background, determining the direction
it moved between successive frames. To perform this task, an iconic memory of the first
frame had to be matched to the second frame. Accuracy at the varying delays thus indicated the extent to which information was still in the icon of the first frame.

Furthermore, the intensities of the dot and background were carefully adjusted so that
in one condition, the rods in the retina had a difficult time separating the dot from the
background. Rods are sensitive to gray but insensitive to red, and they do not differentiate

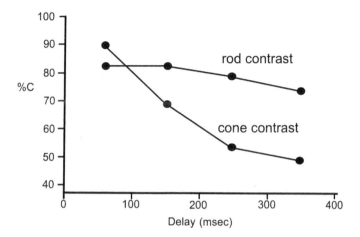

Figure 14.1 Decay of the icon (with the curves smoothed for clarity) under rod and cone contrast conditions.

Source: Original figure.

between colors, so an appropriate high-intensity red and a low-intensity gray effectively blind the rods with respect to any difference. However, the cones can easily tell the difference, so this condition was called the cone contrast condition. In these circumstances, the icon was found to fade rapidly over 400 msec. Presumably the cones in the retina were responsible for the icon in this condition.

In the other half of the experiment, the rod contrast condition, the task was made difficult for the cones by adjusting the intensities so that the dot and background appeared equally bright to the cones. Although the cones could still detect a color difference, this condition made it more likely that the rods would be involved in the task, and it certainly should have been easier for the rods than the cones. In this condition, performance again declined over 400 msec, but to a much smaller extent, implying that the icon was holding information for a much longer time (Figure 14.1).

The beauty of these results is that they accord with what is otherwise known about the duration of the icon. As already indicated, in brightly lit (daytime) conditions it lasts only half a second or less, while in dark (nighttime) conditions it can last up to four seconds. Daytime vision depends on the cones and nighttime vision on the rods, so the difference between the cone contrast and rod contrast conditions accords well with that difference. Most importantly, the results indicate that one component of the icon is located at the retina: The cones are responsible for a rapidly fading version of it, while the rods are responsible for a more slowly fading version. Of course, the usefulness of the rod version is limited by the low resolution of the rod system (Long & McCarthy, 1982).

The central component

On the other hand, there is also strong evidence of a brain-based, central component of the icon. When a visual stimulus is presented briefly and is followed by a *pattern mask* – a second stimulus that spatially covers the first and is composed of jumbled features – recognition of the first is compromised. Furthermore, the mask is most effective not when it is presented immediately after the stimulus, but when it is presented at a short delay of 30–100 msec (Skottun & Skoyles, 2010). This result implies that the mask catches up to the stimulus at a later, time-consuming stage of processing – i.e., feature analysis – and disrupts it (Box 14.1). Most importantly, masking occurs even if the stimulus is presented to one eye and the mask is presented to the other eye. Because inputs to the two eyes do not anatomically converge until they reach the primary cortex, this result strongly implies that there is a central component of the icon located in the brain. Furthermore, a different type of mask has different effects: An unpatterned flash of light is most effective at a zero delay following the stimulus, and only if it is presented to the same eye as the stimulus. Like the study of Sakitt and Long, this implicates a peripheral, retina-based component of the icon (Breitmeyer, 1984).

Other forms of sensory memory that have been investigated include auditory sensory memory, the so-called *echo* lasting about 2–10 seconds (Bijsterveld, 2015), and a somesthetic sensory memory lasting about 1–2 seconds (Bliss, Crane, Mansfield, & Townsend, 1966).

The general interpretation that has been offered for the existence of all of these is that they represent aftereffects of neural activity. With respect to the icon, the cones and rods of the retina remain in a heightened state of activity for a brief period of time following stimulation, preserving a sensory trace of the stimulus. The same is true of visual feature detectors beginning with V1 in the cortex. With respect to the echo, durations toward the lower end of the 2- to 10-second range appear to be associated with residual activity in the primary auditory cortex, while longer durations reflect activity in the association auditory cortex (Lü, Williamson, & Kaufman, 1992).

The evolution of sensory memory

It is evident from the discussion so far that the evolution of sensory memory must be closely tied to the evolution of sensory receptors and a sensory brain. As we have seen

---

Box 14.1   How can a mask catch up to an earlier stimulus?

A pattern mask is most effective when it is briefly delayed relative to the stimulus, a result interpreted as the mask "catching up" to the stimulus at a time-consuming stage of processing. But how is that possible? Logically, it seems that a delay should be maintained throughout the processing system, so that the stimulus always stays ahead of the pursuing mask.

The answer lies in a two-phase response of the visual system. The *magnocellular response* is a rapid response to stimulation that preserves only fuzzy detail but serves to alert the system to the appearance and location of the stimulus. The *parvocellular response* is much slower but contains detailed information that allows feature extraction and identification of the stimulus. The two responses are transmitted by different but roughly parallel systems of cells beginning in the retina.

It is the faster speed of the magnocellular system that allows a pattern mask to "catch up" to a stimulus. The fast magnocellular response to the mask reaches feature analyzers in the cortex at the same time as the slower parvocellular response to the stimulus, even though the mask was presented after the stimulus. Consistent with this explanation, if the task is simply to indicate the location of the stimulus without identifying its features, a delayed pattern mask is not effective. In that case, the magnocellular response to the stimulus does indeed stay ahead of the pursuing mask (Öğmen, Breitmeyer, Todd, & Mardon, 2006).

(Chapter 7), visual sensation has an ancient history, with cnidarians likely having the first eyes that connected to neurons, as well as rhodopsin-based photoreceptors. Thus the retinal component of the icon probably traces back about 605 million years.

Emergence of a brain component, on the other hand, presumably waited on the evolution of nervous system centralization, possibly as early as the emergence of flatworms some 535 million years ago. It is also possible, however, that our central component emerged with the evolution of the primary cortical area V1, which, as we have seen, was a mammalian innovation dating back a little more than 170 million years ago. Certainly masking occurs in Old World monkeys, for example with their perception of animal pictures disruptable by a pattern mask delivered 50 msec after stimulus presentation (Cauchoix, Crouzet, Fize, & Serre, 2016).

Short-term memory

Short-term memory has a much smaller capacity but a much longer duration than sensory memory. Whereas sensory memory preserves a copy of all stimulation for a few seconds at most, short-term memory holds a maximum of about nine items for 20 seconds or so. However, it may be refreshed indefinitely by rehearsal, for example when vocally repeating a phone number from a contact list (Baddeley, 1986; Klatzky, 1980).

The capacity of nine items is a generalization that greatly depends on the type of item held in memory. Capacity is measured by requiring that varying numbers of items be recalled in the order presented. The number of items that can consistently be recalled in correct order is the *memory span*. Strings of random digits (e.g., 3, 6, 1, 7) are subject to the "seven, plus or minus two" dictum made famous by George Miller (1956). Five to nine digits can be remembered in order, depending on the individual and the conditions of the experiment. Cavanagh (1972) reviewed the literature and found that average memory span varied depending on the type of item: For example, about 8 for colors, 5–6 for random words, and 3 for nonsense syllables.

Thus short-term memory does not have a fixed number of "slots" but rather a fixed capacity for information. Items requiring more detailed memory (e.g., random words and nonsense syllables) have fewer slots than items requiring less detailed memory (e.g., digits and colors; Alvarez & Cavanagh, 2004; Cavanagh, 1972). Conversely, stretching capacity over ever-increasing numbers of items results in a progressive degradation of memory content, indicating that a limited information resource is being stretched too thin (Bays, 2018).

The informational view is lent weight by an orderly relationship between memory span and search time through short-term memory. When several items are held in memory (the *memory set*), RT to decide whether a subsequent probe was among the memory set increases linearly with the number of items in the set. For digits, RT increases about 33 msec per digit, implying that it takes that amount of time to compare the probe to a single memory item. As it turns out, plotting the comparison time for a particular kind of memory item against the memory span for that item results in a linear relationship. For example, colors on average show a 38 msec comparison time and a 7.1-item memory span, random words a 47 msec comparison time and a 5.5-item memory span, and nonsense syllables a 73 msec comparison time and a 3.4-item memory span (Cavanagh, 1972). In other words, memory span systematically decreases as comparison time increases. This result supports the informational view of short-term memory because it implies that the greater the detail required to remember an item, as reflected in the memory span, the more time is required to compare the probe to an item in memory.

At least if held in auditory form, information in short-term memory generally lasts 10–30 seconds if it is not refreshed through rehearsal. That makes 20 seconds a reasonable estimate of the duration of short-term memory (Klatzky, 1980).

Comparative studies of short-term memory

Wright and colleagues have reported that pigeons, rhesus macaques, and humans all have short-term memories. They presented sets of four pictures as memory items, followed after a delay by a single probe. Monkeys and humans moved a lever to the right or left, while pigeons pecked right or left colored disks, depending on whether or not the probe was in the memory set. It was found that all three species showed a *recency effect* in which the most recent item or items in a list were recalled better than the list items immediately preceding (Wright, Santiago, Sands, Kendrick, & Cook, 1985). This was an indication of the existence of short-term memory, for the most recent items were presumably those most active in it when the probe was presented.

Most importantly, Wright et al. found that the recency effect disappeared if the probe was sufficiently delayed, implying that information had been lost from memory. This effect varied across species. For pigeons the recency effect was present with a probe delay of 2 seconds, but absent at delays of 6 seconds or more. For monkeys the recency effect was still present at 10 seconds but gone by 20 seconds, while for humans it was still present at 60 seconds but gone by 100 seconds (Wright et al., 1985). The implication is that the species differed in terms of (a) the duration of short-term memory, (b) the ability to rehearse information in short-term memory, or (c) both. Thus humans showed the longest-duration short-term memory and/or the most effective rehearsal process, while pigeons showed the shortest duration memory and/or the least effective rehearsal process. But all showed evidence of short-term memory.

However, if there are species differences in short-term memory duration, they may not follow a phylogenetic progression. Lind, Enquist, and Ghirlanda (2015) compared multiple nonhuman animals by drawing on more than 90 delayed matching-to-sample studies. In these studies a stimulus, typically visual, was followed after a delay by the same stimulus and an accompanying distractor, and the task was simply to choose the match to the first stimulus. Across studies, the duration of memory was found to be disordered across species. For example, pinyon jays and capuchins showed an average duration of 39 seconds, while for chimpanzees it was only 19 seconds. It is possible, however, that methodological differences between studies were responsible for disorderliness in the results (Lind et al., 2015), a supposition that could be tested in future research by using a consistent methodology across species.

Greater insight into the evolutionary origins of short-term memory may come from considering its neurological underpinnings. Reviews of PET and fMRI studies conducted with humans implicate a small number of cortical areas that vary somewhat depending on whether auditory-verbal or visual-spatial short-term memory is considered (Andre, Picchioni, Zhang, & Toulopoulou, 2016; Cabeza & Nyberg, 2000). For the former, most studies show activation in the left inferior parietal cortex (Brodmann area 40; see Figure 10.4) and the left or bilateral premotor cortex in the frontal lobe (areas 6 and 44). For visual-spatial short-term memory, most have shown activation in the bilateral superior parietal cortex (area 7) and the bilateral premotor cortex (area 6). In addition, auditory-verbal and visual-spatial short-term memory both appear to involve left area 10.

Of these, areas 40 and 7 are most consistent with the storage component of short-term memory, with left area 40 storing phonological representations of memory items, and bilateral area 7 storing visual-spatial representations (Cabeza & Nyberg, 2000).

But what about the area 6, 44, and 10 activations? These may best be understood by considering the concept of working memory developed by Alan Baddeley (1986). Working memory incorporates the storage aspects of short-term memory by proposing the existence of an "articulatory loop" in which information may be held through the process of verbal rehearsal (i.e., area 40), and a "visuo-spatial sketch pad" in which information may be held in a visual format (i.e., area 7). Both are subordinate to a reasoning, active component of working memory called the "executive control system" that can consciously manipulate their contents. By this view, the area 6 activation corresponds to activation in the executive control system. Support for this role comes from research linking area 6 activity to problem solving (Fincham, Carter, van Veen, Stenger, & Anderson, 2002), reasoning (Monti & Osherson, 2012), and short-term information updating (Nee & Brown, 2013). Area 6 is part of a larger frontal lobe network involved in executive control, for as we have seen, area 10 appears to evaluate the contents of working memory (see Chapter 11). Left area 44, on the other hand, constitutes a portion of Broca's area. Activation there likely corresponds to the verbal rehearsal component of working memory (Cabeza & Nyberg, 2000).

In contrast to area 6 and 44 activations in humans, macaque working memory activations are generally in areas 9 and 46 (Petrides, 2005; Petrides & Pandya, 1999). Frontal lobe homologies between humans and macaques are close (Figure 14.2), so these really appear to be different areas of activation than those most frequently observed in humans. However, areas 9 and 46 do activate in a fair number of human working memory studies (Cabeza & Nyberg, 2000). It seems possible that they comprise an evolutionarily older substrate for working memory, dating back to at least a macaque-human common ancestor 32 million years ago, and that during human (or possibly ape) evolution, adjacent areas (i.e., 10 and 44) and nearby areas (i.e., 6) have been recruited.

In any case it is clear that in monkeys and humans, short-term memory is based in the neocortex. Birds, including pigeons, do not have a structure homologous to the neocortex of mammals (Krauzlis, Bogadhi, Herman, & Bollimunta, 2018). Thus short-term memory in birds on the one hand, and primates on the other, is best regarded as a matter of convergent evolution.

Intermediate term memory

Mark Rosenzweig and colleagues have called attention to an intermediate term memory in certain animals (Rosenzweig, Bennett, Colombo, Lee, & Serrano, 1993). For example, when chicks are trained to avoid pecking a normally attractive shiny bead, their avoidance dips at about 60 minutes post-training, then recovers. The result is consistent with the phasing out of an intermediate term memory, succeeded by long-term memory (Rosenzweig et al., 1993).

Somewhat similar findings have been reported from the nautilus, a deep-water shelled mollusk. The animals were conditioned to extend their tentacles following presentation of a signal light, and they were subsequently tested at intervals. Memory declined smoothly over one hour, but dramatically rebounded after six hours (Crook & Basil, 2008). Intermediate memories with varying temporal characteristics have also been proposed for fruit flies, honeybees, roundworms, snails, and sea hares, a type of sea slug (Braun & Likowiak, 2011).

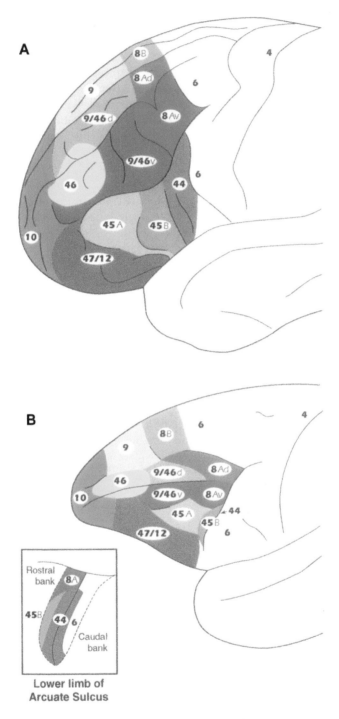

Figure 14.2 Brodmann areas in human frontal cortex (A), and corresponding areas in macaques
(B).

Source: Architectonic map of the human and macaque monkey prefrontal cortex, CC BY-SA 4.0, M. Petrides,
D.N. Pandya.

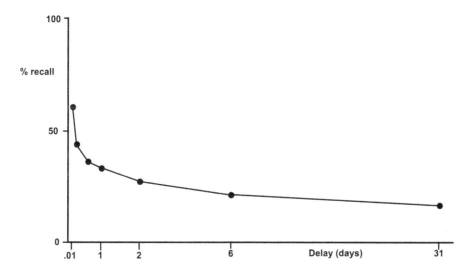

Figure 14.3 Example of the human forgetting function, showing a smooth decline of percent correct recalls over time.

Source: Original figure.

The idea of an hours-long memory intermediate to what in humans is a seconds-long short-term memory and a decades-long long-term memory is certainly appealing. However, to this point there is no evidence for it in people. Human memory follows a smoothly declining trajectory (Figure 14.3), to the extent that detecting any discontinuity at all, let alone a dramatic rebound, is somewhere between difficult and impossible (Rubin, Hinton, & Wenzel, 1999; Wickelgren, 1973). Also, it smoothly declines regardless of whether the memory items are nonsense syllables, words, pictures, or geometric forms (Gilliland, 1948; Wixted & Ebbesen, 1997). That means that the human-animal difference cannot be attributed to the predominant use of verbal memory items in human research on forgetting.

Therefore, it appears that intermediate term memory has no parallel in humans. Perhaps the necessary mutations never occurred, or perhaps they occurred and were subsequently selected against. Presumably a smooth forgetting function improves estimates of when events occurred and in what order, an important capability for a species dependent on experiential recollection. That could be a possible reason for deselection of an intermediate term memory.

Long-term memory and its divisions

If association is what most people think of when they think of memory, long-term memory is its primary repository. In this form of memory, knowledge and experiences can be retained more or less permanently. Models of human cognitive processing hold that information enters long-term memory from short-term memory through rehearsal. *Maintenance rehearsal* (rote repetition) suffices to increase a sense of familiarity of memory items, but *elaborative rehearsal* is required to embed memory items firmly enough in long-term memory to allow their recall. Elaborative rehearsal involves processing the meaning of the items, thus associating them to the existing knowledge structure (Klatzky, 1980).

Long-term memory actually refers to a family of memory systems. Most comparative research, however, has explored two divisions of long-term memory called *episodic memory* and *semantic memory*. Episodic memory is literally memory for episodes or particular events in the past, while semantic memory is memory for the facts that make up our knowledge structure.

A distinction between these systems is strongly indicated on neurological grounds. Amnesic patients who have suffered damage to the hippocampus and neighboring areas of the brain show severely impaired encoding of new information into episodic memory, while simultaneously showing a relative sparing of the acquisition of new semantic information (Rosenbaum, Gilboa, & Moscovitch, 2014). For example, a patient with amnesia may not remember a trip to Florida taken last year, while simultaneously recalling that the capital of Florida is Tallahassee, a fact learned during the trip.

Brain imaging and lesion studies also indicate a difference between episodic and semantic memory. For episodic memory, encoding a new event most often activates the medial temporal cortex, where the hippocampus is located, while retrieving an event most often involves area 10 of the frontal lobe (Cabeza & Nyberg, 2000). Lesions affecting episodic memory also typically involve the medial temporal lobe, especially the hippocampus (McCormick, Ciaramelli, De Luca, & Maguire, 2018).

In contrast, semantic memory retrieval involves the lateral temporal cortex. It also involves area 45 of the frontal lobe in both encoding and retrieval; however, this may be specific to tests that are verbal in nature, calling on language-related structures (Cabeza & Nyberg, 2000).

Lateral temporal involvement was confirmed in a study relating activation in normal brains responding to semantic tasks, to the site of brain *atrophy* (shrinkage) in *semantic dementia*. In semantic dementia, use of semantic knowledge is impaired while episodic memory is spared. The study found the activation and atrophy sites to be the same: The anterior lateral temporal lobe, in both hemispheres. More specifically, the activation data implicate the anterior portions of the middle and superior temporal gyri (see Figure 10.3; Rogers et al., 2006). Others have confirmed that this area is among those affected in semantic dementia (Landin-Romero, Tan, Hodges, & Kumfor, 2016).

Episodic memory

Do animals have episodic memories? Hampton and Schwartz (2004) have emphasized how difficult they are to demonstrate. Because animals lack language, memory is generally tested with recognition tasks in which previously seen items are picked from an array of such items intermixed with distractors. Because the items were learned as an episode, it is true that episodic memory could be responsible for correct recognitions. Unfortunately, mere familiarity could also support recognition. Good performance might therefore result not from recall of the learning episode, but from a vague feeling that items have been seen before.

A second problem was noted by Schwartz, Hoffman, and Evans (2005) in regard to previous research by Menzel (1999). In Menzel's study, a chimpanzee indicated the nature and location of hidden food to human caregivers, having observed it being hidden up to 16 hours previously. While that could certainly be an indication of episodic memory, it could also be that semantic memory was employed, and that the chimpanzee merely indicated what it knew to be true (i.e., that there is a particular food item hidden over *there*), without any specific recall of the episode of the food being hidden.

Therefore, turning this observation on its head, one way of demonstrating episodic memory might be to require the recall of something that was once true and was learned through a single observation, but which should be known by the recaller to no longer be true. For example, as I write this, I can recall being at the library earlier in the day and seeing a particular book on a shelf; it was true then but is no longer true, because I checked it out from the library. My memory must be of an episode, since it is not a truth represented in semantic memory.

This kind of strategy has been employed in a series of studies of a lowland gorilla named King. In the initial study, King was trained to respond to the English question, "What did you eat?" by handing over a card symbolizing one of five kinds of fruit. Subsequently, he was given the corresponding piece of fruit, and after either a short delay (5–65 minutes) or a long delay (24–96 hours), he was asked what he had eaten. At both delays he was 70–80% correct, well above chance performance. In a second study, he was able to indicate both the food eaten and the person who had given it to him, at both short and long delays. Thus in both studies, King recalled information learned in a single trial, that he should have known to no longer be true: He had eaten the food, so it existed only as a memory (Schwartz, Colon, Sanchez, Rodriguez, & Evans, 2002).

Of course, it is always possible that King based his responses not on recalled episodes, but rather on a sense of familiarity evoked by the most recently presented fruit or the most recent human benefactor. Familiarity was perhaps most convincingly ruled out in one of a later set of studies (Schwartz et al., 2005). King was given three pieces of fruit one at a time, at five-minute intervals, e.g., apple-grapes-pear. During training, after five minutes, he had to produce the three cards that symbolized the fruit, in reverse order, i.e., pear-grapes-apple, and if successful, he was given raisins. Subsequently, once this procedure had been learned, King was tested at longer intervals (5–23 minutes) and was successful (Schwartz et al., 2005). Mere familiarity could not have been the basis for his performance, because it could not plausibly have provided sufficient information to correctly order multiple cards. Thus King appears to have an episodic memory.

Different researchers have adopted different criteria for demonstrations of episodic memory, but the requirements that learning (a) be achieved in one trial, (b) reflect circumstances the animal should know are no longer true, (c) persist over intervals of at least several minutes, and (d) produce recognition that cannot plausibly be attributed to familiarity alone, together would seem sufficient to convince most observers that episodic and not semantic memory is involved. Unfortunately, it is a very difficult combination to satisfy.

A revealing study, but one that only partially satisfied the four conditions, was conducted by Lewis, Call, and Berntsen (2017) using a relatively large sample of 33 apes (chimpanzees, orangutans, and bonobos). The apes witnessed an experimenter showing them food that was not part of their regular diet, and then they observed the experimenter hiding it in a concealed location after climbing a ladder. The animals were given immediate access to the enclosure, and those that succeeded in climbing up to and finding the food in less than five minutes were included in the subsequent study. In that study, at a delay of 2, 10, or 50 weeks, the ape entered the enclosure and discovered the same kind of food on the ground below the hiding place. The question was whether the animal would then climb the ladder to search the previous hiding place. The answer was a resounding yes, but with an apparent delay effect such that the longer the delay, the fewer the climbs (Figure 14.4; note, however, that the delay effect was nonsignificant). Furthermore, control animals that had not been exposed to the initial food hiding did not climb the ladder.

The study certainly satisfied condition (a), one-trial learning, and condition (c), persistence. It probably satisfied condition (b), because the animals had previously eaten the food stored in the hiding place. It did not satisfy condition (d), however, because

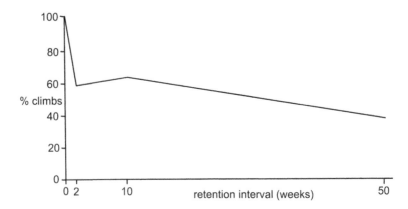

Figure 14.4 Percentage of climbs in the hidden food study, as a function of retention interval.

Source: Original figure.

familiarity alone could plausibly have led animals to climb the ladder. In spite of that, the results are revealing because one-trial learning persisted at a respectable level even 50 weeks after the episode. In addition, something resembling the human forgetting function was seemingly obtained.

Scrub jays show evidence of memory satisfying all four of the episodic memory criteria, as well as correctly recalling "what", "where", and "when" information about hidden food (Hampton & Schwartz, 2004). The "when" aspect is particularly intriguing because the jays avoid looking for food hidden sufficiently long ago that it should be spoiled.

In one of the best designed of these studies, the birds were first taught that cached worms (a preferred food) degraded over time, while nuts remained intact (Clayton, Bussey, & Dickinson, 2003). They then were allowed to cache worms and nuts, on separate sides of a tray and on separate days. Later they were required to choose a side, with the food removed. At short delays the task was to choose the side in which the worms had been cached, but at long delays it was to choose the side in which the nuts had been cached. The jays were able to perform this task successfully, seeming to satisfy the criteria developed earlier. Thus caching a particular food was a one-time act; preferentially avoiding worm locations after a long delay indicated an understanding that circumstances had changed (i.e., the worms had degraded); the delay of hours-to-days showed long persistence; and familiarity could not account for the results since the worm and nut locations were equally familiar. Such memories indeed seem episodic.

Subsequent to the research on jays, a similar experiment was conducted with a mixed group of bonobos, common chimpanzees, and orangutans. Frozen juice and grapes were the perishable and nonperishable foods, with the juice irretrievably melting through a net over the space of an hour. After training, the animals were shown both being hidden, and were then allowed to choose and access one location at 5- or 60-minute delays. It was found that a strong preference for the frozen juice location at the short delay was substantially reduced at the longer delay, similar to the scrub jay results. Interestingly, there was a relationship between delay effect and the animals' age, with only those of middling age (10–14 years) showing a substantial reduction in frozen juice choices at 60 minutes (Martin-Ordas, Haun, Colmenares, & Call, 2010).

Somewhat similar studies have been carried out in rats, and they implicate the hippocampus in animal episodic memory (Eichenbaum & Fortin, 2005; Naqshbandi, Feeney, McKenzie, & Roberts, 2007). Thus Eichenbaum and Fortin (2005) reported studies in which rats were exposed to five odors, then were required to discriminate which of two test odors had been presented prior to the other. Normal rats succeeded, but rats with hippocampal lesions failed. Importantly, both the normal and lesioned rats were able to perform the simpler task of recognizing which of two odors had been presented in the five-odor series, with performance declining for odors presented earlier in the series. This result indicates that both groups of rats had access to a memory trace that faded over time, but only the normal rats were able to use the trace in episode-like fashion: They correctly picked the earlier-presented odor in a pair even though more forgetting had occurred for it. While the degree of control over familiarity is not as good in this paradigm as in the jay paradigm – after all, the task could conceivably be performed by selecting the *less* familiar alternative – the involvement of the hippocampus provides reasonably convincing evidence that episodic, or at least episodic-like, memory is involved.

In fact, "following the hippocampus" may be a way to explore the evolutionary origins of episodic memory. The assumption would be that if a hippocampus is present, then episodic memory, or important elements thereof, may also be present. Such a strategy avoids the complicated behavioral methods necessary to identify episodic memory. However, it is potentially problematic for exactly the same reason: It may be possible for an animal to have a hippocampus without showing all the behavioral features of such a memory.

Birds have a hippocampus, part of the *paleocortex* (ancient cortex), even though they do not have a structure homologous to the mammalian neocortex (Clayton et al., 2003; Vargas, Bingman, Portavella, & López, 2006). Even fish have a hippocampus, or more accurately, a brain structure called the *lateral pallium* that is anatomically, genetically, and functionally homologous to the avian and mammalian hippocampus (Vargas et al., 2006). Fish use it to represent memory of the spatial environment, and it incorporates timing information (Broglio et al., 2005; Vargas et al., 2006). Indeed, as we saw in Chapter 12, the hippocampus is one site of grid cells that play an important role in representing environmental locations even in bats, rodents, and humans. That is certainly episodic in that the cells determine position and timing relative to the present environment, even though the environmental representation itself is built up over many different exposures and thus has semantic qualities as well.

These observations suggest that key components of episodic memory originated with the first appearance of bony fishes about 400 million years ago. However, whether these components have always supported the kind of recollective experience we think of as episodic memory is doubtful. It will take a great deal of clever experimentation to determine whether or not the episodic memory evident in gorillas, rats, and scrub jays shows continuity over nearly 80% of the time span of vertebrate evolution.

Semantic memory

How episodic memory gives way to semantic memory seems a bit mysterious. Logically, every piece of knowledge that we possess began as an episode. For example, there was a specific time and place that we first learned that 8 times 6 equals 48. At one time in our lives, soon after our first exposure to that fact, we might well have remembered the specific circumstances under which we learned it, an indication of reliance on episodic memory. Now, many years later, we do not. Somehow, multiple episodes have become something we simply know to be true.

One possibility is that episodic memory is required for encoding into semantic memory, so that in due course, multiple episodes build our knowledge structure. However, this does not seem to be the case in that some amnesiacs have severely impaired episodic memory yet still acquire new semantic memories (Rosenbaum et al., 2014; Verfaellie, Koseff, & Alexander, 2000). The alternate possibility, which appears more likely, is that the two memory systems exist at least partially in parallel. Faulty operation of one would therefore not necessarily affect the other. As we have seen, there is anatomical evidence for this view, with episodic memory calling on the hippocampus and semantic memory on the anterior temporal cortex.

Superficially, semantic memory should be much easier to demonstrate than episodic memory, seeming to require only that information be recalled in a way that reflects some understanding of meaning, and at a sufficient delay to represent long-term memory. However, the situation is more complex than that, because conditioning also produces behavior that appears meaning-laden. Thus in the classic Pavlovian study, when a dog salivates on hearing a bell, it appears that the dog understands the meaning of the bell: Meat is coming. But does it? Remember than conditioning is a form of learning that is found in flatworms and likely derives from the first emergence of nervous systems 540 or more million years ago. A strengthening of simple neural connections through repeated association seems more plausible than "understanding" at such an evolutionary remove.

One possible way around the conundrum may be to adopt an anatomical criterion. Semantic memory involves the lateral temporal lobe, and thus its evolution might be revealed by tracing the emergence of lateral temporal mechanisms of memory. Functional imaging evidence suggests this may be a workable solution. Thus while semantic memory involves the lateral temporal lobe, conditioning involves widely scattered brain areas (Cabeza & Nyberg, 2000), as might be expected from its arbitrary, associational nature.

The use of the lateral temporal cortex in semantic memory *may* have originated in humans. In monkeys, the ability to combine visual features in order to assign objects to classes, a kind of semantic operation, depends not on the lateral temporal cortex but rather on the perirhinal cortex (literally, cortex "near the nose") located in the medial temporal lobe (Bussey, Saksida, & Murray, 2002). Even so, Rogers et al. (2006) pointed out that this is not quite the same thing as the complex understanding of meaning represented by human semantic memory. They view the human lateral temporal cortex as coordinating information flow between modality-specific areas, lending it a modality-free character that seems closer to complex understanding.

One strong possibility is that the perirhinal cortex was the evolutionary "kernel" of a semantic memory system that now primarily involves the lateral cortex in humans. Semantic dementia affects medial as well as lateral temporal areas (Landin-Romero et al., 2016), although medial damage alone does not produce the full syndrome. For that, lateral damage is required (Rogers et al., 2006). In humans, relative to monkeys and chimpanzees, lateral temporal areas are greatly expanded in size and have become connected to frontal lobe language areas, consistent with the input of meaning (semantic memory) into language (Rilling et al., 2008).

However, it must be stressed that semantic memory is in part dependent on "local" levels of the cortex where perceptual processes are conducted. Thus damage to particular areas of the brain can cause an inability to perceive particular classes of things. For example, fusiform cortex lesions can affect face recognition (Iidaka, 2014), occipital lesions color recognition (Bouvier & Engel, 2006), occipito-temporal lesions the recognition of living things (Humphreys & Riddoch, 2003), superior temporal lesions music recognition (Sihvonen et al., 2016), and so on. In a broad sense, such perceptual deficits can be viewed as semantic: The meaning of a class of objects is lost because they cannot be recognized.

The lateral temporal cortex should therefore be viewed as only one level, although the most general level, of a broader semantic memory system. The implication is that other animals have the local elements of such a system even though they may lack the more general component. Or in evolutionary terms, human semantic memory may be much older in its local levels than it is in its general one.

Conclusion

Truly comparative research in memory is surprisingly meager given that animals have figured prominently as memory research subjects. Nevertheless, habituation, a decline in response to repeated stimulation, can be regarded as a universal memory phenomenon, with origins tracing to single-cell organisms. Association in the form of conditioning is nearly as universal, extending as far back as the origin of flatworms.

A case for universality, at least for vertebrates, can also be made for sensory memory. Sensory memory is a short-lived trace that preserves much of the physical form of stimuli. It lasts from less than half a second up to 10 seconds depending on the sensory modality. In the case of human visual sensory memory, there is strong evidence of both a peripheral retinal component and a central brain component, although in both cases the "icon" is best viewed as an aftereffect of nervous system activity. Because other vertebrates have similar sensory modalities and receptors, it is reasonable to believe they have the peripheral component of sensory memory.

However, the universality assumption may not be valid in the case of the central component, at least for the visual modality, because the brain area involved (V1) is a late mammalian innovation dating to only a little over 170 million years ago. Rhesus monkeys appear to have the central component, as indicated by the existence of delayed masking. Unfortunately, we do not have data on earlier-diverging taxa.

Comparative research is also sparse with respect to short-term memory, a small-capacity form of memory that holds information for about 20 seconds in the absence of rehearsal. Short-term memory capacity and search time are linked, such that fewer "slots" are available for items requiring more memory, while search time is longer. Limited research indicates that pigeons, monkeys, and humans all show evidence of a recency effect, in which the most recently encoding items show increased recall, indicating the widespread existence of short-term memory. However, its duration is haphazard across taxa, and it appears that its neocortical mechanisms as found in primates are absent in birds. Thus primate and bird short-term memories appear to represent convergent evolution.

For the present, there is no evidence of intermediate memory in humans. It has been found in birds and several invertebrates, with a measured duration on the order of one to six hours depending on the species.

Long-term memory has been a more fruitful area for comparative study. Episodic memory, which is memory for particular events, has been demonstrated in the lowland gorilla King. King can reproduce the order of closely spaced events at intervals well beyond the duration of short-term memory. His performance appears to demonstrate one-trial learning of circumstances he knows are no longer true, that persist for a substantial length of time, and that cannot be attributed to familiarity alone, together indicating the presence of episodic memory.

Scrub jays and apes also exhibit episodic memory by searching for either of two types of food cached in single events, contingent on delay, which affects the integrity of the food. Studies of rats likewise suggest the existence of episodic memory, and implicate the hippocampus, the same anatomical substrate of episodic memory that is found in primates including humans. Because all of these species, including birds, have a hippocampus, there may well be a common evolutionary origin for episodic memory. However, it is unclear to what extent it has supported a recollective quality of memory throughout its history.

Finally, semantic memory (memory for facts) may have had its origins in local brain areas where particular classes of stimuli are recognized. If so, its history is likely to prove very complicated, as each local area may have had a separate evolutionary trajectory. However, the most general level of semantic memory may only have emerged in humans, involving modality-free lateral temporal cortex mechanisms. It is possible that this was an outgrowth of primate processing of multiple visual features in the nearby perirhinal cortex.

15 Language

One day in 1995, a *New York Times* reporter visited a primate research facility in Georgia and witnessed an interesting interaction. A bonobo named Panbanisha, walking with her trainer Dr. Sue Savage-Rumbaugh, abruptly pressed three symbols on a portable keyboard, multiple times in different orders. The symbols signified "Mad", "Fight", and "Austin".

Savage-Rumbaugh surmised that there had been a fight at the residence of Austin, a chimpanzee at the facility, which Panbanisha confirmed. Subsequently, it was discovered that there had indeed been a fight between two chimpanzees over the use of computer equipment. It had apparently been heard by Panbanisha in a different building some 200 feet distant (Johnson, 1995).

Copious anecdotal evidence of a wide variety of cognitive skills in general and language skills in particular have been provided by apes trained in the use of symbols or sign language. From the Panbanisha anecdote, it seems straightforward to infer the existence of episodic memory (recollection of a previous incident), semantic memory (identification of Austin through sound alone), representational capacity (mapping the concept of Austin on to a symbol), vocabulary (employment of a set of symbols to convey meaning), and possibly even *syntax* or *grammar* (organization of symbols in orders that succeed in conveying meaning). To say that studies of ape language have revolutionized our thinking about the mental capacities of our nearest relatives would be an understatement.

But how much of the inferred mental repertoire of apes is based in science, and how much in wishful thinking? As we have seen, there is reason to believe that apes do have episodic memories. At minimum they have elements of semantic memory in being able to assign meaning to stimuli, even if they may lack the most general, modality-free aspect of human semantic memory. In this chapter, the remaining inferences from the Panbanisha anecdote are examined, and we consider the extent to which apes show language capabilities in the form of representational capacity, vocabulary, and grammar. We then consider how these capacities led to human language, first considering evolutionary changes in the brain, and then the chronology of changes – both physiological and behavioral – that created modern language.

Representational capacity

The ability to mentally represent the world in a realistic, literal way has been termed *primary representation*. It refers to a direct semantic relationship between what is in the head and what is in the environment (Leslie, 1987). Human infants show evidence of primary representation from a very early age. By 4½ months they reach for suspended objects, indicating that they use boundary information to separate objects from the surrounding environment. At this age infants appear to extract a considerable amount of information about objects, in that distance, size, and movement all affect reaching. Even earlier, by six weeks of age, infants can detect the difference between an object that is partially hidden behind a second object, and one with the hidden part actually missing (Spelke, 1982).

Primary representation is clearly a prerequisite for language in that an object must be mentally represented before it can be named. However, the ability to perceptually separate objects from the environment, and even understand their meaning, is an ancient one and as such sheds little light on language evolution. More complex representations, on the other hand, yield more interesting relationships to language.

Secondary representation

Secondary representation refers to the ability to separate a primary representation from its environmental reference for hypothetical purposes (Suddendorf & Whiten, 2001). For example, using a banana as a play telephone implies secondary representation because in the real world, bananas are not telephones.

The relevance of secondary representation to language is that it allows names to be used without their environmental references being present (Suddendorf & Whiten, 2001). I can say "I have a telephone in my office" without being in my office, a displacement in space, or "I used to have a telephone", a displacement in time, or "I need to find a telephone", a reference to the entire universe of telephones. I can even use the name metaphorically, as in "I played telephone", the child's game in which statements become progressively distorted as they are transmitted in a chain from one individual to another.

Several phenomena indicate fairly strongly that apes are capable of secondary representation. As we have seen in previous chapters, chimpanzees can plan the creation of tools, selecting an appropriate twig from which the leaves are stripped to become a "fishing" tool, or a stick whose end is sharpened to become a spear. Secondary representation must be present to allow the raw materials to be visualized as tools, which before their creation are not literally present in the environment. It is also present when home-reared chimpanzees play with dolls, bathing them or giving them something to "eat" (Tomasello & Call, 1997). Wild chimpanzees, especially female ones, have been observed treating a log or stick as a baby, for example making a nest for it, or carrying it for periods of several hours (Kahlenberg & Wrangham, 2011).

The ape ability to perform hidden displacement tasks has also been interpreted as indicating secondary representation. In one such task, an object is hidden in a small box, which is then concealed in a larger box. The smaller box is then removed but no longer contains the object. On discovering this absence, if test subjects look for the object in the larger box, they are considered to show secondary representation because the hidden object has been inferred to have a hypothetical existence inside the larger box. Human one-year-olds fail this test, but two-year-olds typically pass. Chimpanzees, gorillas, and orangutans also pass (Collier-Baker, Davis, Nielsen, & Suddendorf, 2006; Jaakkola, 2014; Suddendorf & Whiten, 2001).

Some of the most striking evidence of secondary representation comes from studies of mirror-image recognition, in which apes demonstrate that they recognize themselves in a mirror by using it to view parts of their body they otherwise cannot see. For example, many chimpanzees will reach up to touch a mark placed on their brow, having used the mirror to discover the mark (Anderson & Gallup, 2015). This is secondary representation because the animal recognizes that the image is not itself an actual chimpanzee, but rather a hypothetical chimpanzee representing them.

Is there evidence of secondary representation in more distant relatives of humans, such as monkeys? Both New World and Old World monkeys generally fail hidden displacement tasks, with their few apparent successes perhaps attributable to low-level strategies such as "pick the last-touched box" and not to secondary representation. The same may be true of lesser apes (Jaakkola, 2014). Monkeys generally, including capuchins, squirrel monkeys, macaques, and baboons, do not pass the mirror "mark" test, nor do lesser apes as a rule (Anderson & Gallup, 2015). Thus among primates, secondary representation appears to be largely a great ape capacity.

Vocabulary

The creation of a secondary representation is basically the creation of a symbol. A banana is not an actual telephone, but it can be a symbol of a telephone. Indeed, it has been argued that even so simple a cognitive act as creating a mental image is symbolic, because the image is not itself what it represents (Russon, 2004). In effect, some *thing* has been parsed from its environment, isolated, and given independent status as a mental entity.

Conceptually, it seems a small step to give such entities names and begin building a vocabulary. Evolutionarily, however, it is a huge step mastered by only one living species and dabbled in, with prodding, by a handful of others. A little thought gives some idea as to why the step is so large. First, a concrete means of reference must become established, both to transmit symbols and to receive them. Will the symbols be transmitted visually, for example through hand gestures? Or will they be transmitted auditorily, for example through vocalizations? Also, how is such a system bootstrapped from nothing, so that a symbol is both known by its transmitter to be producible, and by its receiver to be meaningful? Finally, there must be sufficient mental capacity on the part of both the transmitter and the receiver that multiple symbols can be combined to convey actions and to flexibly cover new situations. For example, it might do no good to tell hunters to "throw" if they are not sure when to throw, or what, or where. Thus communication is likely to be ineffective if both the transmitter and the receiver have difficulty handling multiple symbols.

Of these varied parts of the larger problem, reception of symbols seems much the easiest to achieve. A wide variety of species – including, it is safe to say, all primates – can readily learn that one symbol represents a food reward and another does not. Discrimination learning has been claimed to be a fundamental ability of all vertebrates (Moore, 2004). However, Russon (2004) pointed out that the symbolic abilities of apes go well beyond mere associative processes. For example, some chimpanzees understand how a scale model or photograph of a room relates to its full-scale counterpart, and they can use it to find a hidden soda. Others, however, fail to master this task (Kuhlmeier, Boysen, & Mukobi, 1999). These results suggest that as a species, chimpanzees are on the edge of understanding that a representation of a space is symbolic of it, an ability acquired in humans' third year of life (Russon, 2004).

Then, too, the use of multiple internalized symbols seems within the capabilities of chimpanzees. For example, the idea of using a hammer to hit a nut placed on an anvil leveled by a wedge represents a hierarchy in three levels, which chimpanzees routinely master. In contrast, monkeys have not demonstrated a capacity to hierarchically represent multiple levels (Russon, 2004).

From the standpoint of great apes including hominins, therefore, the biggest problems in establishing a vocabulary do not appear to be receiving or combining multiple symbols. Instead, they are the establishment of a transmission system, and its bootstrapping so that the transmitter knows the vocabulary is producible and the receiver knows its meaning.

Ape language studies

In ape language studies, both the establishment and bootstrapping of a transmission system are neatly sidestepped. Human tutors provide the system (e.g., sign language) and a producible vocabulary (e.g., individual signs) to teach meaning to the receivers. Given such substantial intervention, the process cannot closely resemble how language initially emerged in hominin prehistory. Nevertheless, the outcomes are likely to be informative concerning early limits in the growth and use of vocabulary.

How do scientists teach an ape to use sign language? First, they show the ape a sign in an appropriate context, molding the animal's hand to produce it. As they do this, they give a reward. Over the next month, caretakers watch to see if the ape produces the sign. If two or more caretakers see it used on at least 15 days of the month, it is concluded that the ape has acquired it. In three years of such training, an average of 130 signs are learned regardless of whether the learners are chimpanzees, gorillas, or orangutans (Blake, 2004).

Use of the signs is often spontaneous and sometimes involves creative combinations such as "eye drink" for contact lens solution, "candy drink" for watermelon, and "cookie rock" for a stale Danish. Signs are also sometimes invented. Generalization often occurs when a specific sign is not known: For example, calling a cookie a cracker or a cow a dog, or using the sign for "open" to request that water be turned on. As the Panbanisha anecdote at the beginning of the chapter implies, and as has been more formally observed, signs can be used to refer to things displaced in time and space (Blake, 2004; Gardner & Gardner, 1998; Savage-Rumbaugh, McDonald, Sevcik, Hopkins, & Rubert, 1986). However, in most cases, signs are used to request either objects or actions, such as "flower" or "chase" (Rivas, 2005).

Panbanisha herself, of course, learned visual symbols called *lexigrams* rather than sign language, which she used by pressing either a keyboard, touchscreen, or card-based array. Panbanisha's half-brother Kanzi learned lexigrams spontaneously without his arms being placed in position and without reward. Both bonobos also consistently responded to human speech as long as the words were in their vocabulary, for example with Kanzi reacting appropriately to the request "Will you take some hamburger to Austin?" (Savage-Rumbaugh, 1992).

Some of these studies give an indication of the vocabulary size that can ultimately be achieved by apes. Koko, a lowland gorilla, reportedly learned 600 signs in about 16 years, with her ultimate acquisition projected to be up to 2000 signs. However, the number of commonly used ones was projected to asymptote at around 500 (Clark et al., 1990). Kanzi, the bonobo, has been said to use somewhat fewer than 400 lexigrams (Paulson, 2007). However, these counts are informal and may be inflated. In more formal, controlled counts, Kanzi and Panbanisha averaged about 80% correct on tests using 256 lexigrams, suggesting a vocabulary size of about 200 (Lyn, 2007). Even so, it could be that fewer lexigrams or signs are commonly used. In a count based on 22 hours of videotaped interactions, involving five chimpanzees, only 88 different signs were observed (Rivas, 2005). In comparison, there are approximately one million English words, and human authors use a corpus numbered in the several to many thousands – for example, 7000 in the case of Jane Austin and 20,000 in the case of James Joyce (Macintyre, 2007). Educated adults have receptive vocabularies of about 20,000 base words and may know several forms of each, such as "legal", "legalese", and "legalize" (Nation, 1993).

It seems clear that living apes, and by likely extension early hominins, are capable of learning signs and symbols referring to objects, conspecifics, attributes, and actions. They also appear to learn and use them appropriately to represent emotion, as in "mad", "happy", "scared", and "hurt", although such uses are infrequent (Lyn & Savage-Rumbaugh, 2013).

Yet with these capabilities come important limitations. Acquisition is slow even with human intervention, between three and four symbols per month. In addition, while sign combinations are frequently semantically meaningful, e.g., "drink coffee", at least as frequently they are not, e.g., "drink gum". Longer combinations of three or four signs are also repetitious about half the time, e.g., "flower gimme flower" (Rivas, 2005).

Nevertheless, it seems undeniable that an important kernel for the evolution of language is present in our closest relatives. Even if utterances do largely represent requests (Rivas, 2005), communication is attempted. Indeed, it is reasonable to suppose that requests predominated early in the history of human language.

But could an incipient language, once started, find a life of its own and be transmitted to others, providing the bootstrapping necessary for the establishment of language? The answer seems to be "yes" in apes. Kanzi learned his first lexigrams not through direct intervention, but by casual observation of human attempts to teach them to his mother (Savage-Rumbaugh et al., 1986). The chimpanzee Loulis learned about 50 signs from other chimpanzees, although at a slow pace of a little less than one sign a month (Fouts, Fouts, & Van Cantfort, 1989; Fouts, Jensvold, & Fouts, 2002). The implication is that the ape-human common ancestor was probably capable of transmitting a small vocabulary-based language system from individual to individual.

If transmission could occur, then the remaining major step in bootstrapping such a system would have been the invention of the vocabulary itself. In the case of sign language, it is not hard to imagine how that might occur. Some signs would very likely have started out as natural communication gestures. A point might mean "that" or "there", and a waving gesture toward the body might mean "gimme" (Rivas, 2005). Individual chimpanzees and lowland gorillas have been observed to develop repertoires of about 10 to 30 natural communication gestures (Corballis, 2003), with several times that number existing across a population (A. I. Roberts, S. G. B. Roberts, & Vick, 2014). Groups of orangutans also have communication gestures numbered in the dozens (Cartmill & Byrne, 2010). Having studied a community of wild chimpanzees in Uganda, Hobaiter and Byrne (2014) identified 36 natural gestures that were employed to achieve specific outcomes, a third of which were considered to have tight as opposed to loose or ambiguous meanings. Once received and acted appropriately upon, such gestures could quickly have become established as an initial vocabulary in an early hominin ancestor.

Grammar

In my undergraduate classes, after describing some of the vocabulary-related findings of sign and lexigram studies with apes, I ask my students to debate whether the capabilities reflect language. Usually I get no takers on the "not language" side of the debate. Typically, every student – or at least every student willing to express an opinion – endorses the conclusion that apes can learn and use a rudimentary language.

But then, virtually all of my students are psychology students. To many psychologists meaning is everything, or a big chunk of it. If a chimpanzee signs "give X" when "X" is available, and the chimpanzee seems satisfied on receiving it, the intent for communication seems clear, the meaning definite, and the use of vocabulary in a language-like way undeniable.

Linguists, in contrast, are often much more skeptical of the language nature of these capabilities. To a linguist, the hallmark of language is grammar, not meaning. Grammar is what allows speakers to communicate (and listeners to distinguish) the meaning of sentences that use similar wording. For example, I might say, "The dog that the child is chasing is wet", or "The dog is chasing the child that is wet". The two sentences use identical words but have very different interpretations. Using grammar, we can generate meanings that are not present in the vocabulary alone, greatly magnifying both the complexity and subtlety of our communications.

On the grammar criterion, apes fall short. This first became apparent in the signing of the chimpanzee Neam Chimpsky, "Nim" for short, playfully named after the famous linguist Noam Chomsky. In his first four years, Nim learned 125 signs. Initial analysis of Nim's two-sign combinations showed some evidence for the consistent ordering that would be expected if ape signing followed grammatical rules. For example, "more X" and "give X" combinations outnumbered "X more" and "X give" combinations by nearly a 2:1 ratio. However, this trend toward ordering broke down completely with combinations of three or more signs, leading to the conclusion that grammatical rules were not present (Terrace, Petitto, Sanders, & Bever, 1979).

Thus was ignited an academic controversy. On the pro-language side, Greenfield and Savage-Rumbaugh (1991) produced data showing that Kanzi ordered lexigrams in a grammatical way. For example, two-sign combinations of action-agent (e.g., "carry person") and action-object (e.g., "keepaway balloon") outnumbered agent-action and object-action combinations by nearly a 6:1 ratio. Subsequent data from two bonobos, including Kanzi, and one chimpanzee confirmed frequent ordering preferences in two-lexigram combinations, and in combinations of a lexigram with a gesture (Lyn, Greenfield, & Savage-Rumbaugh, 2011).

Furthermore, in the 1991 data, three-sign combinations of action-action-agent (e.g., "chase bite P", where "P" was the designation of a human caretaker) outnumbered other orderings by a 7:1 ratio. Unfortunately, however, this particular observation itself proved problematic, as these were the only three-sign combinations sufficiently frequent to be analyzable. Even then, only eight instances had been observed among all combinations of three or more signs, which themselves constituted only 10% of Kanzi's corpus of 13,000 utterances (Greenfield & Savage-Rumbaugh, 1991).

In truth, the average length of ape utterances remains at about two vocabulary items, or a little less, regardless of length of training (Lyn et al., 2011; Miles, 1990; also Terrace et al., 1979, idealized data in Figure 15.1). This renders arguments over grammar rather

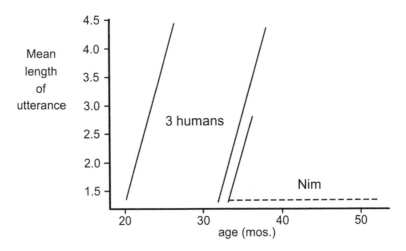

Figure 15.1 Mean length of utterance over time, comparing Nim to three human children using oral language. Nim remained "stuck" at fewer than two words per utterance, unlike human children.

Source: Original figure.

sterile; it may not make sense to argue whether grammar is observed in multiple-sign combinations, when utterance length itself falls so short of human capabilities.

Clearly, apes have minimal access to complex grammatical structures. Nevertheless, the simpler trends in two-sign combinations may be important, particularly given parallels observed in primate gesture and vocalization. In their natural gestures, chimpanzees communicate actions that also imply an actor, a simple two-element grammar (Fouts & Waters, 2003). With respect to vocalization, in both bonobos and common chimpanzees, a kind of *phonological syntax* (ordering of meaningless sound elements) is observed in long-distance calls. A "pant hoot", for example, may be altered by inserting a "bark" into the sequence of vocalizations. Such variations may be related to status, territoriality, and the need to maintain contact with others (Ujhelyi, 1998). As in signing, and in natural gestures, they give a hint of the beginnings of grammar. However, the grammatical structures are not complex.

The role of frontal brain areas

In contrast to apes using sign or lexigram forms of language, humans using language usually vocalize. Nevertheless, our species shows some similarity in how vocalization is controlled. Specifically, regulation of the larynx in part involves area 6 in our frontal lobe (see Figure 14.2). The same area constitutes the *laryngeal motor cortex* (LMC) in New World and Old World monkeys, and in chimpanzees (Kumar, Croxson, & Simonyan, 2016; Simonyan, 2014).

Yet we also differ from other primates. While we do make some use of area 6, the primary motor cortex (area 4) is more involved in vocal control, and these two areas taken together comprise our LMC (Kumar et al., 2016; Simonyan, 2014).

Accompanying the anatomical distinction between species are differences in connectivity. A major difference is that human LMC has vastly stronger connections than macaque LMC with primary somatosensory cortex and the inferior parietal lobe. The strong connections presumably exist to coordinate sensory feedback with phonemic information so that speech can be adjusted on the fly by way of the LMC (Kumar et al., 2016).

Furthermore, connections of the LMC with laryngeal motoneurons in the brainstem are direct in us, but indirect in monkeys. As a result, the LMC produces responses at the larynx that are about three times faster in humans than in monkeys (Simonyan, 2014). That surely was an important development in our evolution, as it enabled rapid, direct cortical control over vocalization.

Broca's area

Broca's area is another frontal area involved in language. In Chapter 10 on praxis and handedness, the roles of Broca's area in human speech production and object manipulation were highlighted. Both have grammatical qualities, requiring the rule-based sequencing of component movements. Brain lesions in the vicinity of Broca's area often result in *agrammatism*, the disordering of grammar. Thus speech becomes telegraphic, for example with the intended sentence "Yesterday I went to the movies" possibly becoming "Yesterday go movie". Also, the decoding of grammatically complex sentences fails. For example, upon hearing the sentence, "The boy that the girl is chasing is tall", agrammatism patients are typically at a loss in judging which person is doing the chasing, and which person is tall (Miceli, 1999).

Recent work suggests that area 44, comprising a portion of Broca's area, is particularly involved in the production of the complex phrase structures characteristic of natural human language. When syntactical processing is complicated, the posterior temporal cortex is also recruited by way of connecting tracts that include the arcuate fasciculus. These connections are much stronger in humans than in other primates, potentially accounting for the weak to nonexistent grammar of ape utterances (Friederici, 2017; Rilling et al., 2008).

As noted in Chapter 10, primate area F5 is homologous to the area 44 portion of Broca's area. F5 contains mirror cells that participate in the copying of sequential actions modeled by others, implicating it in gesture. Furthermore in chimpanzees, asymmetry favoring the left inferior frontal gyrus (the location of F5) correlates to the strength of right-handedness in manual gestures, but not to handedness as evidenced in reaching (Hopkins & Cantalupo, 2004; Taglialatela et al., 2006). Significantly, the correlation also extends to the insertion-turning behaviors we examined in the handedness chapter (Hopkins et al., 2017). Those behaviors can be viewed as having a sequential, grammar-like structure (e.g., grasp an object, insert it into an opening, turn it, and pull it toward oneself). Furthermore, a link to speech is suggested by the fact that chimpanzee communication gestures are more right-handed when accompanied by vocalization than when emitted alone (Hopkins & Cantero, 2003; Hopkins et al., 2005). Together, these findings strongly suggest that in the chimpanzee, vocalization and gesture are both at least partially controlled by F5 in the left hemisphere.

The implication is that the involvement in grammar of what in humans is Broca's area has roots tracing back to F5 in our common ancestor with chimpanzees. But should we attribute the origin of those roots to gesture, or to vocalization? The answer is of critical importance, because the difference is between a gestural origin or a vocal origin of human language. Taglialatela, Russell, Schaeffer, and Hopkins (2008) argued that lesioning the Broca's area homolog in monkeys does not affect their vocalizations, an observation that could be taken as supporting an evolutionary role for the area specifically in gesture. However, monkeys are not chimpanzees, and it is possible that the area's role in vocalization emerged in the immediate ancestors of chimpanzees and not before.

In truth, the most likely scenario for the origin of language is that it was both gestural and vocal. In comparison to using only one modality, using both improves signal detection, disambiguates meaning, and increases the amount of information transmitted (Wilke et al., 2017). Bonobos and common chimpanzees often pair gestures and vocalizations when communicating, and pairing improves the chances of obtaining the desired end compared to the use of either alone (Pollick & de Waal, 2007; Wilke et al., 2017). A further indication of the intimate link between gesture and vocalization is that when chimpanzees make fine motor movements with their hands, their mouths often make accompanying movements (Wacewicz, Zywickzynski, & Orzechowski (2016).

Interestingly, in contrast to common chimpanzees and bonobos, the pairing of vocalization and gesture in gorillas and orangutans appears to be rare, even though each behavior alone is fairly common (Poss, Kuhar, Stoinski, & Hopkins, 2006). This suggests that co-production of vocalizations and gestures may have originated, in large part, with the chimpanzee-human common ancestor. Thus it seems reasonable to suppose that at the very beginning of human language, we made use of all of the communication tools at our disposal in order to make ourselves understood. Gesturing and vocalization together were very likely those tools.

The role of perisylvian brain areas

Producing vocalizations, of course, assumes that the intended recipients can decode the sounds and understand their meaning. In this regard, *perisylvian* brain areas (areas located around the Sylvian fissure, also called the lateral fissure; see Figure 15.2) are

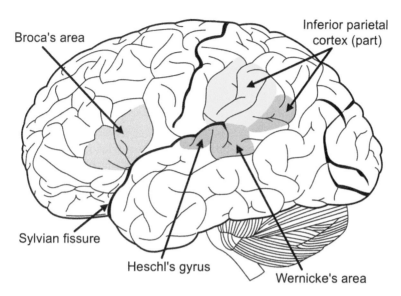

Figure 15.2 The Sylvian fissure in relation to Wernicke's and Broca's areas.

Source: Adaptation of Brain Surface Gyri.SVG, CC BY-SA 3.0, James.mcd.nz.

critically important in modern humans. The primary auditory area, where sound first enters the cortex, is on the superior surface of the temporal lobe in *Heschl's gyrus*. On the left side just posterior to this in the superior temporal gyrus is *Wernicke's area*, implicated in the decoding of speech sounds. Above the Sylvian fissure is the inferior parietal cortex, which is involved in the ongoing storage of *phonology* (language phonetics), necessary to understanding connected speech.

The planum temporale

Comparative studies of the superior temporal areas involved in language have centered around the *planum temporale*, or temporal plane, a small surface of cortex located lateral and posterior to Heschl's gyrus (Meisenzahl et al., 2002). This is generally held to contain secondary auditory cortex as well as a portion of Wernicke's area. Its main advantage as a unit of analysis is that, unlike Wernicke's area itself, it is anatomically easily definable and easily measured (although there are slightly varying, competing definitions; see Meisenzahl et al., 2002).

Anatomical studies have shown that in humans, the planum temporale has a larger area in the left hemisphere than in the right hemisphere (Gough et al., 2018; Meisenzahl et al., 2002). In turn, planum temporale asymmetries, in area or volume, are found to correlate with lateralization of language as assessed during anesthetization of each hemisphere in epilepsy patients (Foundas, Leonard, Gilmore, Fennell, & Heilman, 1994; Oh & Koh, 2009). In the larger of these studies, of patients with both asymmetric planums and asymmetric language, 45 of 46 with larger left planums had left hemisphere language, and 7 of 9 with larger right planums had right hemisphere language, an impressively high agreement rate of 95% (Oh & Koh, 2009). Others have indicated that the size of the planum temporale on one side or the other correlates to language lateralization as assessed by brain imaging (Josse, Mazoyer, Crivello, & Tzourio-Mazoyer, 2003;

Tzourio-Mazoyer, Crivello, & Mazoyer, 2018). Together these studies suggest that the area plays some role in language, and that left hemisphere predominance in language is reflected in a larger left planum temporale.

The left-larger pattern is seen in about 65–80% of humans showing asymmetry in either direction (Gough et al., 2018; Tzourio-Mazoyer et al., 2018). Thus it is a matter of interest that about 80% of tested great apes show it as well (e.g., Cantalupo, Pilcher, & Hopkins, 2003; Hopkins et al., 2016). The asymmetry appears to pertain to orangutans as it does with the other great apes (Cantalupo et al., 2003), suggesting that it originated at least 17 million years ago, although the number of tested animals is small.

Analysis suggests that a similar asymmetry is found in macaques (Gannon, Kheck, & Hof, 2008) and baboons (Marie et al., 2018), potentially extending its evolutionary emergence back to at least 32 million years ago. Accompanying the anatomical asymmetry is a behavioral one in which macaques hear "coo" and "scream" vocalizations better in the right ear than in the left, implicating the left hemisphere in their perception. Also, left superior temporal gyrus cells selectively respond to species-specific calls. However, there may not be a strict homology to the human Wernicke's area, which is in the posterior portion of the gyrus while its macaque equivalent appears to be in the anterior part (Ghazanfar & Hauser, 2001; Poremba et al., 2004).

Inferior parietal cortex

Above the Sylvian fissure, the human inferior parietal cortex is involved in phonological coding and phonological working memory (Liebenthal, Sabri, Beardsley, Mangalathu-Arumana, & Desai, 2013). Its anterior portions in particular are active in word rhyming tasks (Kircher, Nagles, Kirner-Veselinovic, & Krach, 2011; Seghier et al., 2004). Intriguingly, the anterior portions are also involved in comprehending and producing hand gestures (Jax, Buxbaum, & Moll, 2006). It is therefore an area of particular interest if early language was both gestural and vocal in nature.

The evolutionary history of the inferior parietal cortex is divided into two phases, an initial one in which a left hemisphere size advantage was established, and a later one in which the supramarginal and angular gyri first appeared. In modern humans a left hemisphere size advantage is visible in terms of a Sylvian fissure that is longer in the left hemisphere than in the right, especially in the portion posterior to the central sulcus (i.e., bordering the inferior parietal lobe). The asymmetry is also found in great apes, including orangutans, and traces to an extended anterior portion of the left inferior parietal lobe (Cantalupo et al., 2003).

Could it have occurred even earlier than the orangutan divergence? Initial studies of Sylvian fissure asymmetry in monkeys emphasized the lateral aspect of the fissure and produced inconsistent results (Heilbroner & Holloway, 1988). However, Hopkins and colleagues subsequently discovered that the leftward asymmetry is much more pronounced at medial locations of the fissure, and that it exists for apes, Old World monkeys, and New World monkeys about equally (Cantalupo et al., 2003; Hopkins, Pilcher, & MacGregor, 2000). Furthermore, it is the more medial asymmetry that is attributable to an extended anterior parietal lobe (Cantalupo et al., 2003).

The initial phase of the evolution of the inferior parietal cortex, therefore, seems to have occurred prior to the New World monkey divergence, at least 45 million years ago. In all likelihood the function of the area during this phase was concerned less with communication than it was with reaching and with spatial attention, as in living monkeys (Battaglia-Mayer, Mascaro, & Caminiti, 2007; Quraishi, Heider, & Siegel, 2007). These functions were subsequently displaced away from the left inferior parietal lobe of modern humans, to more dorsal parietal locations in the case of reaching, and toward the *right* hemisphere in the case of spatial attention (see Chapter 12).

It is to the later phase of inferior parietal evolution that we must look for involvement in early language. In Chapter 11, it was pointed out that the supramarginal and angular gyri, comprising Brodmann areas 40 and 39, were human innovations that began with the genus *Homo* (Tobias, 1995). They are not found in macaques at all (Figure 15.3). *Homo habilis* endocasts show an accompanying feature not shared by australopithecines and apes: A pronounced "rounded fullness" of the inferior parietal lobe (Tobias & Campbell, 1981). It is this area that is involved in grasping independent of reaching, copying hand postures, phonological coding, and phonological working memory. Thus the supramarginal gyrus (area 40) plays a role in all of these (Frey et al., 2005; Jax et al., 2006; Liebenthal et al., 2013), while the angular gyrus (area 39) aids the copying of hand postures by differentiating between fingers and discriminating right from left (Denes, 1999). They are clearly structures one might expect evolution to have altered if early language was both gestural and vocal in origin, and evolution has done so.

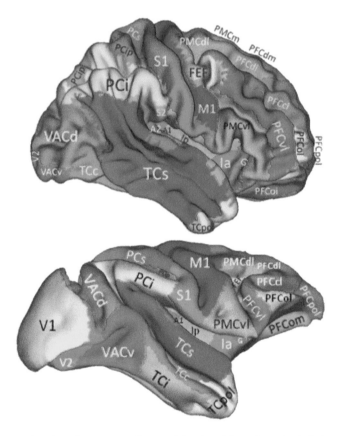

Figure 15.3 Human (top) and macaque (bottom) brains. Note that the inferior parietal cortex (PCi) of the macaque lacks gyri present in the human.

Source: The RM was used in order to construct the connectivity matrices of the two species, CC BY-SA 4.0, A. Goulas, M. Bastiani, G. Bezgin, H.B.M. Uylings, A. Roebroeck, et al.

Other temporal lobe areas

Finally, the role of other temporal lobe areas in language comprehension should be mentioned. In humans the middle and inferior temporal gyri are massively connected to the region around and including Broca's area. Again, the arcuate fasciculus is responsible for much of this, and the connections do not exist in similar magnitude in macaques and chimpanzees. Furthermore, they are larger in the human left hemisphere than in the right hemisphere. Their function appears to be to support lexical-semantic processing (Rilling, 2014). That, of course, is consistent with the lateral temporal lobe's role in semantic memory, as we saw in Chapter 14.

The hominin evolution of language: language area changes

Thus far the evidence argues for a joint gestural and vocal origin of language, with each reinforcing the other. At the moment of the chimpanzee-human divergence, the human ancestor very likely had chimpanzee-like capabilities. Those included a primary capacity to mentally represent objects in the environment; a secondary representational capacity to refer to them metaphorically or when they were otherwise not present; the ability to represent objects with a small symbolic vocabulary of gestures and sounds, and to generalize, differentiate, and combine their meanings; highly developed learning mechanisms that allowed the transmission of vocabulary; and an ability to order, in a very basic way, at least two such symbols.

Neurologically, area F5 and its surroundings in the frontal lobe had become specialized, by virtue of mirror cells, in copying and sequencing actions modeled by others. The temporal lobe was involved in recognizing vocalizations, and the region around the planum temporale had long ago evolved the left-sided asymmetry that would serve as a substrate for further language development. On the other side of the Sylvian fissure, the parietal cortex had long been involved in mirror cell activity (Bonini, 2017), and the left inferior parietal cortex had expanded.

Broca's cap

Following on that beginning, it can probably be assumed that attaining speech-like utterances depended at least in part on the conversion of F5 to more resemble the modern Broca's area. An indirect indicator of Broca's area is *Broca's cap*, an outward bulge of cortex in that vicinity. It first appears with consistency in *Homo habilis* endocasts, having had variable expression in apes and in australopithecines (Broadfield et al., 2001; Tobias & Campbell, 1981). Recently, it was discovered that a well-defined Broca's cap in endocasts counterintuitively depends on the cap in the left hemisphere having shrunk relative to the corresponding area in the right hemisphere, bringing its borders into better relief (Balzeau, Gilissen, Holloway, Prima, & Grimaud-Hervé, 2014). Thus, if the cap is first clearly visible in *Homo habilis*, the shrinkage may have begun by 2.4 million years ago. At about the same time, the gyri more directly marking Broca's area may have first made impressions on endocasts (Falk, 1983).

Falk (2014) reported that chimpanzees and humans have different frontal lobe sulcal patterns in the region of areas 44 and 45. Also, while both species can have a Broca's cap, the human one includes areas 44 and 47 but not 45, while the chimpanzee one includes area 44 and sometimes a portion of area 45. Furthermore, there were possible indications in an endocast of the 2-million-year-old species *Australopithecus sediba* of a transitional state between chimpanzees and modern humans. Taken together, these results suggest reorganization of the region in hominins following the last chimpanzee-human common ancestor.

Another discontinuity across species is apparent when the area of the third frontal convolution is measured as an approximation of Broca's area, a quantity that can be assessed using endocasts. In early hominins, the area was smaller in smaller brains, but beginning with late *Homo* species (*rhodensiensis/heidelbergensis, neanderthalensis,* and *sapiens*), the size of Broca's area has been the same irrespective of brain size, even in relatively small-brained individuals (Balzeau et al., 2014). This seems reminiscent of the preservation of area 10 during the dwarfing of the brain of *Homo floresiensis*, which was interpreted in Chapter 11 as suggesting the area's critical importance to that species. Thus, in completion of the analogy, Broca's area may have assumed critical functional significance by the onset of the later *Homo* species about one million years ago.

By 370,000 years ago, according to a molecular-clock estimate, an important genetic change occurred that further enabled language. A gene known as FOXP2, located on chromosome 7, has been linked to language capacity via an extended family with a defective allele. Affected members of the family cannot properly combine component oral movements, such as protruding the tongue while closing the lips, and have trouble repeating even short word sequences (Lieberman, 2007). Imaging indicates that their Broca's area has reduced gray matter but increased activation, in the left hemisphere specifically (Vargha-Khadem et al., 1998). The significance is that changes to the gene preceded the split between Neanderthals and modern humans, although it postdated the chimpanzee-human divergence. Thus DNA analysis indicates that Neanderthals shared our form of the gene (Krause et al., 2007), placing its origin somewhere in excess of 370,000 years ago (Noonan et al., 2006). To the extent that it is a "language gene" (a cautionary note is that intelligence is affected as well; Lieberman, 2012), the result implies the existence of language in that time frame.

Wernicke's area

Although changes to Wernicke's area in the temporal lobe are difficult to date, at some point that area's cellular microstructure changed. In chimpanzees, the cell bodies of neurons in the planum temporale line up in *minicolumns* that are about 36 micrometers (0.036 mm) wide in both hemispheres. In contrast, in humans there is a noticeable asymmetry, with wider columns existing in the left hemisphere (50 micrometers) than in the right (43 micrometers). Thus there appears to have been a "rewiring" of the planum temporale, presumably related to the left hemisphere comprehension of language (Buxhoeveden, Switala, Litaker, Roy, & Casanova, 2001). Unfortunately, there is no evidence bearing on exactly when this change occurred.

The hominin evolution of language: peripheral changes

Alterations to anatomy in the periphery – i.e., the body and skull – were also necessary to fully support language. Relevant information begins about three million years ago with the hominin *Australopithecus africanus*. A *hyoid* (a horseshoe-shaped bone in the neck) from a juvenile of that period reveals a bubble of bone upon it that is characteristic of chimpanzees and gorillas but that does not at all resemble modern humans. The significance is that the bubble probably reflects the presence of air sacs in the larynx (Alemseged et al., 2006). They increase acoustic energy at low frequencies, allowing calls to extend further in forest and woodland. However, simulations suggest that these effects make human speech sounds indistinct (de Boer, 2012). The implication is that australopithecine air sacs resulted in the production of sounds more like ape vocal calls than modern human speech.

When was this feature lost? Unfortunately, hyoids are not common among fossil assemblages. However, two exist from *Homo heidelbergensis* specimens dating to 530,000 years ago. Both lack the bubble and resemble the modern human hyoid (Martínez et al., 2008). Of course, the loss of the bubble, and thus presumably the shift from vocal calls to more speech-like utterances, could have occurred anytime in the 2.5 million years between the australopithecine and *heidelbergensis* examples.

The range of speech sounds that could be uttered may also have increased with changes to the face that began occurring between *Homo habilis* and *Homo ergaster*. Specifically, the face began to flatten and the jaw to shorten. The tongue also shortened and reoriented, with the result that it could produce greater leverage, increasing its flexibility and ability to articulate sounds (MacLarnon & Hewitt, 2004).

The lungs

One substantial species distinction is that humans can control vocal air flow to a much greater extent than apes. Our abilities are easily self-observed by monitoring breathing while speaking out loud. Clearly, we can produce either sound bursts or continuous vocalizations at will. This allows the sequencing of phonetic information, to the extent that human sequencing of vocalizations is 10 times faster than that of apes (MacLarnon & Hewitt, 2004).

An important element in the control of air flow is the control of breathing by the lungs themselves. While speaking, exhalations become extended in time in order to produce the air movement required for modification by the throat and mouth. Air pressure can also be regulated by abdominal muscles controlling the lungs, to produce either loud or soft vocalizations. In contrast, other primates have much less control over breathing and cannot produce vocalizations of anything like human complexity. In them, the regulation of vocalization much more involves the diaphragm that it does the abdominal muscles, the opposite of humans (MacLarnon & Hewitt, 2004).

Nevertheless, in humans, some control over the breathing musculature is exerted by the chest. It involves certain nerves exiting from the spinal cord at the *thoracic vertebrae*, those at the level of the thorax. The nerves partially comprise the spinal cord, and so the *vertebral canal* (the opening in the vertebrae through which the cord passes) must be wide enough to accommodate them. This observation leads to a simple prediction: Species with the fine breathing control needed for rapidly sequenced vocalizations should have enlarged vertebral canals at the thoracic level, because of the increased size of nerves needed to control the lungs. Of course, when looking for enlargement, it is necessary to take body size into account, because larger species will have larger canals for that reason alone.

Putting this prediction to the test, MacLarnon and Hewitt showed that when thoracic canal size is plotted against body size, all nonhuman primates track very closely to the same line. In other words, when body weight is accounted for, they all have similar-sized canals. However, modern humans place significantly above the line, indicating an enlarged canal. Most critical to present purposes, though, were measurements of the size of the thoracic canal in fossil hominin vertebrae. Neanderthals and early modern humans likewise tracked above the primate line, while *Australopithecus afarensis*, *Australopithecus africanus*, and a single *Homo ergaster* specimen fell almost exactly on it. The results strongly suggest that precise control over breathing was achieved sometime between 1.6 million years ago and 100,000 years ago, the age limits of the *Homo*

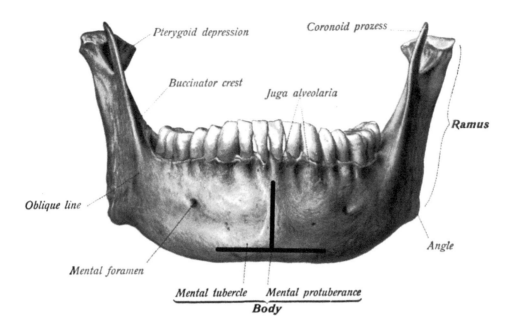

Figure 15.4 A human chin with an "inverted T" (added for emphasis) formed from the mental protuberance in the center, and the mental tubercle on each side.

Source: Adaptation of Sobo 1909 90.png, public domain, Dr. Johannes Sobotta.

ergaster specimen and the more modern hominins. Unfortunately, at present we do not have sufficient vertebral remains of intervening hominins to narrow this range further (MacLarnon & Hewitt, 1999, 2004).

The jaw

Turning to the jaw, unlike *Homo erectus* we have a protruding, flat-surfaced chin, sometimes described as an "inverted T" (Figure 15.4). Analysis of a series of jaws of modern *Homo sapiens*, Neanderthals, and earlier specimens often attributed to *Homo heidelbergensis* suggest that it emerged about 90,000 years ago in modern humans (Schwartz & Tattersall, 2000).

The protruding chin has long been an evolutionary enigma, as its appearance does not appear related to evolutionary changes in chewing. However, analysis of forces exerted by the tongue during *speech* do indicate that protruding chins reduce stress (Ichim, Kieser, & Swain, 2007). Thus its appearance about 90,000 years ago may be a sign of increased articulatory ability.

The vocal tract

A relatively late evolutionary change is the "verticalization" of the vocal tract (Figure 15.5). At least as early as *Homo heidelbergensis* (530 kya), and as late as Neanderthals (30–200 kya), the vocal cavities in adults were oriented more horizontally, as in apes, rather than vertically as in modern humans (Lieberman, 2012; Martínez et al., 2013). Modeling the sounds that such an airway can produce suggests that vowels would

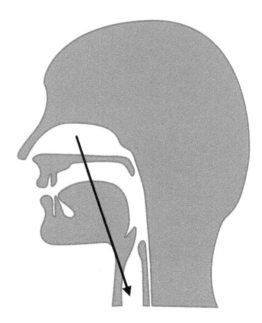

Figure 15.5 The "verticalization" of the modern human vocal tract, which is much more horizontal in apes, *Homo heidelbergensis*, and Neanderthals.

Source: Adaptation of Midsagittal diagram unlabeled.svg, public domain, Wugapodes.

have had reduced intelligibility compared to modern speech (Lieberman, 2012). Brown and Golston (2006) have argued that evolving a vertical vocal tract was also essential to produce a full range of consonants. Either way, the full modern range of sounds would not have become possible until the vocal cavities became fully vertical in modern *Homo sapiens* (Barney, Martelli, Serrurier, & Steele, 2012; Lieberman, 2007). The date has been suggested as somewhere between 40,000 and 90,000 years ago (Lieberman, 2006). At the same time, the root of the tongue descended deeper in the throat, becoming thicker and more rounded, and better able to control air flow during vocalization (Lieberman, 2007).

The hominin evolution of language: behavioral changes

Some theorists have argued that language developed rapidly only with the proliferation of tools and decorative artifacts some tens of thousands of years ago (e.g., d'Errico, Henshilwood, Vanhaeren, & van Niekirk, 2005; Zilhão, 2007). In their eyes, the creation of artifacts is a symbolic act that implies the use of language for purposes of representation and instruction.

Proposing a link between language and the proliferation of artifacts does seem a leap of faith, however. After all, it is not hard to imagine sufficiently intelligent but language-less beings using *spatial* perceptual, memorial, and motor processes both to create artifacts and to transmit knowledge of how they were created. If decorative artifacts are indicative of language, then any decorative artifact of any age must be assumed to imply the existence of language. In that case, are we really willing to assume that language is implied by stone tools decorated with red ocher, dating from over 160,000 years ago (Marean et al., 2007)? Or by a shell etched with a geometric design, dating from about

500,000 years ago and attributed to *Homo erectus* (Joordens et al., 2015)? It doesn't seem unreasonable to require additional evidence before drawing such conclusions.

"Proto-World" language

However, there is additional behavioral evidence – though highly controversial – of a development in language on the tens-of-thousands-of-years scale. It emerges from the comparative study of languages, suggesting that all existing language families may partly derive from an original language spoken about 45,000 years ago (Ruhlen, 1994b). Specifically, tracing the common roots of words has led to a claimed convergence on a small vocabulary reflective of a hunter-gatherer society, with words that mean "ashes", "bone", "dog", "earth", "water", and so on. A reconstruction of 27 of the original words has been proposed (Ruhlen, 1994a). For example, "kuan" is taken to mean "dog" and is proposed as the root of the word "hound" in English, while "aq'wa" is thought to mean "water" and is proposed as the origin of the Latin and English word "aqua" (Bradshaw, 1997; Ruhlen, 1994a).

This hypothetical "Proto-World" language is quite controversial. Evidence of it has been called subjective, poorly repeatable, and even unscientific (McMahon, 2004). Notable problems are that apparent correspondences between languages can be coincidental, and that words not only can be inherited from an earlier language, but can also be borrowed from unrelated ones (Nettle, 1999). It also is not clear how a 45,000-year-old language would have spread across the globe.

However, it is worth pointing out that the Proto-World hypothesis places something of an upper limit on the emergence of modern language. If it is untrue, then modern languages have even more recent origins. Either way, it must be the case that earlier languages, assuming they existed, were less successful. The reason why that might be can only be speculated. However, the 45,000-year-old date roughly coincides with the lower date limit on the verticalization of the vocal tract, suggesting that modern language may have depended on evolving a novel capability to produce a wide range of sounds.

The evolutionary process

In this description of language evolution, the evolutionary process itself has received short shrift. However, for the most part, little need be said. Language has such clear adaptive value, in its ability to recruit the help of others and in teaching the young and transmitting culture, that it seems certain that once started, natural selection would ensure its continuance. The main caveat concerns whether the most recent elaborations of language were too recent to have been much affected by natural selection (Box 15.1). Otherwise, the universal nature of language is a testament to the power of evolutionary processes to produce profound change.

Box 15.1   Did reading evolve?

Writing systems date back only about 5500 years, yet reading is a nearly universal human capability. The short time frame in which reading emerged, and its rapid spread, imply that it is not a direct product of natural selection.

Instead, reading must have emerged as a co-option of other capabilities. Dehaene and Cohen (2007) have suggested that one of the brain sites involved in reading occupies an intersection of cortical areas sensitive to object identity and foveal stimulation. The implication is that this *visual word form area* (VWFA), in or near the occipito-temporal sulcus, is located where it is because recognizing words presented visually calls for sensitivity to particular objects (words) at a particular location (the fovea). In other words, reading has co-opted preexisting capabilities, and a preexisting cortical map, for new purposes.

Even so, however, co-option requires the preexisting circuitry to be altered during development. Consistent with this, the predominant role of the left hemisphere in reading does not emerge until several years after reading begins. Thus visual field studies of word recognition show no visual field difference – and therefore no hemispheric difference – from the beginning of reading at about age 5 to around ages 8–10, when a transition to increasing left hemisphere representation begins. By age 11, the leftward shift is complete (Boles et al., 2008). Similarly, reading-related evoked potentials also shift leftward during development as reading finds its "optimal location" in the left ventral visual system (Dehaene & Cohen, 2007). Presumably it is optimal because other language processes are located in the left hemisphere as well.

A second reading-relevant location is the left angular gyrus (Brodmann area 39), involved in recoding visual words into phonological forms (Boles, 2002; Pugh et al., 2000). The angular gyrus adjoins Wernicke's area, yet how "left hemisphered" an individual is in reading, fails to positively predict how left hemisphered that individual is in recognizing auditory language. In fact, there is actually a slight negative relationship between the two, so that a person with an especially strong left hemisphere involvement in one tends to have a reduced left hemisphere involvement in the other (Boles, 1998). This pattern is consistent with the co-option view: The more left parieto-temporal cortex is co-opted by either reading or auditory language comprehension, the less is available for the other.

It may be more appropriate to ask, "*Will* reading evolve?" Given its importance to education and economic security, and the impact those have on reproduction, the answer is almost certainly "yes". However, as with natural selection more generally, there is no way to predict the time frame or the precise nature of future adaptations.

Conclusion

Language did not spring forward in one great leap, like Athena from the forehead of her father Zeus in Greek mythology. Instead, it followed the course of other complex abilities, starting crudely and acquiring increased sophistication by degrees over many millions of years (Table 15.1). It was prefigured by the appearance of frontal lobe mirror cells prior to the Old World monkey divergence 32 million years ago. These assisted the copying of others' actions, an operation that would later prove critically important in learning communication gestures.

However, deeper cognitive components would also be necessary. Secondary representational capacity, appearing sometime prior to the orangutan-human common ancestor 17 million years ago, allowed understanding of displacements in time and space. It also

Table 15.1 Milestones in the evolution of language. mya = millions of years ago

Date	Achievement	Examples/notes
Before 32 mya	Frontal lobe mirror cells	In living monkeys: Activity during grasping, etc., and observations of same by others.
	Asymmetric planum temporale	In living monkeys: Larger on the left side.
	Superior temporal gyrus involvement in recognizing vocalizations	In living monkeys: Greater left-side involvement.
	Inferior parietal cortex asymmetry	In living monkeys: Longer Sylvian fissure on the left, indicating a larger left side.
Before 17 mya	Secondary representation (displacements and metaphors)	In living apes: Tool manufacture; hidden displacement; mirror-image recognition.
	Symbolic capacity	In living apes: Play objects; hierarchical planning; learning of externally provided languages (signs and lexigrams).
	Simple grammar	In living apes: Tendency to order two-element sequences in externally provided languages and in natural gestures; vocalization insertions.
	A "vocabulary" of natural communication gestures	In individual living apes: Repertoires of 10–30 gestures.
~ 10 mya	Skilled handedness, possibly indicating left hemisphere area F5 development	In living gorillas and chimps: Population right-handedness in tube task, hammering.
Before 7.5 mya	F5 co-production of communication gestures and vocalizations	In living chimps: Activation of F5 during co-occurrence of right-handed communication gestures and vocalizations; mouth movements accompanying fine hand movements.
After 6 mya (otherwise uncertain)	Increased laryngeal motor control	In hominins: Laryngeal motor cortex recruitment of area 4, with stronger parietal connections and direct connections to the brainstem.
	Shift in location of Broca's cap	In hominins: Inclusion of area 47 and exclusion of area 45, relative to chimps.
	Increased connections between Broca's area and the temporal lobe, especially on the left	In hominins: Relative to chimps.
	Asymmetric minicolumns in Wernicke's area	In hominins: Found in humans but not chimps.
After 3 mya	Loss of air sacs, allowing speech-like vocalizations	In hominins: Hyoid bones suggest air sacs in *Australopithecus africanus* 3 mya, not in *Homo heidelbergensis* 0.5 mya.
2.4 mya	Appearance of a well-defined Broca's cap and gyri marking Broca's area (homologous to F5)	In hominins: First found in *Homo habilis*.
	Appearance of supramarginal and angular gyri; "rounded fullness" of inferior parietal lobe	In hominins: First found in *Homo habilis*.

(Continued)

Table 15.1 (Continued)

Date	Achievement	Examples/notes
~1.9 mya	Face flattening and jaw shortening begin; shortening and reorienting of tongue	In hominins: First occurred between *Homo habilis* and *Homo ergaster*.
~1 mya	Stabilization of size of Broca's area	In hominins: Size of third frontal convolution stabilized beginning with *Homo heidelbergensis*.
0.1–1.6 mya	Increased breathing control	In hominins: Enlarged thoracic vertebral canal.
Before 0.37 mya	FOXP2 allele affecting Broca's area	In hominins: Shared by Neanderthals and *Homo sapiens*.
~ 0.09 mya	Protruding, flat-surfaced chin supporting speech stresses	In hominins: First appeared in modern humans.
0.04–0.09 mya	Verticalization of vocal tract; full range of vocalizations	In hominins: First appeared in modern humans.
0.045 mya?	Appearance of modern vocabulary?	

provided mental stand-ins that could be dealt with metaphorically in place of actual objects. Mirror-image recognition became possible, too, with an image standing independently as a metaphorical representation of the self. The importance of this capacity cannot be overestimated, because the units of language – gestures and words – would need to be understood as metaphorical references and not as objects in and of themselves. Thus was born a symbolic capacity.

A small "vocabulary" of natural communication gestures likely provided the first true grist for the language evolution mill. Dozens of gestures are found in living orangutans, gorillas, and chimpanzees. That pushes a natural gestural "vocabulary" back beyond 17 million years ago. However, a fuller exploitation of it probably depended on developments in frontal lobe area F5, dating back about 10 million years ago, as evidenced by population right-handedness in gorillas and chimpanzees performing skilled tasks. This was a marker for the F5 ability to sequence motor acts – for example, communication gestures – another critical necessity in the evolution of language.

Before 7.5 million years ago, a Rubicon was reached: In the chimp-human common ancestor, and possibly not before, F5 could co-produce communication gestures and vocalizations. Now two communication modalities could be mutually reinforcing, allowing disambiguation of messages that might be obscured in only one. Sometime after three million years ago, the loss of air sacs made speech-like vocalizations possible.

By the appearance of *Homo habilis* two million years ago, F5 had evolved into a discernible Broca's area, appearing as a notable protrusion (Broca's cap) and as gyri known from the traces of endocast sulci. Possibly at this time, language of a crude, limited type was being produced, perhaps both gestural and vocal in nature. To accommodate it, the inferior parietal lobe developed its "rounded fullness" as well as the supramarginal and angular gyri, presumably playing something like its modern role in comprehending gestures and speech sounds. It may have been at this point, though the dating is uncertain, that minicolumn asymmetry appeared in Wernicke's area in the temporal lobe, and the arcuate fasciculus increased in size and connections to improve transmissions between Wernicke's area and Broca's area.

The remaining hominin language adaptations all appear to have had the function of allowing more precise and discriminable vocalizations, so that after this point language became heavily dominated by speech as opposed to gesture. Face flattening and jaw shortening commenced, and the tongue began shortening and reorienting. As early as 1.6 million years ago, but certainly by 100,000 years ago, the thoracic vertebral canal enlarged to pass nerves to the chest that allowed increased breathing control during speech. Before 370,000 years ago, the FOXP2 gene mutated, affecting the development of Broca's area and presumably profoundly influencing speech.

The increased mechanical stress of speech-related oral movement resulted in selection of a protruding, flat-surfaced chin by about 90,000 years ago, and around that time, or perhaps as late as 40,000 years ago, the vocal tract became vertical, supporting a full range of vocalizations including a full set of consonants and discriminable vowels. By 45,000 years ago, a modern vocabulary may have emerged, which some believe has left its imprint on all of today's language families.

Thus language has at least a 32-million-year background, divided into roughly four periods. In the first period, the foundational capacities of imitation, secondary representation, symbolism, and a natural vocabulary of communication gestures, emerged in monkeys and ancestral apes. The second period, dating from perhaps 17 million years ago to shortly before 7.5 million years ago, saw an increasing specialization of area F5 in producing skilled motor acts, to the point that it could co-produce gestural and vocal communications. By the end of the third period, about two million years ago, F5 had become Broca's area, and the inferior parietal lobe had further developed. Crude, limited speech may have been produced and understood. Finally, in the fourth period, from about 1.9 million years ago to about 40,000 years ago, there were a number of adaptations of the face, jaw, tongue, vocal tract, and chest nerves that resulted in more discriminable speech and, perhaps at the end of the period, the beginning of modern vocabulary.

16 Consciousness

Our human cognitive nature produces a number of complex behaviors such as toolmaking and language. However, in the opinion of many, a non-behavior, consciousness, is the supreme achievement of the human mind.

Daniel Dennett (1981) likened the popular view of consciousness to an "inner light" that is either on or off, and that is obvious *from the inside*. The problem, Dennett noted, is identifying consciousness *from the outside*. Are animals conscious? Could a machine be conscious? Now we must define external criteria that denote consciousness.

Dennett's description, however, only scratches the surface of the mystery. Many psychologists closely identify consciousness with attention, taking the view that whatever we are attending *to*, we must be conscious *of* (e.g., Posner, 1978). However, that assumes a unitary nature of attention that may not exist. Attention can be split between information sources (Scharlau, 2004) and, according to some theories, between tasks (Boles & Dillard, 2015). But are we really splitting consciousness when we split attention?

Even so, attention is only a beginning point in the universe of thought about consciousness. In ever-expanding circles, consciousness has been equated with intellect (Ornstein, 1972), the sum total of cognitive processes (Titchener, 1909), matter (Miller, 1962), reality (Eccles, 1966), and even unity with God (White, 1974). Some have linked it to quantum physics (Mukhopadhyay, 2014). Small wonder, then, that others have argued that consciousness cannot be defined (Granit, 1977).

The approach taken in this chapter is to examine behaviors that have been interpreted as implying consciousness. This is a functional approach to the topic, emphasizing what consciousness brings to behavior. Even so, some readers will demur. Thus, while supporting functional approaches, Allen and Bekoff (2007) noted that some philosophers consider consciousness an epiphenomenon of other mental processes, itself having *no* effect on behavior.

It is likely that some of the behaviors that have been interpreted as implying consciousness will be viewed as more compelling than others. But perhaps somewhere in the darkness, we can catch a glimpse of the "inner light".

Sensory awareness

An initial function of consciousness to consider is that of sensory awareness, the subjective experience that we have sensed something external to us. At first glance this may seem too simplistic and self-evident to lend value to evolutionary analyses. However, certain failures of sensory awareness in fact prove informative.

Blindsight and deaf hearing

Blindsight is a phenomenon in which an observer claims to be blind, yet shows unexpected residual visual abilities. It results from partial or complete destruction of the primary visual cortex (area 17; see Figures 10.4 and 16.1), producing blindness for part or all of visual space. The basic demonstration of blindsight begins with the presentation of a visual stimulus in the blinded area, to which the patient fails to respond. On prompting, the patient states that they are blind and see nothing. However, if they are asked to point toward the place where the stimulus was shown, they often are able to do so with surprising accuracy. Some patients can even make discriminations of shape (e.g., squares from rectangles), identify line drawings of animals or other objects, or discriminate colors, while simultaneously stating that they are guessing (Ajina & Bridge, 2017; Trevethan, Sahraie, & Weiskrantz, 2007).

A similar phenomenon, *deaf hearing*, likewise occurs due to destruction of the primary sensory cortex, this time Heschl's gyrus or Brodmann area 41 bilaterally (see Figures 10.4 and 16.2). The patient denies hearing anything but nevertheless orients to loud sounds like hands clapping or fingers snapping (Garde & Cowey, 2000).

Blindsight and deaf hearing demonstrate that while sensory information entering the brain through the primary cortex routinely reaches consciousness, when that route is blocked it does not. The major visual route is the *geniculostriate route*, which proceeds from the eye to the lateral geniculate nucleus to the primary visual cortex (area 17). It is this route that produces visual awareness and is disrupted in blindsight. However, there is an alternate *tectopulvinar pathway* that proceeds from the eye to the superior colliculus, then to the *pulvinar* (a nucleus in the thalamus), and finally to the secondary visual areas of the cortex (see Figure 16.1). This route bypasses area 17 altogether, providing cortical visual input in the absence of area 17 function.

Is it the tectopulvinar pathway that supports the residual capabilities of blindsight? Though long thought so, confirmatory evidence emerged only relatively recently. Leh, Mullen, and Ptito (2006) studied hemispherectomy patients in which an entire cerebral

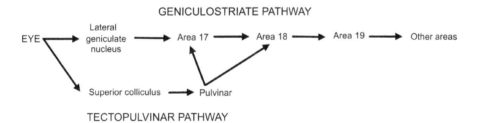

Figure 16.1 The geniculostriate and tectopulvinar visual pathways to the cortex.
Source: Original figure.

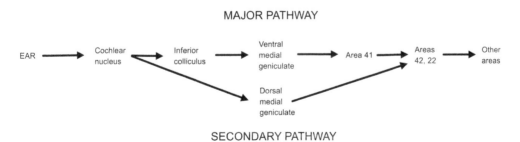

Figure 16.2 The auditory pathways to the cortex.
Source: Original figure.

hemisphere had been surgically removed, producing profound blindness in one visual field. They found that responses to visual stimuli in the sighted field were speeded by the redundant presentation of the same stimuli in the blind field, a demonstration of blindsight. However, the speeding of responses occurred only if the redundant stimuli were achromatic (i.e., in shades of white, gray, and black) and thus visible to the medium- and long-wavelength cones that have inputs into the superior colliculus. If they were in colors distinguishable only to short-wavelength cones, which have no input into the superior colliculus, no speeding occurred. A subsequent fMRI study reported activation of the visual cortex beyond V1 and V2, when the redundant stimulus was achromatic, but not when it could be seen only by short-wavelength cones (Leh, Ptito, Schönwiesner, Chakravarty, & Mullen, 2010). Together, these results argue strongly that the intact pathway in blindsight is routed through the superior colliculus, as is true of the tectopulvinar pathway. This route, of course, does not appear to provide visual awareness.

In the auditory modality the major pathway is from the ear to the inferior colliculus, and then to the ventral portion of the medial geniculate nucleus, before ending up in Brodmann area 41, the primary auditory cortex (see Figure 16.2). Destruction of area 41 on both sides can result in the deaf hearing phenomenon. The apparent cause is a secondary pathway leading from the inferior colliculus to the dorsal portion of the medial geniculate nucleus, then bypassing area 41 before ending up in the secondary auditory cortex (area 42). No auditory awareness appears possible through this route, although as indicated, the patient orients to loud sounds (Garde & Cowey, 2000).

What can blindsight and deaf hearing tell us about the neural substrate of consciousness? On one hand, it seems clear that under normal circumstances, stimulation of primary cortex is a prerequisite for sensory awareness. If such stimulation doesn't occur as in blindsight and deaf hearing, sensory awareness does not occur. The most obvious possibility, then, is that stimulation of the primary cortex directly generates awareness. However, this can quickly be rejected on the basis of a study of macaques by Nakamura and Mishkin (1980). They surgically isolated the visual cortex in one hemisphere – striate, inferotemporal, and other higher visual areas – from all other cortexes in that hemisphere. They also blinded the other hemisphere by disconnecting optic nerve input to it, and they disconnected the hemispheres by severing the cerebral commissures. The result was behavioral blindness. The monkeys could not perform a previously trained visual discrimination, bumped into objects, and failed to react to fear-inducing objects. In other words, even though the visual system seemed to be fully intact in one hemisphere – though disconnected from the rest of the cortex – the monkeys showed no evidence of visual awareness. Awareness therefore is not a *property* of visual cortical stimulation alone.

A second possibility is that information from the primary cortex travels to some other brain structure that itself produces awareness. While this is difficult to rule out – a definitive test would require ruling out all other brain structures, one by one – it seems unlikely. That's because in blindsight and deaf hearing, the residual sensory capabilities enter the cortex in secondary areas just barely removed from the primary cortex. It seems very likely that they have access to all the brain structures that primary cortex can access, yet awareness does not result. Note that this also means the secondary areas, taken alone, do not produce awareness.

If primary and secondary sensory areas are not alone responsible for consciousness, what is? There appears to be something unique about primary cortex *in interaction* with other areas of the brain that results in awareness.

The role of reentry

A promising idea is that neurologically speaking, *reentrant* connections support such interactions. In other words, the primary cortex stimulates higher-order cortical sites that in turn send connections back to primary cortex. Perhaps that is somehow what results in sensory awareness (Lamme, 2003; Rees & Frith, 2007).

There is in fact direct experimental evidence that supports the role of reentry. When Transcranial Magnetic Stimulation (TMS) to V1 is used to evoke an experience of *phosphenes* (imaginary points of light) in intact brains, awareness of the phosphenes can be prevented by V1 restimulation at a lower intensity that does not itself generate phosphenes, 30 msec or so after the initial stimulation (Pascual-Leone & Walsh, 2001). This strongly suggests that if reentry is prevented using subthreshold TMS, there is no awareness. Delayed TMS has also been found to degrade V1 function compared to no delay, again consistent with reentry, although any effects on subjective awareness were not examined (Camprodon, Zohary, Brodbeck, & Pascual-Leone, 2010).

Of course, the reentry solution does not completely resolve the problem, for we still have little idea of how it is that reentry results in sensory awareness (Lamme, 2003). In particular we have little understanding how we generate the ineffable sensory qualities that baffle even the most dedicated introspective observers (e.g., why is yellow yellow, or sour sour?). However, reentry does at least allow us to begin an evolutionary analysis of the origins of sensory awareness, by tracing the history of reentrant connections.

In this regard, reentrant connections into V1 certainly exist in macaques, where they have been most extensively studied (Hupé et al., 2001). They also exist in squirrel monkeys, a New World species (Rockland & Knutson, 2000), as well as in rats (Johnson & Burkhalter, 1996), ferrets, and cats (Cantone, Xiao, McFarlane, & Levitt, 2005). Indeed, feedback connections are a fundamental characteristic of the mammalian brain (Johnson & Burkhalter, 1996). Thus their evolutionary history dates as far back as the origin of mammals about 195 million years ago.

However, it is not clear whether they have supported sensory awareness across that entire time span. Although cats have residual visual abilities following destruction of the primary visual cortex (Payne, Lomber, MacNeil, & Cornwell, 1996), which are consistent with blindsight, we do not know whether the animals experience reduced visual awareness. What we do know, however, is that *monkeys* experience blindsight, while simultaneously believing themselves blind. Thus Cowey and colleagues have found that macaques can localize stimuli in their blind field. Yet if the same stimuli are shown in the same blind field, and the choice is whether or not a stimulus was presented, the monkeys choose to respond that none appeared (Cowey & Alexander, 2012). Thus the monkeys appear to inform us that they believe themselves blind. Monkeys can also be trained to respond to the location of stimuli in a blind area even though they do not spontaneously shift their eyes to them (Moore, Rodman, Repp, & Gross, 1995). These performances are very similar to those of human blindsight patients who deny that a stimulus is present but who can localize it when encouraged to.

Putting these observations together, it seems clear that sensory awareness dates back at least to the human-macaque common ancestor about 32 million years ago. It may date back to the origin of mammals 195 million years ago, if reentrant connections from higher to lower brain areas are diagnostic of sensory awareness.

Attention

William James is widely credited with establishing a close correspondence between attention and consciousness. In his book *Psychology* (1892), he maintained that attention accentuates and emphasizes perceptions that become the objects of consciousness, and that it ignores (and thus causes consciousness to ignore) the rest of the many stimuli that constantly besiege us. Later authors equated consciousness with awareness and likewise emphasized the role of attention in creating it (Posner, 1978; Taylor & Fragopanagos, 2007).

Problems with attention as consciousness

However, at least two issues arise with using attention as a proxy for the existence of consciousness. The first is whether attention even implies consciousness. It can be argued that any selection of one of multiple stimuli attributes impinging on the nervous system of an organism is a form of selective attention; yet this capacity is so basic that it undoubtedly exists in all animals with nervous systems. Thus chickens can make use of spatial cues to attend to a location, while ignoring other locational cues, and zebrafish can quickly reverse their "attentional set" if a previously unrewarded color suddenly becomes rewarded (Krauzlis et al., 2018). Without the ability to select stimuli or attributes significant to survival, and to ignore unimportant ones, an animal's chances of survival and propagation would be compromised. Thus selection should be (and presumably is) a universal nervous system characteristic, but we should hesitate before claiming that the existence of a nervous system necessarily implies the existence of consciousness.

A second issue with using attention as a proxy for consciousness is that attention is not a unitary concept. Posner (1992) identified three distinct components. First, there is a selective attention component, which for spatial selection at least depends on the parietal lobe, primarily but not exclusively in the right hemisphere. Second, there is a component that recognizes the meaningfulness of stimulation and that involves the anterior cingulate cortex. Third, there is an alertness component, maintaining wakefulness and readiness to respond to stimulation, that is dependent on a right frontal mechanism. All of these components also involve subcortical areas such as the thalamus.

But which component should we identify with the existence of consciousness? The selective component is what William James had in mind, but we also habitually speak of being conscious (awake) versus unconscious (asleep), roughly corresponding to Posner's third component. Most of us should have no problem viewing his second component as consciousness-related, too. Thus, if we receive a stimulus that suddenly provokes a recognition of meaning, we often say that the meaning has become conscious. But if we accept that all three represent consciousness, we not only make the task of identifying consciousness in animals a much harder proposition, but we open the possibility that a particular animal might be conscious in some ways but not others.

Unfortunately, there is also no consensus as to what neural structures support attention-as-consciousness. Posner's work appears to implicate roles for the frontal and parietal lobes, and for the cingulate cortex, but Lamme (2003) suggested that attention plays only a supporting role in consciousness. He argued that while we are conscious of many things, attention is merely what allows us to report one of those things, and that consciousness itself results from recurrent interactions between sensory areas and widespread areas of cortex (frontal, prefrontal, parietal, and temporal), that brings information into register with needs, goals, and context. Merker (2007) has even questioned whether consciousness depends on the thalamic-cortical connections involved in attention. He noted that children with hydroanencephaly, lacking a cortex but not in a persistent vegetative state, show indications of consciousness by reacting to environmental events.

Also troubling are the results of studies using TMS. In humans, TMS applied to the parietal cortex disrupts both selective attention, and consciousness as defined as the ability to report stimuli – but with no correlation across participants in the sizes of these disruptions (Babiloni et al., 2007). The result suggests that the relationship between attention and consciousness is at best a loose one.

Historically, the relationship between attention and consciousness has been an important one. However, the issues surrounding how best to conceptualize that relationship, and how to relate it to behaviors or to neural structures whose evolution can be traced, appear too fraught at present to take us far in using attention to trace the evolution of consciousness.

Metacognition

One possible function of consciousness is that it permits *metacognition*, the awareness of one's own cognitive states and processes. For example, it might involve a judgment of how much one knows in a particular circumstance. This aids decision making by distinguishing between situations where action can be taken because knowledge is sufficient, and situations where it should be withheld because knowledge is insufficient. Such "feelings-of-knowing" are vague and error-prone, as are other products of metacognition like judgments of how easily one will learn something, or of the quality of one's retrieved answers (Nelson, 1996). They can therefore be viewed as a type of "fringe consciousness" that is constructed from the outcomes of unconscious processes (Norman, 2017).

Even young children have metacognitive skills. In one study, picture-number pairs such as a dog and the number "2" were presented for learning. Then in the testing phase, a number was presented and the child could either choose to decline answering or to view an array of pictures with the aim of choosing the correct partner – e.g., the dog if a "2" had been presented. If correct, a highly rewarding visual display was shown. If incorrect none was shown, while declining to respond received a less rewarding display. The results showed that 4.5- to 5-year-old children declined to respond to pictures that were difficult to remember 40% of the time, but those that were easy to remember only 21% of the time. However younger children, 4–4.5 years old, did not do this (34% versus 33%). The implication is that "feeling of knowing" is a metacognitive faculty that develops beginning at about 4.5 years of age (Liu, Su, Xu, & Pei, 2018).

In a second study, children always responded by trying to pick out a studied picture from a larger array, but could indicate afterward whether or not they felt confident in their decision. Children age 4–5 showed much higher accuracy when they felt confident than when they did not (82% versus 28%), a difference that wasn't found in children age 3–4 (56% versus 46%, a nonsignificant difference). Thus the "confidence in retrieved answers" form of metacognition (Nelson, 1996) appears to begin developing around age 4 (Liu et al., 2018).

Monkeys, too, show evidence of metacognition. Shields and colleagues tested two rhesus macaques on their confidence in retrieved answers. A rectangle that was filled either densely or sparsely with pixels was presented, and the task was to respond by pressing an "X" to a dense rectangle or "O" to a sparse rectangle. Difficulty was varied by changing the density within each of these categories. Thus a dense rectangle could be relatively sparse, or a sparse rectangle relatively dense, with either of these representing a difficult trial. After learning the task, the monkeys were next introduced to confidence ratings, learning to press a blue bar if they had confidence in their just-made response or a pink bar if not. Food rewards and time-out punishments were arranged to appropriately incentivize the monkeys. It was found that accuracy was higher when the monkeys expressed confidence in their judgment that when they did not, by 14%. This was quite similar to the outcome of a parallel experiment using adult humans, also finding a 14% differential (Shields, Smith, Guttmannova, & Washburn, 2005).

Are New World monkeys also aware of what they know? Tests administering a version of the sparsity task to capuchins produced ambiguous results. Monkeys were allowed to make an opt-out response when presented with the rectangle, with the expectation that they would make the response more frequently in the most difficult trials. Five monkeys completely failed to do so in spite of appropriate incentives. However, a sixth monkey did so after extensive experience with the task (Beran, Smith, Coutinho, Couchman, & Boomer, 2009). A later set of studies used the opt-out procedure with a different discrimination task involving picking the largest of a set of squares. Macaques were found to regularly use the opt-out response, reaching 40% of the most difficult trials. Capuchins were much less likely to use it, opting out of only 10% of the most difficult trials. A second experiment shrunk this difference between species but did not eliminate it (Beran, Perdue, & Smith, 2014).

Other studies using varying methods have supported the existence of metacognition in all of the great ape species (including humans) and in Old World monkeys (i.e., macaques). Evidence for it is mixed in New World monkeys (i.e., capuchins), with some studies failing to find it, and others finding either reduced proficiency or proficiency only in some individuals and not others (Beran et al., 2014). While including other monkey species in future research would be desirable, the tentative conclusion is that New World monkeys are "on the cusp" with respect to metacognitive abilities – and thus on the cusp with respect to "fringe consciousness" as well. Thus metacognition may have barely begun to evolve in our common ancestor with New World monkeys 45 million years ago.

Self-recognition

Another possible function of consciousness is that it allows self-recognition. Gordon Gallup (1968) first reported self-directed behaviors in a chimpanzee following brief exposure to a mirror:

> The chimp's initial responses to the mirror consisted of the typical other-directed behaviors such as lip smacking, vigorous head bobbing, a few threats, and intense inspection of the mirror image. [Afterward,] however, the other-directed orientation was abruptly displaced. . . . He was observed to stick out his tongue at the mirror, and shake his brass neck tag in front of the mirror. Perhaps the most significant evidence of self-directed behavior was the fact that the chimp would actively groom various parts of his body otherwise not visually accessible except in the presence of a mirror (for example, the forehead and eyebrows).

Humans, of course, have no problem recognizing themselves in mirrors. Nevertheless, at this point in his research, Gallup interpreted self-directed behaviors in animals less in terms of higher-order cognition than in terms of conditioning. The mirror, he suggested, might disrupt the stimulus-response pairings governing social interactions, and lead to the learning of new contingencies between what the mirror shows and movement of the body (Gallup, 1968).

All of that would soon change. By 1970, Gallup was able to report the consequences of placing a mirror with each of four chimpanzees. The initial responses were as if directed to another animal. However, around the third day, self-directed behaviors came to the fore. Subsequently, and most importantly, after the chimps were anesthetized and a red dye was applied to their eyebrow ridge and ear, they almost immediately began touching the affected areas on reexposure to the mirror. The dye could be seen by them *only* in the mirror, so clearly they used it to perceive and explore the markings. That in turn implied that they understood the image was of themselves. In contrast, four macaques showed no self-directed behaviors with similar treatment, even after prolonged exposure to the mirror. Based on this difference between species, Gallup abandoned the conditioning explanation, arguing instead that mirror self-recognition requires advanced intellect. Indeed, he indicated, it implies "a concept of self" (Gallup, 1970).

In 1977, consciousness entered the argument. Gallup noted that the research of others found no evidence of self-recognition in spider monkeys, capuchins, baboons, and gibbons, among other primate species (Gallup, 1977). This led him to conclude:

> The unique feature of mirror-image stimulation is that the identity of the observer and his reflection in a mirror are necessarily one and the same. The capacity to correctly infer the identity of the reflection must, therefore, presuppose an already existent identity on the part of the organism making this inference. Without an identity of your own it would be impossible to recognize yourself. And therein may lie the basic difference between monkeys and great apes. The monkey's inability to recognize himself may be due to the absence of a sufficiently well-integrated self-concept.

In other words, chimpanzees, who unlike monkeys can self-recognize, have a self-concept and are thus self-aware. In turn, self-awareness was taken as implying consciousness (Gallup, 1977).

Individual differences in self-recognition

Subsequently, Gallup and his colleagues reported on a long-term study of chimpanzee self-recognition that revealed individual differences. A retesting of 12 animals first examined eight years previously showed that, while most individuals who could self-recognize earlier continued to self-recognize (N = 6), and most that could not self-recognize earlier continued not to (N = 2), others showed an unmistakable decline in self-recognition with age (N = 4 individuals age 18–24 years). Thus not all of the chimpanzees had ever been able to self-recognize, and there was an age-related decline in the ability among those that could (de Veer, Gallup, Theall, van den Bos, & Povinelli, 2003).

The aging effect confirmed the result of an earlier study, using cross-sectional design, finding that while 75% of adolescent and young adult chimpanzees could self-recognize, only 26% of older ones could (Povinelli, Rulf, Landau, & Bierschwale, 1993). In the words of the authors, there appeared to be "a precipitous decline" beginning at 16–20 years of age. On the other end of the lifespan, there were indications that mirror self-recognition may not begin in chimpanzees until 4–5 years of age (Custance & Bard, 1994; Povinelli et al., 1993), in contrast to beginning at 1½ to 2 years of age in most human children (Schulman & Kaplowitz, 1977). Another important finding of the study by Povinelli et al. was that the "mark test" of self-recognition sometimes gave results that were incongruent with observations of other self-directed behaviors in a mirror. Thus some animals failed to touch the red dye yet manipulated their genitals or lips using the mirror (Povinelli et al., 1993).

A number of these observations suggest that chimpanzees are on the evolutionary cusp of self-recognition – and thus potentially of consciousness. Many, but by no means all, pass the mark test. Some who fail the mark test nevertheless can use mirrors for self-inspection and self-manipulation. Mirror self-recognition follows a developmental time-course longer than that of human children. It is also vulnerable to the effects of aging. In each case, these conclusions suggest that chimpanzees are not far over an evolutionary divide between species that can or cannot self-recognize.

Self-recognizing species

So what species can self-recognize? Several orangutans have passed the mark test (Swartz, Sarauw, & Evans, 1999). Gorillas were the subject of early controversy that has been resolved in favor of self-recognition, including passage of the mark test (Posada & Colell, 2007; Swartz et al., 1999). Finally, common chimpanzees have been the most extensively studied species in this regard, with 42 of 97 animals (43%) having passed the mark test as of 1999 (Swartz et al., 1999). It is unclear whether there is any significant variation across great ape species in ability to pass the test, because sample sizes have been small, and there has been insufficient systematic control over such variables as the age of the subject. Nevertheless, all great ape species appear capable of self-recognition.

Among lesser apes, gibbons use mirrors to examine parts of their bodies, but they fail the mark test (Hyatt, 1998; Ujhelyi, Merker, Buk, & Geissmann, 2000). One human-reared siamang, however, passed it (Ujhelyi et al., 2000).

On the other hand, Old World monkeys such as macaques and baboons generally fail the test (Gallup, 1970, 1977). Gallup reported leaving mirrors with macaques for periods up to 17 years without ever eliciting self-recognition, although the monkeys could use the mirrors to orient to a person entering their room (Gallup, 1998). Attempts to shape the behavior of macaques to produce self-touching of mirrored marks have had greater success (Chang, Fang, Zhang, Poo, & Gong, 2015; Thompson & Boatright-Horowitz, 1994). However, it is unclear whether success following conditioning has the same meaning as spontaneous self-touching (Anderson & Gallup, 2015; Gallup, 1982). Thus, while Chang et al. (2015) were able to rule out the possibility that conditioned self-touching of a dye mark was stimulated by a mark on *any* monkey's face, it seems possible that monkeys learned to self-touch in response to a *particular* face, i.e., their own, without really recognizing that the face represented them.

On the other hand, Boccia (1994) reported that one of 14 pigtail macaques did pass the mark test even without training, first wiping at the mark and then inspecting the hand. Because this was such a rare and transient response, she suggested that the behavior be considered "incipient recognition or protorecognition rather than as full self-recognition" (Boccia, 1994, p. 358).

Another indication that some self-recognition may be present in macaques comes from a study in which rhesus monkeys were fitted with electrode implants for other studies. The implants used highly conspicuous blue acrylic blocks, and the monkeys used mirrors to inspect them. The monkeys then also used the mirrors to inspect body parts such as the genitals (Rajala, Reininger, Lancaster, & Populin, 2010). However, in interpreting the results of this study, it should be kept in mind that Gallup's original aim was to use marks that could not be felt or smelled (Gallup, 1970). This aim was not met in the case of the implants, which certainly afforded tactile if not olfactory stimulation.

With respect to New World species, Gallup indicated that spider monkeys and squirrel monkeys fail the test. Capuchins were also reported to consistently fail the mark test (Gallup, 1977, 1982; Roma et al., 2007). Yet subsequent research has found that they treat mirror images differently from how they view strangers of their species. While found in both sexes, the effect is clearest in females, who show greater eye contact and less anxiety in the mirror condition (de Waal, Dindo, Freeman, & Hall, 2005). It seems possible, therefore, that while New World species are not generally capable of self-recognition, some notice that something is "different" about mirror images.

The results appear to indicate that true mirror self-recognition, performed on a strictly visual basis, exists among primates only in apes, and possibly only in *great* apes. Lesser apes and macaques may show partial signs of self-recognition that depend greatly on the individual or on conspicuous, multimodal stimuli. It may be that providing a tactile stimulus such as an implant can cause a macaque to recognize a relationship between a mirror image and its own body, without triggering true self-recognition.

Several other unrelated species have also been found to pass the mark test. A possible encephalization effect is suggested by reports that dolphins and elephants do so. Reiss and colleagues have found that bottlenose dolphins engage in self-directed behaviors while viewing their marked bodies in mirrors. For example, they stretch the neck upward when marked under the chin, or orient a marked side toward the mirror (Morrison & Reiss, 2018; Reiss & Marino, 2001). Of course, this may not be the quite the same thing as self-touching, for as we have seen, it is possible for an animal to use a mirror to inspect body parts while failing the mark test.

For their part, Asian elephants were initially reported to fail the mark test (Povinelli, 1989). However, in that study the mirror was small and out of reach. Plotnik, de Waal, and Reiss (2006) tested three Asian elephants using a larger, accessible mirror and found that one passed. The others failed but nevertheless used the mirror for self-inspection.

Finally, the mark test has seemingly been passed by some corvids, the bird family that includes crows, ravens, jays, magpies, and jackdaws, among others. Prior, Schwarz, and Güntürkün (2008) found that some individual European magpies attempt to use their beaks, and succeed in using their feet, to remove yellow marks visible only in a mirror. However, the result was called into question by Soler, Pérez-Contreras, and Peralta-Sánchez (2014), who found that while jackdaws similarly removed marks visible in a mirror, they did the same in front of nonreflective cardboard. Apparently, the birds felt the presence of the stickers used to mark their feathers, the same method that had been used in the magpie experiment. To be sure, certain outcomes of the Prior et al. experiment are difficult to explain in that way. For example, the successful individuals mostly withheld action toward the marks unless a mirror was present and the marks were colorful. However, the ideal experiment using a dye with no tactile properties has not been performed to date. Most recently, another corvid, the Clark's nutcracker, has been reported to pass the mark test, though, oddly, more clearly did so when a blurry rather than a regular mirror was used. Again, though, stickers were used (Clary & Kelly, 2016).

The role of encephalization

The diversity of species showing self-recognition – the great apes, Asian elephants, and possibly siamangs, dolphins, and corvids – poses a challenge for phylogenetic interpretations. However, there is a sufficient correlation to encephalization to raise the question of whether self-recognition emerges out of raw brain power. As we have seen, great apes including humans have the highest encephalization quotients of all primates. Siamangs are also rather high, though no more so than non-self-recognizing gibbons, baboons, and spider monkeys (see Table 11.1).

Extending an encephalization analysis to mammals outside the primate order is problematic, because as we have seen (Chapter 5), the power function exponent that best relates brain size to body size in primates is not the same exponent that best fits mammals in general. Nevertheless, if we replace the 0.28 primate exponent with a 0.67 mammalian exponent, the results are instructive as long as we recognize that it produces disordered results within primates. What the results show is that Asian elephants have a relatively high encephalization quotient, extending into the range of apes. Bottlenose dolphins have a quotient even higher than that (Table 16.1).

But what about corvids? Attempting to include members of a different order along with mammals in an encephalization analysis is truly dicey, especially when there is evidence of greater neural density in bird brains than in the brains of mammals generally and of primates specifically (Olkowicz et al., 2016). However, it is possible to place magpies relative to other birds. Based on the tabled brain volumes and body weights of nearly 1500 bird species (Iwaniuk & Nelson, 2003, appendix A1), a best-fitting exponent of 0.57 may be calculated. In turn, applying that in an EQ equation places magpies at the 92nd percentile. That is certainly a high value, although it is not truly exceptional in that 8% of the tabled bird species have higher EQs. However, it is possible members of some of those species would show self-recognition if tested. Of the tabled species with higher EQs than magpies, about half are of the family *Psittacidae* (parrots), about an eighth are of the family *Cacatuidae* (cockatoos) and another eighth are fellow corvids (family *Corvidae*). Many of the higher-ranked species have a reputation for intelligence, yet it appears that none have been subjected to the mark test.

These results seem generally consistent with an encephalization effect on self-recognition. However, the relationship is by no means perfect. At present it is not possible to assess whether it makes more sense to view self-recognition as an emergent property of increased intelligence, or as the product of the convergent evolution of more specific brain mechanisms supporting it.

Nevertheless, the observations that mirror self-recognition is most prevalent among the great apes but may have precursors in mirror-based self-inspection in gibbons

Table 16.1 Encephalization values for mammals including elephants, based on a 0.67 exponent.

Species	EQ
pig	0.27
sheep	0.54
rat	0.84
European cat	1.14
fox squirrel	1.43
ring-tailed lemur	1.45
gorilla	1.54
African elephant	1.67
Asian elephant	2.14
chimpanzee	2.32
bottlenose dolphin	3.93
human	7.51

Notes: EQ = brain mass / (0.12 × body mass$^{0.67}$), with masses measured in grams. All values except bottlenose dolphins from Shoshani, Kupsky, and Marchant (2006). Dolphin EQ calculated from brain and body sizes provided by Lefebvre, Marino, Sol, Lemieux-Lefebvre, and Arshad (2006). These values cannot be directly compared to EQ values cited elsewhere in the book, because of the use of a different exponent as well as a multiplier in the divisor.

and macaques and in perceiving "something different" about mirror images in capuchins, strongly suggest a long-evolving capacity among primates. If self-recognition reflects consciousness, then consciousness may have evolved over many millions of years as well.

Theory of mind

As originally formulated, the term theory of mind referred to a proposed ability of chimpanzees to impute the existence of mental states in themselves and others (Premack & Woodruff, 1978). Though conceptually similar to metacognition, which involves assessing the status of task-related mental processes, theory of mind is considered a separate ability because it involves assessing states rather than processes, e.g., desires, thoughts, and beliefs (Papaleontiou-Louca, 2008).

Gallup, Anderson, and Shillito (2002) asserted that having a theory of mind depends on self-awareness, for without it, there is no basis for imagining the mental states of others. This position predicts that animals that have no self-awareness as assessed by the mark test should also have no theory of mind. However, the relationship may not be so clear-cut. For example, a basic theory of mind might be inferred from the ability to follow the gaze of others, which indicates knowledge that others may be viewing something important. Yet gaze following is an ability that extends across many different species, including ones in which there is no evidence of self-awareness. For example, members of some lemur species, representatives of the earliest-branching primate group, the strepsirrhines, follow one another's gaze toward humans, food sources, and other lemurs (Ruiz, Gómez, Roeder, & Byrne, 2009; Sandel, MacLean, & Hare, 2011; Shepherd & Platt, 2008). Gaze following has also been reported in ravens (Bugnyar, Stöwe, & Heinrich, 2004), dolphins (Tschudin, Call, Dunbar, Harris, & van der Elst, 2001), dogs (Call, Bräuer, Kaminski, & Tomasello, 2003), and goats (Kaminski, Riedel, Call, & Tomasello, 2005). It should not be surprising then, that among primates, gaze following occurs not just among lemurs but also among mangabeys and macaques (Drayton & Santos, 2017; Tomasello, Call, & Hare, 1998) and all great apes (Bräuer, Call, & Tomasello, 2005; Kano & Call, 2014).

Skepticism that gaze following denotes a theory of mind has been expressed by Povinelli and Bering (2003). In their view, the ability is one among many that are associated with highly social species – deception, holding grudges, and organized hunting being others – that emerge far in advance of the ability to interpret mental states. Simpler attentional, motivational, perceptual, and other abilities were regarded by Povinelli and Bering as underlying gaze following in animals.

This could be a valid point if gaze following was as simple as merely shifting the eyes in response to eye shifts in others. Even simple stimulus-response association could account for the behavior, since shifting is likely to lead to more positive outcomes than not shifting. However, Tomasello and colleagues found that chimpanzees will effortfully approach and look around barriers in order to see what human observers are gazing at (Tomasello, Hare, & Agnetta, 1999), an observation that seems resistant to a simple stimulus-response association explanation.

That chimpanzees understand something about what others see comes from research on food competition. When a subordinate has the opportunity to view food hidden from a dominant individual, the subordinate selectively retrieves that food rather than food both can see (Bräuer, Call, & Tomasello, 2007; Tomasello, Call, & Hare, 2003). The implication is that the subordinate understands that the competitor cannot see the food, and strategically targets it for retrieval.

Referential pointing

Also relevant are observations of referential pointing in apes. Frans de Waal stated, "there is no point to pointing unless one understands that the other has not seen what you have seen" (de Waal, 2003, p. 23). Anecdotally, the majority of caged chimpanzees will point to food visible to them outside the cage that is seemingly unnoticed by human caretakers, clearly in an attempt to obtain the food. This behavior is untrained. There are also anecdotes from the wild of chimpanzees using pointing to draw attention to other chimpanzees (de Waal, 2003).

An experimental approach was taken by Cartmill and Byrne (2007), using orangutans. In their study a human caretaker withheld half or all of a treat and then remained unresponsive to food entreaties. When half the treat was withheld, the orangutans often repeated behaviors – including pointing – that had gotten them half the treat. However, when all of it was withheld, they tended to switch to other behaviors.

These results may indicate that the animals understood the difference between incomplete understanding and a total lack of understanding on the part of the caretaker, prompting a behavioral adjustment. If so, they are consistent with the existence of a theory of mind. However, an explanation that doesn't involve theory of mind seems equally plausible: When something partially worked, it was repeated, but if it didn't, it wasn't. Perhaps a similar explanation can account for the ability of certain domesticated species like dogs and horses to use human cues to find food (Call et al., 2003; Proops, Rayner, Taylor, & McComb, 2013).

In any case, the Cartmill and Byrne study is a good illustration of the surprising difficulty inherent in distinguishing from behavior alone, between an internal cognitive state (theory of mind), and an internal precognitive state (e.g., stimulus-response association). This is much the same problem, of course, that Povinelli and Bering (2003) highlighted, and that was pointed out at the beginning of the chapter as a thorny issue when inferring conscious states more generally.

Helping and deception

Helping behaviors and deception have also been cited as suggesting a theory of mind. Elephants are often observed helping their disabled fellows, for example by attempts to lift a prone body or to pull a calf from a hole (B. L. Hart, L. A. Hart, & Pinter-Wollman, 2008). Dolphins and whales sometimes respond to distress in conspecifics by maintaining proximity or by supporting the injured animal at the surface. With regard to deception, species as diverse as hyenas and dolphins have been observed hiding objects from conspecifics (Kuczaj, Tranel, Trone, & Hill, 2001).

Whether one accepts any of these behaviors as indicating a complex state like theory of mind may depend on whether the behavior is itself sufficiently complex to render a less complex state implausible. Simple gaze following, repeating versus switching behaviors depending on the level of success, pointing, cue use to find food, helping behaviors, and deception may not rise to the standard required. In each case, it is not hard to imagine instinctual or associational learning mechanisms that could produce the behavior. Many of us may be more convinced when a chimpanzee looks around barriers to see what a human is gazing at, or selectively targets foods that a dominant rival cannot see. In these cases, the complexity of the behavior seems to imply a more complex state we can label a theory of mind.

However, if our main requirement is that behaviors be complex before we entertain them as evidence, we must be prepared to accept the possibility that nonprimate species can also have a theory of mind. Scrub jays, for example, show surprisingly complex behaviors when food-caching. When observed by other jays, individuals appear to guard against pilfering by caching food in locations (a) farther rather than closer to the observer, (b) hidden from the observer by a barrier rather than in plain view, and (c) less illuminated rather than more illuminated. These distinctions are not made when caching occurs in private. Furthermore, when caching has been observed, those caches are retrieved first when a private opportunity arises, and they are often moved to new locations (Emery, 2006). These behaviors seem sufficiently complex to indicate the existence of theory of mind. But if at least some birds have it, then the viewpoint that it is a primate characteristic (Brüne & Brüne-Cohrs, 2006) is obviously problematic.

Once again, we seem to have little more than the encephalization hypothesis to fall back upon in judging which species are likely to possess a consciousness-related construct. If we accept that only complex behaviors can constitute acceptable evidence of theory of mind, we are left with relatively complex – i.e., big-brained – animals that have it. Chimpanzees do, given that they look around barriers to see what others are looking at, and they selectively target food that more dominant individuals cannot see. Scrub jays do, too, given the increase in complexity of their food-hiding behaviors when being observed. Apes and corvids, as we have seen, are both relatively large-brained. The situation is much the same as exists in mirror self-recognition, except that fewer species have been investigated with respect to complex behaviors that imply theory of mind.

Conclusion

There are many conceptions of consciousness, ranging from an identification with attention, to considering it a byproduct of large-scale systems like the totality of cognitive processes or even of reality itself. The comparative study of consciousness introduces further uncertainties because external behaviors are used to infer the functions of consciousness, and disagreements often arise over the meaning of such behaviors.

Sensory awareness, while a seemingly simplistic function of consciousness, in fact is quite informative when impaired, as in blindsight and deaf hearing. Patients who deny seeing anything but who can localize stimuli, and others who deny hearing anything but who orient to loud sounds, have in common a destruction of primary sensory cortex accompanied by a lack of sensory awareness. Experiments using TMS in humans, and disconnection of visual cortex from higher-order cortex in monkeys, suggest that sensory awareness is the result of reentrant connections from higher-order cortex back to primary cortex. These connections appear to be characteristic of mammals in general, and thus have a 195-million-year history. While it is unknown whether they supported sensory awareness that long ago, macaques experience blindsight, putting a lower limit of about 32 million years on the origin of sensory awareness.

In contrast, the idea that attention connotes consciousness is more problematic, in part because it is not a unitary concept. Attention as selection, as recognition of meaning, and as alertness are all fair characterizations, but such a subdivision greatly complicates the task of tracing the evolution of consciousness. One of these, selection, seems especially problematic in that it may have its origins in the first appearance of nervous systems.

Metacognition, the awareness of one's own cognitive states and processes, includes such phenomena as "feelings of knowing" and the ability to judge the quality of one's retrieved answers. Because these feelings and judgments are vague and error-prone, they can be viewed as a kind of "fringe consciousness" constructed from unconscious processes. Human research suggests that some forms of metacognition begin developing between 4 and 5 years of age. All great apes as well as an Old World monkey (the rhesus macaque) show evidence of metacognition, but a New World species (the capuchin monkey) is "on the cusp", with studies finding either no proficiency, reduced proficiency, or proficiency only in some individuals.

Self-recognition seems a fairly straightforward function of consciousness, in the sense that it is easily demonstrated through self-directed responses using mirrors. It has been found in apes and Asian elephants, and perhaps in dolphins and corvids. Apes, which have been most extensively tested, show individual differences in that about a quarter of adolescents and young adults, and three quarters of older adults, fail to self-recognize. This effect appears to be at least partially developmental in that some animals that self-recognize earlier in life fail to do so later. Many individuals never succeed in recognizing themselves. These observations suggest that apes may be barely over an evolutionary divide that excludes some individuals as well as Old World and New World monkeys. Encephalization may well play a role, in that all self-recognizing species have relatively high encephalization quotients.

Theory of mind is often taken as a manifestation of self-awareness, on the grounds that imputing a state of knowledge in others presupposes understanding similar states in oneself. Some simple behaviors that might be taken as implying a theory of mind, such as gaze following, helping behaviors, and deception, are widespread among animal species but are also potentially interpretable in stimulus-response or instinctual terms. More complex behaviors, such as approaching and looking around a barrier to see what a human is staring at, or selectively targeting foods that a dominant rival cannot see, seem more convincing of a theory of mind. Certainly chimpanzees perform these, but scrub jays also engage in surprisingly complex alterations of behavior when food-caching, depending on whether or not they are being observed by other jays. This again suggests a possible link to encephalization.

Thus in the end, evolutionary analyses appear to confirm the multi-faceted nature of consciousness. Selective attention may be as ancient as nervous systems, and sensory awareness possibly traces to the first mammals. But in any case, sensory awareness certainly originated earlier than self-recognition.

17 A summary in nine firsts

The roots of cognition date nearly from the origin of life four billion years ago. When the first cell responded to interior pressure and expelled waste products, a primitive type of sensation was born. Emerging about the same time was the ability to pass useful substances into the cell while excluding less useful ones. Soon the addition of flagella or cilia allowed movement away from less productive to more productive areas. Together, internal mechanical sensing, external chemical sensing, and the ability to move constituted the first building blocks of cognition.

Over 600 million years ago, hair-based mechanoreceptors were incorporated into the stinging cells of jellyfish and other cnidarians. Neurons may have evolved from them, with neural nets emerging by 605 million years ago. A little later, early chordates had sensing hair cells and a nervous system. The nervous system in turn enabled the selection of one of multiple stimuli, an early form of selective attention. Early nervous systems, however, depended on primary sensory cells with sensors and an axon on opposite ends, limiting their architectural complexity.

Metazoan vision likely appeared first in cnidarians. Whether vertebrate eyes descended directly from them is unclear, but all eyes use variations of vitamin A–based photopigment. An additional photopigment class, the flavins, originated even before cnidarians and was later used to regulate circadian rhythms.

Paired, pigmented, though likely lensless cup eyes probably emerged with the first bilateria about 580 million years ago. Early chordates likely had only one eye, used for sensing direction. Optical reversal in the eye resulted in a nervous system crossover, with the left side of the world represented on the right side of the nervous system, and vice versa. It seems likely that this scheme was simply duplicated when eyes again paired in vertebrates.

One "backward" aspect of the visual system, namely the slowing rather than initiation of neurotransmitter flow in response to light, was probably in place in chordates. Multiple cone types likely date to the vertebrate emergence.

The ability to move is thought to have been characteristic of early metazoa as early as 665 million years ago, specifically in swimming sponge larvae. It was accomplished with beating cilia, propelling free-swimming forms through water, or, a little later, in benthic species, over a mucus-lubricated trail. Ancestral flatworms both crept on the sea bottom and used muscles for full-body undulation when swimming, prior to 540 million years ago.

The ability to remember was probably present even in early single-cell organisms, in the form of habituation to repeated stimuli. Visual sensory memory, produced as an aftereffect of neural activity, likely originated with the first cnidarian eyes and in all probability was present in chordates.

These characteristics – receptors linking to nervous systems enabling sensory selectivity; the ability to move through water either in suspension or on the ocean bottom; and the beginnings of sensory memory – were the birthright of chordates. From those beginnings sprang numerous cognitive mechanisms and cognitive-influenced behaviors. These may be loosely organized under nine "firsts".

The first vertebrate (520 million years ago)

The first vertebrate was a jawless fish with a cartilage-like notochord extending to the head, which housed a brain. Close descendants had hard teeth formed of dentine. A nervous system was present that included the beginnings of the cerebral hemispheres. Sensory cells with sensory and synaptic ends allowed a more complicated neural architecture than previously possible.

Sensory capabilities included a vestibular sense based on one semicircular canal within an inner ear. A second canal would soon form in most vertebrate descendants. There were probably olfactory nerves, an olfactory bulb, and external nostrils. Bitter taste receptor genes had evolved, and taste buds would soon emerge. Vision probably used paired, optically reversing eyes as well as a retina containing both rods and multiple cone types.

Crossover existed in a portion of the motor system, and as that strengthened it brought each side of the world into visual and motoric alignment in the opposite cerebral hemisphere. Meanwhile, partial crossover coordinated movement between the body's sides. An evolving cerebellum in most descendants, initially composed of only a few cells, soon contributed to consistent mapping of the sensory and motoric worlds.

Conditioning was a form of memory, the product of a central nervous system organized to allow the arbitrary pairing of stimuli and responses. With it came discrimination learning, so that beneficial and noxious stimuli could be distinguished and responded to appropriately.

The first tetrapod (365 million years ago)

The first tetrapod was an amphibian resembling the closely related lungfish. In place of the lungfish's fleshy, lobed body fins were crude limbs and digits that propped the head up in shallow water, allowing air breathing.

Three semicircular canals in the inner ear provided a fine sense of balance. Also present was a basilar membrane embedded with basilar papillae, for transducing sound. However, the middle ear contained only one ossicle, limiting the high frequencies that could be heard. Olfactory genes allowed a range of odors to be detected, and would soon vastly expand in number.

Several changes occurred during the transition from direct tail-based propulsion in water to the use of limbs on land. "Rear-wheel drive" evolved from earlier "front-wheel drive", with the tail used to anchor muscles connected to the rear limbs. Muscles and bones enlarged to provide body support, and the limbs shifted laterally, allowing a sprawling gait with symmetric alternation of sides.

A hippocampus was present, having expanded from the few cells in bony fish ancestors. Although its functional significance at that time is unclear, one function of the present-day hippocampus is to support episodic memory.

The first eutherian (170 million years ago)

The first eutherian weighed less than an ounce, but had a relatively large brain for its size. Its middle ear ossicles had separated from the jaw bone, and it probably had fur, both characteristics identifying it as a mammal.

All three ossicles were in place, allowing the impedance of air to be matched to that of fluid in the inner ear. A fully coiled basilar membrane also existed. As a result, sound could be heard with considerable sensitivity, with an upper-extended frequency response and with improved sound localization.

Tactile sense developments included Meissner's corpuscles in the skin to detect texture, and the primary and secondary somatosensory cortex in the parietal cortex.

The olfactory bulbs had undergone multiple expansions. Bitter taste pseudogene formation had increased relative to earlier tetrapods, perhaps due to a decreased dietary reliance on plants.

Visually, two of four cone types had been lost as unnecessary metabolic expenses, in that early nocturnal mammals relied more on rod-based night vision than on cone-based day vision. In the brain, areas V1 and V2 had formed, respectively reflecting increased processing of visual features and contours, and of figure-ground segregation. The after-effects of neural activity in V1 provided a brain-based component of visual sensory memory.

Area VP/V3 had evolved for increased processing of visual form. Reentrant connections between forward areas of the brain and V1 may have supported sensory awareness.

Movement involved a largely but not fully erect body posture, with weight distributed more directly over the legs. An asymmetrical gait, with different timing cycles for the front and rear limbs, likely supplemented symmetrical ones, resulting in higher speeds. These and other movements no doubt benefitted from the recently evolved primary motor cortex.

The first primate (83 million years ago)

Although primates left no fossils until nearly 20 million years after their origin, the earliest known ones resembled large squirrels in size and body plan. They had mobile, grasping hands and feet, with relatively flat nails on some digits, and opposable thumbs and big toes. Weights had increased to 1 kg (2 pounds) and beyond.

A reduction of the olfactory bulbs would soon begin in most of the order, indicating less reliance on smell and an increasing dependence on vision. Early primates probably had large corneas for improved light-gathering, reflecting their nocturnal lifestyle. They had evolved, or would soon evolve, a tapetum lucidum to further increase light sensitivity. A fovea supporting high-resolution vision would soon develop, although it would be ill-defined and partially occluded by blood vessels. There was substantial binocular convergence, providing good depth perception for insect predation.

Movement involved an ambling, diagonal gait with extended strides, appropriate to a fine-branch environment. The limbs were compliant and highly mobile. The tail was large and heavy, and it was used to help maintain balance on branches. In the brain, supplementary, premotor, and parietal motor areas had evolved and become interconnected for greater movement control. There may also have been a weak population left-handedness in manual actions.

The equivalent of human area MT/V5 had evolved to support motion perception. Early primates likely also had the ability to follow gaze. Patches of cortex recognized classes of objects, including face patches that were configuration-based and orientation-specific. If episodic memory had not already evolved with tetrapods, it probably existed in early primates. By this time the hippocampus also possessed place cells and grid cells, which were used to represent the spatial environment.

The first ape (32 million years ago)

The first ape undoubtedly resembled *Aegyptopithecus*, the possible monkey-ape common ancestor. If so, it weighed about 6 kg (13 pounds) and had a brain 30 cubic centimeters in size, resulting in an EQ of a little under 3 when calculated using a primate-appropriate exponent. It had the heavy limbs of a slowly moving arboreal quadruped. It had a tail, but that would soon be lost by its descendants, possibly because it served no useful function in countering the movements of a heavier body.

The olfactory bulbs had further decreased in size, accompanied by increased olfactory gene deactivation, and the vomeronasal nerves had been lost. These circumstances point to a reduced reliance on odors.

The visual system showed substantial change. A third cone type had re-evolved for better color vision. The tapetum lucidum had been lost, and the cornea and lens had reduced in size, so that the eye had less light-gathering power but produced a sharper image. The unoccluded fovea was well-defined, and there was an enlarged optic nerve. All of this is consistent with a visually dependent, diurnal, probably frugivorous lifestyle.

In the brain, the visual areas had enlarged, and depth was perceived with considerable accuracy aided by increased binocular convergence. The ventral parietal cortex was in place that was sensitive to visual flow.

Visual selectivity had increased due to a narrowing of the field of best vision. Accompanying it, sound could be localized with high accuracy. Improved cross-modal integration of sensory information was served by the brain's intraparietal sulcus, which also processed egocentric location. Concomitantly, it was involved in controlling eye movements and in deploying attention.

The inferior parietal lobe was involved in visual enumeration, and a subitizing process based on analog quantities was in place to quickly enumerate small numbers of stimuli. Mental maps were used in navigation, supported by the parietal lobe. The left inferior parietal lobe had especially enlarged and was probably used both in spatial attention and in reaching and grasping.

Ventral visual pathways had also further developed. The first ape may have had an area V4 capable of perceiving curvatures in depth and color, and an IT/LOC area that recognized both two- and three-dimensional patterns with movement. Perceptual invariance was in place. However, there was probably little if any global-over-local bias. Recognition did not exist of mirror images of the self in reflecting pools of water, even though something "different" may have been perceived about them.

Motor planning and movement complexity had increased, aided by premotor cortex that controlled both proximal and distal muscles, and by an enlarged cerebellum. Population right-handedness had likely emerged in unskilled reaching, gesturing, and feeding behaviors. This was possibly accompanied by a leftward asymmetry in the hand area of the primary motor cortex.

Several brain areas had evolved that would figure into the acquisition of language. Broca's area existed in the form of homologs to areas 44 and 45, and area 6 regulated laryngeal vocalization. A larger left- than right-sided planum temporale was in place. There was also a right ear advantage over the left in recognizing vocalizations, although that may have owed less to the planum temporale that it did to the anterior superior temporal gyrus.

A short-term memory lasting a number of seconds existed, in part supported by areas 9 and 46 in the brain. A "local" type of semantic memory was also present, involving the perirhinal cortex. It represented particular classes of objects, but it was not yet modality-free.

Although the first ape showed no inkling of self-recognition, it nevertheless exhibited metacognition by knowing to withhold a response when it judged its knowledge to be insufficient.

The first great ape (21 million years ago)

The first great ape likely resembled *Proconsul* and *Morotopithecus*, with a greatly increased body weight somewhere between 15 and 45 kg (30–100 pounds). Its brain was probably 170–370 cubic centimeters in size, resulting in an EQ in the range of 11 to 18. It was arboreal and frugivorous, although whether it primarily locomoted above- or below-branch is unclear. Either way, it could probably adopt eccentric grasping postures, allowing suspension from multiple limb supports and thus access to fruit at the periphery of trees. Sensory capabilities had not changed much since the first ape.

Population left-handedness in tube-type extraction tasks had likely evolved but, in a few million years, would switch weakly to the right in most of the first great ape's descendants. This upcoming change, largely involving the left hemisphere of the brain, probably co-occurred with the ability to use a simple set of communication gestures similar to the naturally occurring gestures of living great apes. A rudimentary grammar may have evolved at the same time, perhaps initially limited to ordering a very small number of relatively meaningless phonological elements, but coming to be applied in an equally limited way to gestures.

Simple tools were likely created from vegetation, using an ability to form secondary representations of not-yet-existing, hypothetical objects. Planning existed in the form of simple scripts, enabled by the brain's enlarged area 10. Causal relationships were beginning to be understood, but only in a very limited way. The beginnings of a self-concept may also have existed in the form of self-recognition of mirror images by some individuals.

The first hominin (7.5 million years ago)

The first hominin weighed perhaps 45 kg (100 pounds) and had a roughly 370-cubic-centimeter brain, with an EQ of about 18. It probably locomoted using a combination of above- and below-branch arboreality and (most significantly) bipedalism. Bipedal walking was likely hampered to some extent by a modestly splayed hallux, and was accompanied by side-to-side rocking. The trunk, however, was vertical while walking, or soon would be in the line leading to modern humans.

The arms, freed by bipedalism, could carry limited loads. Those included simple tools made from vegetation, such as sticks for food extraction and hunting weapons like spears or clubs. Natural stones could also be carried, to be used as hammers and anvils. In fact, different groups of the first hominins likely had tool cultures that were transmitted within and between groups.

Objects were held in a strong, precision finger-to-thumb grip, possibly pad-to-pad, and to some extent, preferentially by the right hand. Tube-type tasks had become predominantly right-handed, although a complex insertion-turning component of handedness remained mostly left-handed. A second increase in area 10 size and connectivity may have occurred, improving planning.

Visually, a weak global-over-local bias existed that would subsequently strengthen. The first hominin could likely understand spatial representations to some extent, and the enumerating subitizing process had expanded to three to four items. Face patches in the brain showed increased orientation specificity, responding mostly to upright faces.

A rudimentary ability to understand symbolic hierarchies existed, such as that involved in wedge-anvil-nut-hammer constructions. This ability may have assisted the use of sequential gestures and vocalizations used in communication, partially controlled by an area F5/44 that was likely larger on the left side of the brain. Broca's cap may have started to appear in some cases and included area 44 and possibly part of area 45.

Finally, the proportion of individuals able to self-recognize had increased, suggesting increased self-awareness. Concomitantly, it was possible to have a rudimentary theory of the mind of others, as indicated by looking around barriers to see what others could see, or by strategically targeting food competitors could not see.

The first human (2.4 million years ago)

The first human, i.e., the first member of the genus *Homo*, weighed about 35 kg (80 pounds) and had a 610-cubic-centimeter brain, yielding an EQ somewhat above 30. It had a stiffer and less compliant bipedal gait than its ancestors, recovering more energy between steps. It had an arched foot as well as the beginning of endurance running adaptations such as shortened toes. The pelvis had become more cup-like, cradling the internal organs. It also provided attachments for expanded gluteus maximus muscles, enabling a stable, nonrocking stride. The fingers were straighter, accompanying reduced arboreal abilities.

Worked stone tools had existed for some time and would soon show increasing symmetry. All components of handedness showed or would soon show a strong population trend to the right. Toolmaking involved greater planning, aided by new inferior parietal gyri that allowed improvements in copying gestures and understanding physical causality. Butchery was practiced, and wood was either being chopped or soon would be. However, the use of fire lay a million years in the future.

Broca's cap was more consistently in evidence, although the size of Broca's area was not yet fixed and depended on brain size. Many other language-related changes had not yet occurred, so if language was produced, any vocal component had probably not advanced much beyond that possessed by the first hominin. Communication gestures may have done so, though, given the aforementioned inferior parietal lobe developments.

The first modern human (200,000 years ago)

The first modern human weighed about 55 kg (120 pounds) and had a brain around 1350 cubic centimeters, for an EQ of about 64. It had a stiff, noncompliant gait with a minimally splayed hallux, and a cylindrical rather than cone-shaped ribcage that aided endurance running. Most body hair had long been lost, and the first modern human's descendants would soon adopt clothing.

Tools were increasingly sophisticated, and included symmetrical handaxes as well as spears with hafted points for hunting. Tool use was aided by an increased overlap of motion and form processing in the brain.

There had been substantial olfactory pseudogene formation, continuing the trend toward increased reliance on vision. Reflecting that, the field of best vision had narrowed to 1.5°, and auditory localization abilities had followed suit to match.

Nevertheless, the occipital lobe had decreased in size relative to the rest of the brain, accompanied by an extreme expansion of the parietal lobe. There was a strong global-over-local bias in visual processing, and face patches had become widespread and more differentiated.

Numerous adaptations were in place to serve language. Area 4 in the frontal lobe had exerted predominant control of the larynx, and was strongly connected to the parietal cortex as well as to the brainstem for fast control of vocalizations. Broca's cap included both areas 44 and 47, and its size had long stabilized irrespective of brain size. The FOXP2 gene had mutated since the last common ancestor of chimpanzees and humans, affecting the language-related expression of Broca's area. There were massive arcuate fasciculus connections between Broca's area and the middle and inferior temporal gyri, supporting lexical-semantic processing. Accompanying this was probably a greatly expanded vocabulary. Wernicke's area was likely being used in language.

Cranial, throat, and chest anatomy had evolved to support language as well. The face had flattened and the jaw and tongue had shortened, allowing more distinctive speech sounds. The "inverted T" of the chin had evolved, reducing stress on the chin during speech. The vocal cavities had become relatively more vertical, supporting a deeper tongue root and increasing the intelligibility of vowel sounds. At some point the "bubble" on the hyoid bone had also been lost, likewise allowing more distinctive speech. Finally, the thoracic nerves had enlarged for breathing control during vocalization.

Memory had improved in at least two ways. Short-term memory duration had increased relative to that in monkeys, possibly due to participation of brain areas 10, 44, and 6. A modality-free, general semantic memory had also emerged, using lateral temporal cortex that was strongly connected to frontal language areas.

The continuing story

Change, of course, did not end with the first modern human. Indeed, except for limited tool cultures, virtually all cultural developments lay in the future. To name just a few, they included the development of all existing languages, the emergence and elaboration of decorative artifacts, the shift from hunting to agriculture, and the invention of philosophy, religion, and the sciences, as well as all of the technological artifacts that both enrich and clutter our lives. To take one modest example, it took virtually all of the remaining 200,000 years to develop papermaking and crayons, so that a 6-year-old girl could draw a picture of a monkey and banana using the secondary representation capabilities that had evolved by the lifetime of the first great ape.

How much of this explosion in culture was evolutionary? It is, of course, very hard to say. Nevertheless, it is difficult to believe that we passed from Acheulean handaxes to Swiss Army knives without at least some genetically based neurological change. Unfortunately, what we can derive from endocasts is insufficient to address the question. Nor are comparative methods available, because our hominin ancestors did not spin off any living descendants that are genetically isolated from us.

Yet it is important to keep in mind that whatever else we are, we are the product of environment. If natural selection means anything, it means that if our species survives but our environment changes, selection pressures should cause us to change as an inevitable genetic consequence. The environment has already changed radically with the shift to sedentary occupations and lifestyles, the increasing substitution of corporate agriculture for family farming, and the development of small communities and then larger cities that service every need through occupational specialization.

One physical result is that our bones have weakened. The bending strength of the femur and tibia declined as much as 20–30% between the Upper Paleolithic and the Iron Age/Roman periods, i.e., between roughly 22,000 and 2,000 years ago. Ruff et al. (2015) attributed it to a shift from a more mobile to a more sedentary lifestyle as agriculture began. They viewed this as more environmental than genetic, with sedentary lifestyles resulting in a load reduction that limited bone development during the lifespan. However, there is nothing to preclude a possible shift in gene frequencies accompanying the weakening, lending it a partially genetic character. After all, one would expect that any genetic coding for stronger bones would be selected against if they were no longer required (cf. Chirchir et al., 2015, who acknowledged a similar point). Further weakening down to modern times may have continued as a trend, although it has been too small to reach the threshold of statistical significance (Ruff et al., 2015).

The problem is that identifying shifting traits alone does not separate environmental from genetic influences. It may be more useful to examine how traits are affected by changing gene frequencies. An example in northern Europe is the frequency of a gene, designated -13910*T, for *lactase persistence*, i.e., for an enzyme allowing adults to digest the sugar (lactose) present in milk. The emergence of the gene accompanied the shift from a hunter-gatherer lifestyle prior to 8400 years ago to a predominantly sedentary, farming lifestyle after 6000 years ago. Molecular evidence suggests that the gene first appeared roughly in that period, the same one in which milk began to be collected. Presently, only a few thousand years later, over 90% of northern Europeans carry the gene, indicating its frequency must have dramatically increased across time and space under strong selective pressure. Yet even today, 65% of the *world's* people have no lactase persistence. That is likely to continue to change as -13190*T and other persistence genes spread and are selected for (Gerbault et al., 2011).

Gene frequency studies also indicate strong recent selection for reduced pigmentation in western Eurasia. Wilde and colleagues extracted DNA from 63 individuals dating from 4000 to 6500 years ago, mostly from modern-day Ukraine, and compared it to modern Ukrainian as well as more general European populations (Wilde et al., 2014). Alleles associated with lighter skin, hair, and eyes were found to have increased in frequency, with high selection pressure. The effects have been dramatic, showing increases from 4% to 37%, 16% to 71%, and 43% to 97% in the percentage of individuals with three light-pigmentation alleles, between the ancient sample and modern Europeans (with slightly smaller increases compared to modern Ukrainians). Among the reasons proposed by the authors for the shift toward lighter coloration was that it was an adaptation to reduced sunlight at high latitudes. Light-colored skin increased the absorption of sunlight, maintaining vitamin D synthesis under low-light conditions.

Recently, Field and colleagues developed a powerful new technique for detecting relatively recent mutations. It exploits the fact that locations on chromosomes near a recent mutation are relatively unlikely to have sustained recent mutations themselves; conversely, locations near an older mutation are much more likely to have sustained them due to the passage of time. Therefore, as a first approximation, determining the distance between mutations on either side of a third one indexes the amount of time since the third mutation, with longer distances indicating that it occurred correspondingly recently. Finally, the frequency of the third mutation in the population is determined, and that along with the time determination allows an estimate of selection pressure (Field et al., 2016).

Using this strategy, Field et al. confirmed that lactase persistence and pigmentation have been under heavy positive selective pressure. Also under positive pressure are height, infant head circumference, decreased body mass index in males, female hip size, and later sexual maturation in females.

However, it is not clear whether all human populations are showing the same trends (Field et al., 2016). This raises an important point: The mere fact of genetic change does not necessarily mean that the species as a whole is changing. It remains to be seen, for example, whether lactase persistence becomes a universal human trait. Lighter pigmentation almost certainly will not, because people at more tropical latitudes derive benefits from heavier pigmentation, such as a limit on the harmful effects of solar radiation (Jablonski & Chaplin, 2017). In fact, this is clear in the data of Wilde et al. (2014), in which all three low-pigmentation alleles were found to have near-zero frequencies in Asian and African populations.

A better candidate for a future species-wide characteristic, and importantly, one with cognitive implications, might be a bigger head at birth. A larger infant head circumference was one of the characteristics identified by Field et al. (2016) as being under strong selective pressure. Presumably this is universally beneficial due to a larger brain. But to be fully realized, will universal Caesarian birth be required to escape pelvic size limits? That would require the easy availability of the surgical procedure on a worldwide scale, regardless of individual economic circumstances. If that problem is overcome, then over the long run, the brain of our species will be free to evolve greater cognitive complexity.

Our inability to envision whether that is even feasible points to the heart of the problem of predicting evolution. For our capacity to do so, even in general outline, is limited by the feebleness of our insight into the future. All we can be sure of is that as the environment changes, so must we.

Glossary

Acheulean: a hominid tool culture that began about 1.8 million years ago, marked by the rise of handaxes

acritarchs: tiny acid-resistant fossils showing evidence of a cell wall

acuity: the ability to resolve fine visual details

adductors: in the leg, muscles on the inside of the thigh that move it inward

agonist: a drug that temporarily binds to neural receptors and prevents their normal function

agrammatism: a clinical disorder marked by the disordering of syntax (or grammar)

Ahrensburgian: a human tool culture known for extensive organic remains, and for showing evidence of hunting using spearthrowers or bows and arrows

allele: a variation of a gene

allocentric frame of reference: a mental framework allowing judgment of location relative to the environment

allopatric speciation: the evolution of new species by physical separation

amino acids: the building blocks of proteins

amniotes: animals whose embryos develop within a membrane

anaerobic: living without oxygen

anapsids: tetrapods with no openings in the side of the skull behind the eyes

anterior: forward

Anthropoidea: anthropoids, i.e., monkeys, apes, and humans

arboreal: living in trees

archaea: a domain and kingdom of single-cell organism

artifactual: pertaining to an artifact, such as a stone tool

artiodactyls: a clade within mammals, consisting of mammals that are hooved and have an even number of toes

association: the arbitrary pairing of a stimulus with a response in memory

asymmetrical gait: a gait with a movement cycle taking a different length of time for the forelimbs and hindlimbs

atrophy: shrinkage

auditory sense: the sense of hearing

Aurignacian: a human tool culture marked by an increase in microblades and microliths

autosomal: pertaining to the autosomes

autosomes: all of the chromosomes other than X and Y

axon: a long thread-like projection of a neuron

backed blade: a stone blade blunted on one edge

basilar membrane: a membrane with hair cells, located inside the cochlea

basilar papillae: rounded projections in the inner ear, each containing a few hair cells

benthic: bottom-dwelling

bilateral: two-sided

bilateria: bilaterally symmetric animals

bilayers: films

binocular convergence: the state of both eyes imaging the same object at the same time

binocular cue: a depth cue requiring two eyes to see

binocular disparity: a binocular depth cue using the discrepancy between images from the two eyes

binocular field of view (BFOV): the opening angle of the joint field of view of the two eyes

biofilms: cooperative communities of bacteria

biomineralization: the incorporation of minerals into structures such as bones or teeth

bipedal: two-legged; walking on two legs

bipolar cells: a class of neurons located in front of the retina that connect the ganglion cells to the rods and cones

blindsight: a phenomenon in which an observer claims to be blind, yet shows residual visual abilities

brachiation: a form of locomotion characterized by swinging from branch to branch

Broca's cap: an outward bulge of cortex in Broca's area

Cacatuidae: the family of cockatoos

Cambrian explosion: the rapid expansion of metazoan phyla in a portion of the Cambrian period, roughly 520 or 530 to 540 million years ago

catalyze: accelerate, especially a chemical reaction

Catarrhini: "narrow-nosed" monkeys; also known as Old World monkeys

caudal: rear

caudofemoralis longus: in most tailed animals, a muscle located largely in the tail and connected to bones of the rear limb

Cercopithecidae: a family of catarrhines consisting of two subfamilies, the Cercopithecinae and the Colobinae

Cercopithecinae: a subfamily of Cercopithecidae consisting of macaques, baboons, and vervet monkeys

Chordata, or chordates: relatively complex animals with a hollow neural tube behind a notochord

cilia: small hairlike cellular outgrowths

circadian rhythm: a biological rhythm governing our sensitivity to the 24-hour day

clade: a branch on the tree of life defined by a novel characteristic

cladistics: the grouping of species by novel characteristics not present in other, less related species

cnidarians: a phylum of animals composed of jellyfish, hydras, sea anemones, etc.

coalescence time: the latest date at which two species had completely common ancestry

cochlea: a spiral-shaped cavity of the inner ear

Colobinae: a diverse subfamily of Cercopithecidae that includes colobus and proboscis monkeys, doucs, langurs, etc.

complex cells: in the visual system, neurons that respond to orientation over a large area, and often to movement in a particular direction

composite tools: tools made of more than one material

concordant: using the same side, as in two right-handed siblings

cones: a class of receptors in the retina, chiefly responsible for color vision

conspecifics: other members of the same species

contralateral: opposite-sided

convergent evolution: the evolution of similar form or behavior in relatively unrelated species, due to similar environmental pressures

core: in the making of stone tools, the stone from which the tool is made

coronal organ: a vibration-sensing organ in some chordates, located in the oral region

cortical magnification factor: a mathematical relationship between the acuteness of an area of vision and the amount of cortex that processes it

corticospinal tract: a motor tract originating in the precentral gyrus of the brain that crosses over to the opposite side of the body; also known as the pyramidal or dorsolateral tract

Corvidae: the family of corvids, which includes crows, ravens, and jays, among others

cranial nerves: sensory and motor nerves completely located in the head

craniates: animals with skulls

cranium: bony braincase

cryptochromes: a class of photopigments, more recently called flavins

ctenophores: comb jellies

cubozoans: box jellyfish

cultural variation: in ape communities, the extent to which a behavior differs between communities

cutaneous nerves: nerves serving somatosensation, which enter the spinal cord

deaf hearing: a phenomenon in which the patient denies hearing anything but nevertheless orients to loud sounds

dendritic arbor: the treelike branching of dendrites

dental comb: outward-projecting lower canines and incisors, used as a comb while grooming

dentine: bonelike tissue without cells

dentition: the form and patterning of teeth

derived character: a characteristic possessed by one subgroup but not all subgroups within a clade

descent with modification: the intergenerational transfer of genetic instructions that differ between parents and offspring

diapsids: tetrapods with two openings in the skull behind the eye

dichromacy: the state of having two types of cones

diffusion: in ape communities, the social transmission of behaviors across communities

dinoflagellates: ocean-dwelling unicellular organisms with two flagella

distal: farther from the core

diurnal: active in daytime

dizygotic twins: nonidentical (fraternal) twins

dorsal: top

dorsal visual stream: the visual pathway from brain area V1 to the parietal lobe

dorsolateral tract: a motor tract originating in the precentral gyrus of the brain, that crosses over to the opposite side of the body; also known as the corticospinal or pyramidal tract

echo: the auditory form of sensory memory

ectoderm: in vertebrate prenatal development, the cells giving rise to external features like the skin, nails, and hair, as well as the brain and nervous system

egocentric frame of reference: a mental framework allowing judgment of location relative to the body

elaborative rehearsal: processing the meaning of items to be remembered, thus associating them to the existing knowledge structure

encephalization quotient (EQ):   a calculated quantity expressing brain size in a form that takes the brain-body size relationship into account

endocast: a latex rubber molding of the inside of a skull; or more recently, a virtual rendition of the inside of a skull using imaging data

endothermy: a state characterized by the internal generation of heat to maintain body temperature, i.e., "warm-bloodedness"

episodic memory: memory for episodes or particular events in the past

eukaryotes: "true kernels", i.e., cells having nuclei

Eutheria: "true beasts", i.e., placental mammals

exome: the reduced portion of the genome that codes for proteins

extralemniscal system: somatosensory neural pathways that run upward to the secondary somatosensory cortex

extrapyramidal tract: a collective term for motor pathways with only partial crossover between the brain and the body

factor analysis: a statistical technique that groups measures according to their interrelationships

fatty acids: a class of lipids

femur: the thigh bone that forms a ball joint with a socket in the pelvis

figure-ground segregation: in vision, the process of extracting an object from its background

flagella: whip-like cell structures

flakes: in the making of stone tools, the stone fragments struck from the core

flavins: a class of photopigments

flexed: bent

focal length: the distance required for an image to come to a focus

folivores: leaf eaters

foramen magnum: the opening on the bottom of the skull allowing entry of the spinal column

fovea: the central area of maximally acute vision located within the macula

frontal eye field: a location in the frontal lobe of the brain involved in making voluntary saccades

frontal pole: the foremost part of the frontal lobe

frugivorous: eating fruit

functional magnetic resonance imaging (fMRI): a method for imaging brain activity involving detection of electrical currents influenced by proton spin

gait: the cyclical means by which the limbs move the body

ganglia: functional groupings of nerve cells

ganglion cells: a class of neurons located in front of the retina, whose axons form the optic nerve

gating: the controlled passing of substances through a membrane

generator potential: a net positive electrical charge inside a neuron that has increased past a threshold value, opening sodium channels in the cell membrane

genetic recombination: an error during the reproduction of chromosomes resulting in the movement of genetic instructions from one chromosome to another

geniculostriate route: a visual pathway which proceeds from the eye to the lateral geniculate nucleus to the primary visual cortex

genome: the set of genetic instructions; all the chromosomes together

gill arches: thin support structures for the gills, made of cartilage or bone

global precedence effect: in its first sense, faster response to global (broad) than to local (narrow) aspects of a scene; in its second sense, the greater effect of conflicting information at the global level on the local level, than vice versa

gluteus maximus muscles: large muscles in the butt that allow us to walk stably

gnostic cell, or grandmother cell: in vision, a neuron that responds to a specific object, such as one's grandmother

gracile: slender

grammar, or syntax: the rule-based organization of symbols in orders that succeed in conveying meaning

grandmother cell, or gnostic cell: in vision, a neuron that responds to a specific object, such as one's grandmother

Gravettian (called Périgordian in France):   a human tool culture characterized by backed blades and end-scrapers

grid cells: neurons that respond to locations in an overall grid representation of the environment

gustatory sense, or gustation: the sense of taste

habituation: a decline in response to repeated stimulation

hair follicle receptors: nerve endings wrapped around the base of a hair, that respond to its bending

hallux: big toe

Hamburgian: a human tool culture known for extensive organic remains, and for showing evidence of hunting using spearthrowers or bows and arrows

hamstrings: muscles on the back of the thigh that move it backward

Haplorhini, or haplorhines: dry-nosed primates

heading: the direction of movement; bearing

height in the visual field: a monocular depth cue in which the images of objects lower in the visual field are usually judged closer than those higher in the visual field

Heschl's gyrus: a gyrus on the superior surface of the temporal lobe that constitutes the primary auditory area

heuristics: rules of thumb that tend toward the solution of a problem

Homeobox genes: genes "like a box", because the DNA sequence is short and can be enclosed by a box in written form

hominid: in outmoded usage, humans and their immediate ancestors; now used synonymously with Hominidae

Hominidae: a family consisting of all great apes including humans

Homininae: a subfamily consisting of all African apes including humans

Hominini, or hominins: a tribe consisting of all bipedal apes including humans

Hominoidea, or hominoids: the superfamily consisting of apes including humans

homologous: traceable from common origins

homology: a similarity in traits traceable to common ancestry

horizontal gene transfer: the movement of genetic material from one organism to another in the absence of a parent-child relationship

Hox: abbreviation for Homeobox

humerus: the upper arm bone

hyoid: a horseshoe-shaped bone in the neck

hypercomplex cells: in the visual system, neurons that respond to edges having both an orientation and an end; some also respond to curvature

hyperpolarize: to become more negatively charged

icon: the visual form of sensory memory

ilium: each of the bladelike extensions at the back of the pelvis

impedance: the amount of pressure required to produce a given displacement

insectivores: insect eaters

interneurons: small, laterally connecting neurons

interorbital distance: the space between the eyes, measured between the nearest edges

ions: electrically charged atoms

ipsilateral: same-sided

ischial callosities: callused skin patches on the buttocks that serve as padding while sitting

island dwarfism: the state of having reduced size, due to the species' isolation on an island over time

kinesthetic sense: a mechanical sense that assesses joint position and muscle tension

knapping: the act of shaping a stone core by breaking off flakes

lactase persistence: the presence of lactase beyond childhood and into adulthood, allowing the digestion of the sugar lactose, present in milk

laryngeal motor cortex: cortex involved in regulating the larynx

lateral line system: in fish, hair cells arranged in a line from gills to tail, detecting pressure gradients

lateral pallium: a brain structure in fish that is anatomically, genetically, and functionally homologous to the avian and mammalian hippocampus

lesions: areas of physical damage, in the brain usually resulting from strokes or penetrating wounds

Levallois technique: a stone toolmaking technique involving removing many flakes from a core in preparation for striking off one desired flake

lexigram: a type of visual symbol conveying meaning

lipids: fatty or waxy molecules

locomotion: movement from place to place

macula: a small spot of relatively acute central vision, containing the fovea

Magdalenian: a human tool culture that made many types of tools from bone and antler, some decorated with carvings

magnocellular response: a rapid response to visual stimulation that preserves only fuzzy detail but serves to alert the system to the appearance and location of the stimulus

maintenance rehearsal: rote repetition

mechanoreceptors: specialized sensory receptor cells that react to mechanical forces

medial: inner; toward the middle

medial lemniscal system: somatosensory neural pathways that run upward to the medulla of the brain, and then via synapses to the primary somatosensory cortex

Meissner's corpuscles: touch receptors below the outer layers of skin, detecting flutter

memory set: several items held in memory

memory span: the number of items that can consistently be recalled in correct order

memory stores: kinds of memory

Merkel's receptors (or discs): touch receptors below the outer layers of skin, detecting fine details

metabolism: the cellular process by which energy is made available

metacognition: the awareness of one's own cognitive states and processes

metazoa: multicellular animals

microblades: small stone blades less than 5 cm in length

microcephalic: abnormally small-brained

microliths: small stone blades with at least three distinct sides

microvilli: tiny projections of cell membrane

minicolumns: columns of neurons in the cortex, oriented vertically to the cortical surface and extending through most of the cell layers

mirror cells: neurons that rapidly fire when their possessor observes the actions of others, and when he or she makes the same motions

mitochondria: small structures responsible for energy production in a cell

molecular clock: a method of dating based on genetic differences between species

monocular cues: depth cues that can be seen by a single eye

monozygotic twins: identical twins

motile: independently moving

Mousterian: a hominid tool culture associated with Neanderthals

mutation: a copying error within a chromosome resulting in an altered set of instructions

myofibrils: chains of protein in muscle fibers that contract when stimulated

natural selection: the environmental favoring of traits, improving survival and reproduction

negative hand: the outline of a hand held against a rock and painted by blowing pigment through a tube held by the other

neoteny: the retention of juvenile characteristics in adulthood

neural crest: a cluster of embryonic cells arranged in ridge-like formations to the sides of the neural tube

neuromuscular junction: the location where a motor neuron synapses onto muscle fiber

neurons: nerve cells

New World monkeys: monkeys native to the Americas

nocturnal: active at night

notocord: a supporting rod in chordates, located in front of a hollow neural tube

nucleotide: a molecule composed of a nucleobase, a sugar, and a phosphate

occlusion: a monocular depth cue in which an object whose image partially blocks another is judged closer

Old World monkeys: monkeys native to Africa, Asia, or Europe

olfactory epithelium: a layer of neurons lining the upper portion of the nasal passageway

olfactory sense, or olfaction: sense of smell

omnivore: an animal that eats both animals and plants

opposable thumb: a thumb that can rotate along its long axis, allowing its pad to contact the pads of other fingers

opsins: a family of photopigments

optic flow: the experience of objects rushing past in the periphery while the view far ahead changes much more slowly

optic foramen: a small opening in the bone of the eye socket through which the optic nerve passes

orienting: the shifting of eyes, attention, or both toward a source of stimulation

oscula: mouth-like openings

ossicles: three small bones in the middle ear

Pacinian corpuscles: touch receptors, deeply embedded in the skin, that respond to vibration

paleocortex: ancient cortex

parietal eye field: a location in the parietal lobe of the brain that updates spatial information in response to movement

parvocellular response: a relatively slow response to visual stimulation that contains detailed information

pattern mask: a stimulus composed of jumbled features that spatially covers another stimulus

peptides: short chains of amino acids

perceptual invariance: the ability to recognize an object regardless of physical transformations of characteristics like position, depth, size, viewing angle, contrast, and texture

Périgordian (called Gravettian outside France):   a human tool culture characterized by backed blades and end-scrapers

perirhinal cortex: cortex located in the medial temporal lobe, literally "near the nose"

perisylvian cortex: cortex located around the Sylvian fissure

phonological syntax: the ordering of meaningless sound elements

phonology: language phonetics

phosphenes: imaginary points of light

photomotility: movement in response to light

photopigments: pigments that react chemically to light

phylogeny: evolutionary relationships

phytoliths: microscopic mineral particles formed by plants

pinna: outer ear

place cells: neurons that respond to particular locations in the environment

placode: a specialized cell cluster, representing a thickening of the ectoderm during vertebrate prenatal development that develops into a structure

planum temporale, or temporal plane: a small surface of cortex located lateral and posterior to Heschl's gyrus

Platyrrhini: "broad-nosed" monkeys, also known as New World monkeys

"pli-de-passage fronto-parietal moyen parietale" (PPFM): a gyrus in the brain which is buried along the central sulcus, connecting the precentral and postcentral gyri

polydactyly: the condition of having more than five digits per limb

population left-handedness (PLH): a predominance of left-handedness in a population

population right-handedness (PRH): a predominance of right-handedness in a population

positron emission tomography (PET): a method for imaging brain activity that involves injecting a radioactive substance into the bloodstream, followed by the detection of decay products immediately outside of the head

postcranial: pertaining to the skeleton below the head

potassium-argon dating: a geological clock method that assigns dates according to the ratio of potassium-40 to argon-40

praxis: the ability to conceive, initiate, and complete movement

preadaptation: an evolutionary change later built upon by other changes

precision grip: the delicate holding of a small object between the index finger and thumb

precuneus: the medial portion of area 7

prefontal cortex: the forward portion of the frontal lobes

prehensile: able to grasp objects

primary representation: the ability to mentally represent the world in a realistic, literal way

primary sensory cells: neurons with sensors at one end and an axon out the other end

primates: an order of eutherians with highly mobile feet and hands, and opposable digits

primitive character: a characteristic possessed by the earliest-evolving member of a clade

primordial soup: the thin solution of chemicals comprising the early ocean

prokaryotes: cells without nuclei

prosimians: outmoded term for primates "before apes", consisting of the Strepsirrhini and tarsiers

prosopagnosia: the inability to recognize faces presented visually

proximal: nearer to the core

pseudogenes: genes that have become deactivated because of disruptive mutations

pseudoneglect: overattention to one side of space due to an intact attentional mechanism on the opposite side of the brain

Psittacidae: the family of parrots

pulvinar: a nucleus in the thalamus

pursuit eye movements: eye movements that follow a moving object

pyramidal tract: a motor tract originating in the precentral gyrus of the brain, that crosses over to the opposite side of the body; also known as the corticospinal or dorsolateral tract

quadriceps: muscles on the front of the thigh that extend it

quadrupedal: four-legged; moving on four legs

quorum sensing: the ability of some bacteria to detect the presence of increasing numbers of other bacteria, by sensing a building concentration of chemical excretions

radiation: expansion in the number of species

recency effect: a memory phenomenon in which the most recent item or items in a list are recalled better than the items immediately preceding

reentrant connections: connections from higher-order cortical areas back to the primary cortex

relative size: a monocular depth cue in which the images of two similar objects are compared and the larger is judged closer

residual brain size: the amount of brain over and above that predicted by body weight

retouching: secondary flaking along the edge of a stone tool

retrosplenial cortex: cortex immediately posterior to the corpus callosum

right shift theory: the theory stating that human handedness is the product of a bell-shaped distribution in combination with a right-shift allele that is present in some individuals and not others

robust: heavily built

rods: a class of receptors in the retina, chiefly responsible for night vision

root hair plexus: nerve endings wrapped around the base of a hair

rostral: forward

route knowledge: knowledge of landmarks and associated turns, used in learning pathways from one point to another

Ruffini's endings (or corpuscles):   touch receptors, deeply embedded in the skin, that respond to skin stretching

saccades: eye movements toward a source of stimulation

saccule: a fluid-filled cavity in the inner ear with sensors that respond to acceleration

sagittal crest: a ridge of bone running front to back on the top of the skull

savanna: grassland with sparse stands of trees

script: a contextually dependent series of actions leading to a goal

secondary representation: the ability to separate a primary representation from its environmental reference for hypothetical purposes

secondary sensory cells: neurons with a sensory end and a broad synaptic surface on the other end

semantic dementia: a disorder in which use of semantic knowledge is impaired even though episodic memory is spared

semantic memory: memory for facts that make up our knowledge structure

sensory memory: a very short-lived memory trace that preserves much of the physical form of stimuli

sessile: attached, as to the sea bottom

sexual selection: mate selection by desirable qualities, e.g., particular colorations or forms

simple cells: in the visual system, neurons that perform a feature analysis by representing a scene in terms of small segments of contours, and that may incorporate color

social learning: the acquisition of new behaviors by observing others

Solutrean: a human tool culture known for creating the laurel-leaf point through pressure-flaking

somatosensation: the sense of touch, or the tactile sense

spatial attention: the ability to attend to objects or locations independent of eye position

spatial neglect, or visual neglect: difficulty in orienting to objects opposite the side of a brain lesion

speciation: the creation of new species

spike potential: a sudden increase in positivity that travels up a neuron's axon, opening sodium channels

split time: the latest date at which there was any genetic interchange between two species

Strepsirrhini: strepsirrhines, i.e., wet-nosed primates

striking platform: in the making of a stone tool, a small flat surface where a blow landed

stromatolites: mats of sediment formed by bacteria

subitizing: a perceptual process that rapidly enumerates small numbers of items

superficial layer of V1: the layer of V1 cells closest to the surface of the brain, also known as layer 1

suprachiasmatic nucleus: a nucleus of neurons connected to the ganglion cells of the retina, involved in maintaining the circadian rhythm

survey knowledge: knowledge of how landmarks are configured relative to one another

symbiotic: mutually beneficial

symmetrical gait: a gait with a movement cycle taking the same length of time for the forelimbs as for the hindlimbs

sympatric speciation: the evolution of new species within an unseparated population

synapsids: tetrapods with one opening in the skull behind the eye

syntax, or grammar: the rule-based organization of symbols in orders that succeed in conveying meaning

tactile: pertaining to touch

Talairach space: a set of X, Y, and Z coordinates allowing the mapping of brain activity to structures in an "average" brain

tapetum lucidum: literally "silvery carpet"; the reflective layer behind the retinas of certain animals such as cats and dogs

tectopulvinar pathway: a visual pathway that proceeds from the eye to the superior colliculus, then to the pulvinar, and finally to the secondary visual areas of the cortex

termite fishing: a behavior in which a twig is prepared and then inserted into termite mounds, allowing the withdrawal of termites

terrestrial: living on the ground

tetrachromacy: the state of having four types of cones

tetrapods: vertebrates with four feet or legs

theory of mind: the ability to ascribe mental and emotional states to others

thermophilic: heat-loving

theropods: large carnivorous dinosaurs with short forelimbs and large jaws

thoracic vertebrae: vertebrae at the level of the thorax or chest

trace fossils: fossils of behavior as opposed to physical form

trichromacy: the state of having three types of cones

trichromatic vision: three-color vision

tuff: a layer of rock formed from compacted volcanic ash

umami: a meaty or savory flavor

utricle: a fluid-filled cavity in the inner ear with receptors that respond to acceleration

ventral: bottom

ventral visual stream: the visual pathway from brain area V1 to the temporal lobe

Vernier acuity: the ability to determine whether two aligned line segments are continuations of the same line or rather are offset from one another

vertebral canal: the opening in the vertebrae through which the spinal cord passes

vertebrates: animals with backbones

vesicles: sacs

vestibular sense: the sense of balance

visual agnosia: the inability to recognize a class of objects presented visually

Visual neglect, visuospatial neglect, or spatial neglect: difficulty in orienting to objects opposite the side of a brain lesion

visual word form area (VWFA):   a brain area in or near the occipito-temporal sulcus, involved in recognizing words presented visually

visuospatial neglect, visual neglect, or spatial neglect: difficulty in orienting to objects opposite the side of a brain lesion

vomeronasal sense: a chemical sense that detects pheromones using receptors in the nasal passages

Wernicke's area: the posterior portion of the superior temporal gyrus implicated in the decoding of speech sounds

working memory: a form of memory that holds and manipulates information over the short term

References

Afraz, S-R., Kiani, R., & Esteky, H. (2006). Microstimulation of inferotemporal cortex influences face categorization. *Nature, 442*, 692–695.

Ahlberg, P. E., & Clack, J. A. (2006). A firm step from water to land. *Nature, 440*, 747–749.

Aiello, L. C. (2010). Five years of *Homo floresiensis*. *American Journal of Physical Anthropology, 142*, 167–179.

Ajina, S., & Bridge, H. (2017). Blindsight and unconscious vision: What they teach us about the human visual system. *Neuroscientist, 23*, 529–541.

Albrecht, D. G., De Valois, R. L., & Thorell, L. G. (1980). Visual cortical neurons: Are bars or gratings the optimal stimuli? *Science, 207*, 88–90.

Aldridge, K. (2011). Patterns of differences in brain morphology in humans as compared to extant apes. *Journal of Human Evolution, 60*, 94–105.

Alemseged, Z., Spoor, F., Kimbel, W. H., Bobe, R., Geraads, D., Reed, D., & Wynn, J. G. (2006). A juvenile early hominin skeleton from Dikika, Ethiopia. *Nature, 443*, 296–301.

Alexander, J. R., & Challef, S. (2000). Control: An emergent property of biological neurons. *International Journal of Systems Science, 31*, 895–909.

Alexander, R. M. (1992). Human locomotion. In S. Jones, R. Martin, & D. Pilbeam (Eds.), *The Cambridge encyclopedia of human evolution* (pp. 80–85). Cambridge, UK: Cambridge University Press.

Allen, C., & Bekoff, M. (2007). Animal consciousness. In M. Velmans & S. Schneider (Eds.), *The Blackwell companion to consciousness* (pp. 58–71). Malden, MA: Blackwell Publishing.

Allison, T., Puce, A., Spencer, D. D., & McCarthy, G. (1999). Electrophysiological studies of human face perception. I: Potentials generated in occipitotemporal cortex by face and non-face stimuli. *Cerebral Cortex, 9*, 415–430.

Almécija, S., Shrewsbury, M., Rook, L., & Moyà-Solà, S. (2014). The morphology of *Oreopithecus bombolii* pollical distal phalanx. *American Journal of Physical Anthropology, 153*, 582–597.

Almécija, S., Smaers, J. B., & Jungers, W. L. (2015). The evolution of human and ape hand proportions. *Nature Communications, 6*, article 7717.

Almécija, S., Tallman, M., Alba, D. M., Pina, M., Moyà-Solà, S., & Jungers, W. L. (2013). The femur of *Orrorin tugenensis* exhibits morphometric affinities with both Miocene apes and later hominins. *Nature Communications, 4*, article 2888.

Alvarez, G. A., & Cavanagh, P. (2004). The capacity of visual short-term memory is set both by visual information load and by number of objects. *Psychological Science, 15*, 106–111.

Anderson, J. R., & Gallup, G. G. (2015). Mirror self-recognition: A review and critique of attempts to promote and engineer self-recognition in primates. *Primates, 56*, 317–326.

Andre, J., Picchioni, M., Zhang, R., & Toulopoulou, T. (2016). Working memory circuit as a function of increasing age in healthy adolescence: A systematic review and meta-analyses. *NeuroImage: Clinical, 12*, 940–948.

Annett, M. (1998). Handedness and cerebral dominance: The right shift theory. *Journal of Neuropsychiatry, 10*, 459–469.

Annett, M. (2006). The distribution of handedness in chimpanzees: Estimating right shift in Hopkins' sample. *Laterality, 11*, 101–109.

Antcliffe, J., Callow, R., & Brasier, M. (2014). Giving the early fossil record of sponges a squeeze. *Biological Reviews*, *89*, doi:10.1111/brv.12090

Antón, S. C. (2003). Natural history of *Homo erectus*. *Yearbook of Physical Anthropology*, *46*, 126–170.

Arendt, D., Benito-Gutierrez, E., Brunet, T., & Marlow, H. (2015). Gastric pouches and the muco-ciliary sole: Setting the stage for nervous system evolution. *Philosophical Transactions of the Royal Society, B*, *370*, article 20150286.

Arendt, D., Tosches, M. A., & Marlow, H. (2016). From nerve net to nerve ring, nerve cord and brain – evolution of the nervous system. *Nature Reviews Neuroscience*, *17*, 61–72.

Arendt, D., & Wittbrodt, J. (2001). Reconstructing the eyes of Urbilateria. *Philosophical Transactions of the Royal Society of London, B*, *356*, 1545–1563.

Argue, D., Morwood, M. J., Sutinka, T., Jatmiko, & Saptomo, E. W. (2009). *Homo floresiensis*: A cladistic analysis. *Journal of Human Evolution*, *57*, 623–639.

Arias-Martorell, J., Potau, J. M., Bello-Hellegouarch, G., Pastor, J. F., & Pérez-Pérez, A. (2012). 3D geometric morphometric analysis of the proximal epiphysis of the hominoid humerus. *Journal of Anatomy*, *221*, 394–405.

Arias-Martorell, J., Potau, J. M., Bello-Hellegouarch, G., Pastor, J. F., & Pérez-Pérez, A. (2015). Like father, like son: Assessment of the morphological affinities of A.L. 288-1 (*A. afarensis*), Sts 7 (*A. africanus*) and Omo 119-73-2718 (*Australopithecus* sp.) through a three-dimensional shape analysis of the shoulder joint. *PLoS ONE*, *10*(2), e0117408.

Armour, J. A. L., Davison, A., & McManus, I. C. (2014). Genome-wide association study of handedness excludes simple genetic models. *Heredity*, *112*, 221–225.

Arrese, C. A., Hart, N. S., Thomas, N., Beazley, L. D., & Shand, J. (2002). Trichromacy in Australian marsupials. *Current Biology*, *12*, 657–660.

Asara, J. M., Schweitzer, M. H., Freimark, L. M., Phillips, M., & Cantley, L. C. (2007). Protein sequences from mastodon and *Tyrannosaurus rex* revealed by mass spectrometry. *Science*, *316*, 280–285.

Asfaw, B., White, T., Lovejoy, O., Latimer, B., Simpson, S., & Suwa, G. (1999). *Australopithecus garhi*: A new species of early hominid from Ethiopia. *Science*, *284*, 629–635.

Ashley-Ross, M. A. (1994). Hindlimb kinematics during terrestrial locomotion in a salamander (*Dicamptodon tenebrosus*). *Journal of Experimental Biology*, *193*, 255–283.

Astafiev, S. V., Shulman, G. L., Stanley, C. M., Snyder, A. Z., Van Essen, D. C., & Corbetta, M. (2003). Functional organization of human intraparietal and frontal cortex for attending, looking, and pointing. *The Journal of Neuroscience*, *23*, 4689–4699.

Averbach, E. (1963). The span of apprehension as a function of exposure duration. *Journal of Verbal Learning and Verbal Behavior*, *2*, 60–64.

Baab, K. L., & McNulty, K. P. (2009). Size, shape, and asymmetry in fossil hominins: The status of the LB1 cranium based on 3D morphometric analyses. *Journal of Human Evolution*, *57*, 608–622.

Babiloni, C., Vecchio, F., Rossi, S., De Capua, A., Bartalini, S., Ulivelli, M., & Rossini, P. M. (2007). Human ventral parietal cortex plays a functional role on visuospatial attention and primary consciousness. A repetitive transcranial magnetic stimulation study. *Cerebral Cortex*, *17*, 1486–1492.

Backwell, L., d'Errico, F., & Wadley, L. (2008). Middle stone age bone tools from the Howiesons Poort layers, Sibudu Cave, South Africa. *Journal of Archaeological Science*, *35*, 1566–1580.

Baddeley, A. (1986). *Working memory*. Oxford: Clarendon Press.

Bagnara, S., Boles, D. B., Simion, F., & Umiltà, C. (1982). Can an analytic/holistic dichotomy explain hemispheric asymmetries? *Cortex*, *18*, 67–78.

Bajpai, S., Kay, R. F., Williams, B. A., Das, D. P., Kapur, V. V., & Tiwari, B. N. (2008). The oldest Asian record of Anthropoidea. *Proceedings of the National Academy of Sciences*, *105*, 11093–11098.

Baldauf, S. L., Bhattacharya, D., Cockrill, J., Hugenholtz, P., Pawlowski, J., & Simpson, A. G. B. (2004). The tree of life: An overview. In J. Cracraft & M. J. Donohue (Eds.), *Assembling the tree of life* (pp. 43–75). Oxford: Oxford University Press.

Balleza, D. (2011). Toward understanding protocell mechanosensation. *Origins of Life and Evolution of Biospheres, 41,* 281–304.

Balzeau, A., Gilissen, E., Holloway, R. L., Prima, S., & Grimaud-Hervé, D. (2014). Variations in size, shape and asymmetries of the third frontal convolution in hominids: Paleoneurological implications for hominin evolution and the origin of language. *Journal of Human Evolution, 76,* 116–128.

Bard, J. (2017). *Principles of evolution: Systems, species, and the history of life.* London: Garland Science.

Bardack, D. (1991). First fossil hagfish (*Myxinoidea*): A record from the Pennsylvanian of Illinois. *Science, 254,* 701–703.

Barfod, G. H., Albarède, F., Knoll, A. H., Xiao, S., Télouk, P., Frie, R., & Baker, J. (2002). New Lu-Hf and Pb-Pb age constraints on the earliest animal fossils. *Earth and Planetary Science Letters, 201,* 203–212.

Bargalló, A., & Mosquera, M. (2014). Can hand laterality be identified through lithic technology? *Laterality, 19,* 37–63.

Barney, A., Martelli, S., Serrurier, A., & Steele, J. (2012). Articulatory capacity of Neanderthals, a very recent and human-like fossil hominin. *Philosophical Transactions of the Royal Society, B, 367,* 88–102.

Barton, R. A., & Venditti, C. (2013). Human frontal lobes are not relatively large. *Proceedings of the National Academy of Sciences, 110,* 9001–9006.

Bassler, B. L., & Losick, R. (2006). Bacterially speaking. *Cell, 125,* 237–246.

Battaglia-Mayer, A., Mascaro, M., & Caminiti, R. (2007). Temporal evolution and strength of neural activity in parietal cortex during eye and hand movements. *Cerebral Cortex, 17,* 1350–1363.

Bays, P. M. (2018). Reassessing the evidence for capacity limits in neural signals related to working memory. *Cerebral Cortex, 28,* 1432–1438.

BBN International. (2008). *Anthropology – The doctrine of man.* Retrieved February 15, 2019 from https://bbn1.bbnradio.org/english/tools-home/bible-doctrines/bible-doctrines-anthropology.

Bear, D. M., Lassance, J-M., Hoekstra, H. E., & Datta, S. R. (2016). The evolving neural and genetic architecture of vertebrate olfaction. *Current Biology, 26,* R1039–R1049.

Beaumont, P. B. (2011). The edge: More on fire-making by about 1.7 million years ago at Wonderwerk Cave in South Africa. *Current Anthropology, 52,* 585–595.

Begun, D. R. (2010). Miocene hominids and the origins of the African apes and humans. *Annual Review of Anthropology, 39,* 67–84.

Begun, D. R., & Kordos, L. (2004). Cranial evidence of the evolution of intelligence in fossil apes. In A. E. Russon & D. R. Begun (Eds.), *The evolution of thought: Evolutionary origins of great ape intelligence* (pp. 260–279). Cambridge, UK: Cambridge University Press.

Beisel, K. W., Wang-Lundberg, Y., Maklad, A., & Fritzsch, B. (2005). Development and evolution of the vestibular sensory apparatus of the mammalian. *Journal of Vestibular Research, 15,* 225–241.

Bell, E. A., Boehnke, P., Harrison, T. M., & Mao, W. L. (2015). Potentially biogenic carbon preserved in a 4.1 billion-year-old zircon. *Proceedings of the National Academy of Sciences, 112,* 14518–14521.

Benhamou, S., & Poucet, B. (1996). A comparative analysis of spatial memory processes. *Behavioural Processes, 35,* 113–126.

Bennett, M. R., Harris, J. W. K., Richmond, B. G., Braun, D. R., Mbua, E., Kiura, P., . . . Gonzalez, S. (2009). Early hominin foot morphology based on 1.5-million-year-old footprints from Ileret, Kenya. *Science, 323,* 1197–1201.

Bennett, M. R., Reynolds, S. C., Morse, S. A., & Budka, M. (2016). Footprints and human evolution: Homeostasis in foot function? *Palaeogeography, Palaeoclimatology, Palaeoecology, 461,* 214–223.

Bentley-Condit, V. K., & Smith, E. O. (2010). Animal tool use: Current definitions and an updated comprehensive catalog. *Behaviour, 147,* 185–221.

Benton, M. J. (2005). *Vertebrate palaeontology* (3rd ed.). Malden, MA: Blackwell Publishing.

Beran, M. J., Perdue, B. M., & Smith, J. D. (2014). What are my chances? Closing the gap in uncertainty monitoring between rhesus monkeys (*Macaca mulatta*) and capuchin monkeys (*Cebus apella*). *Journal of Experimental Psychology: Animal Learning and Cognition, 40*, 303–316.

Beran, M. J., Smith, J. D., Coutinho, M. V. C., Couchman, J. J., & Boomer, J. (2009). The psychological organization of "uncertainty" responses and "middle" responses: A dissociation in capuchin monkeys (*Cebus apella*). *Journal of Experimental Psychology: Animal Behavior Processes, 35*, 371–381.

Berger, L. R., de Ruiter, D. J., Churchill, S. E., Schmid, P., Carlson, K. J., Dirks, P. H. G. M., & Kibii, J. M. (2010). *Australopithecus sediba*: A new species of Homo-like Australopith from South Africa. *Science, 328*, 195–204.

Berger, L. R., Hawks, J., de Ruiter, D. J., Churchill, S. E., Schmid, P., Delezene, L. K., . . . Walker, C. S. (2015). *Homo naledi*, a new species of the genus *Homo* from the Dinaledi chamber, South Africa. *eLife, 4*, article e09560.

Berlin, J. C., Kirk, E. C., & Rowe, T. B. (2013). Functional implications of ubiquitous semicircular canal non-orthogonality in mammals. *PLoS One, 8*, article e79585.

Bernhardt, H. S. (2012). The RNA world hypothesis: The worst theory of the early evolution of life (except for all the others). *Biology Direct, 7*, 23.

Beste, C., Arning, L., Gerding, W. M., Epplen, J. T., Mertins, A., Röder, M. C., . . . Ocklenburg, S. (2018). Cognitive control processes and functional cerebral asymmetries: Association with variation in the handedness-associated gene *LRRTM1. Molecular Neurobiology, 55*, 2268–2274.

Bianchi, S., Stimpson, C. D., Bauernfeind, A. L., Schapiro, S. J., Baze, W. B., McArthur, M. J., . . . Sherwood, C. C. (2013). Dendritic morphology of pyramidal neurons in the chimpanzee neocortex: Regional specializations and comparison to humans. *Cerebral Cortex, 23*, 2429–2436.

Bijsterveld, K. (2015). Beyond echoic memory: Introduction to the special issue on auditory history. *The Public Historian, 37*, 7–13.

Biro, D., Inoue-Nakamura, N., Tonooka, R., Yamakoshi, G., Sousa, C., & Matsuzawa, T. (2003). Cultural innovation and transmission of tool use in wild chimpanzees: Evidence from field experiments. *Animal Cognition, 6*, 213–223.

Blake, J. (2004). Gestural communication in the great apes. In A. E. Russon & D. R. Begun (Eds.), *The evolution of thought: Evolutionary origins of great ape intelligence* (pp. 61–75). Cambridge, UK: Cambridge University Press.

Blanca, M. J., & López-Montiel, G. (2009). Hemispheric differences for global and local processing: Effect of stimulus size and sparsity. *The Spanish Journal of Psychology, 12*, 21–31.

Bliss, J. C., Crane, H. D., Mansfield, P. K., & Townsend, J. T. (1966). Information available in brief tactile presentations. *Perception & Psychophysics, 1*, 273–283.

Blob, R. W.(2001). Evolution of hindlimb posture in nonmammalian therapsids: Biomechanical tests of paleontological hypotheses. *Paleobiology, 27*, 14–38.

Bloch, J. I., & Wilcox, M. T. (2006). Cranial anatomy of the Paleocene plesiadapiform *Carpolestes simpsoni* (Mammalia, Primates) using ultra high-resolution X-ray computed tomography, and the relationships of plesiadapiforms to Euprimates. *Journal of Human Evolution, 50*, 1–35.

Bloemendal, H., de Jong, W., Jaenicke, R., Lubsen, N. H., Slingsby, C., & Tardieu, A. (2004). Ageing and vision: Structure, stability and function of lens crystallins. *Progress in Biophysics & Molecular Biology, 86*, 407–485.

Boccia, M. L. (1994). Mirror behavior in macaques. In S. T. Parker, R. W. Mitchell, & M. L. Boccia (Eds.), *Self-awareness in animals and humans* (pp. 350–360). Cambridge, UK: Cambridge University Press.

Boeckmann, B., Marcet-Houben, M., Rees, J. A., Forslund, K., Huerta-Cepas, J., Muffato, M., . . . Gabaldón, T. (2015). Quest for orthologs entails quest for Tree of Life: In search of the gene stream. *Genome Biology and Evolution, 7*, 1988–1999.

Boghi, A., Rampado, O., Bergui, M., Avidano, F., Manzone, C., Coriasco, M., . . . Bradac, G. B. (2006). Functional MR study of a motor task and the Tower of London task at 1.0 T. *Neuroradiology, 48*, 763–771.

Boisvert, C. A. (2005). The pelvic fin and girdle of *Panderichthys* and the origin of tetrapod locomotion. *Nature, 438*, 1145–1147.

Boles, D. B. (1984). Global versus local processing: Is there a hemispheric dichotomy? *Neuropsychologia, 22*, 445–455.

Boles, D. B. (1991). Factor analysis and the cerebral hemispheres: Pilot study and parietal functions. *Neuropsychologia, 29*, 59–91.

Boles, D. B. (1997). Multiple resource contributions to training. *Proceedings of the Human Factors and Ergonomics Society*, 41st Annual Meeting, 1176–1179.

Boles, D. B. (1998). Relationships among multiple task asymmetries: II. A large-sample factor analysis. *Brain and Cognition, 36*, 268–289.

Boles, D. B. (2002). Lateralized spatial processes and their lexical implications. *Neuropsychologia, 40*, 2125–2135.

Boles, D. B., Adair, L. P., & Joubert, A-M. (2009). A preliminary study of lateralized processing in attention-deficit/hyperactivity disorder. *Journal of General Psychology, 136*, 243–258.

Boles, D. B., Barth, J. M., & Merrill, E. C. (2008). Asymmetry and performance: Toward a neurodevelopmental theory. *Brain & Cognition, 66*, 124–139.

Boles, D. B., & Dillard, M. (2015). The measurement of perceptual resources and workload. In R. R. Hoffman, P. A. Hancock, M. W. Scerbo, R. Parasuraman, & J. L. Szalma (Eds.), *Cambridge handbook of applied perception research* (Vol. 1, pp. 39–59). New York, NY: Cambridge University Press.

Boles, D. B., & Karner, T. A. (1996). Hemispheric differences in global versus local processing: Still unclear. *Brain and Cognition, 30*, 232–243.

Boles, D. B., & Law, M. B. (1998). A simultaneous task comparison of differentiated and undifferentiated hemispheric resource theories. *Journal of Experimental Psychology: Human Perception and Performance, 24*, 204–215.

Boles, D. B., Phillips, J. B., & Givens, S. M. (2007). What dot clusters and bargraphs reveal: Subitizing is fast counting and subtraction. *Perception & Psychophysics, 69*, 913–922.

Bonini, L. (2017). The extended mirror neuron network: Anatomy, origin, and functions. *The Neuroscientist, 23*, 56–67.

Boogert, N. J., Fawcett, T. W., & Lefebvre, L. (2011). Mate choice for cognitive traits: A review of the evidence in nonhuman vertebrates. *Behavioral Ecology, 22*, 447–459.

Born, R. T., Trott, A. R., & Hartmann, T. S. (2015). Cortical magnification plus cortical plasticity equals vision? *Vision Research, 111*, 161–169.

Botta, O. (2004). The chemistry of the origin of life. In P. Ehrenfreund, W. Irvine, T. Owen, L. Becker, J. Blank, J. Brucato, . . . F. Robert (Eds.), *Astrobiology: Future perspectives* (pp. 359–391). Dordrecht: Kluwer Academic Publishers.

Botella, J., Privado, J., de Liaño, B. G., & Suero, M. (2011). Illusory conjunctions reflect the time course of the attentional blink. *Attention, Perception, and Psychophysics, 73*, 1361–1373.

Boussau, B., Blanquart, S., Necsulea, A., Lartillot, N., & Gouy, M. (2008). Parallel adaptations to high temperatures in the Archaean eon. *Nature, 456*, 942–946.

Bouvier, S. E., & Engle, S. A. (2006). Behavioral deficits and cortical damage loci in cerebral achromatopsia. *Cerebral Cortex, 16*, 183–191.

Bowers, J. S. (2017). Parallel distributed processing theory in the age of deep networks. *Trends in Cognitive Sciences, 21*, 950–961.

Bowmaker, J. K., & Hunt, D. M. (2006). Evolution of vertebrate visual pigments. *Current Biology, 16*, R484–R489.

Bradshaw, J. (1997). *Human evolution: A neuropsychological perspective*. East Sussex, UK: Psychology Press.

Bramble, D. M., & Lieberman, D. E. (2004). Endurance running and the evolution of *Homo*. *Nature, 432*, 345–352.

Brandler, W. M., Morris, A. P., Evans, D. M., Scerri, T. S., Kemp, J. P., Timpson, N. J., . . . Paracchini, S. (2013). Common variants in left/right asymmetry genes and pathways are associated with relative hand skill. *PLoS Genetics, 9*, article e1003751.

Bräuer, J., Call, J., & Tomasello, M. (2005). All great ape species follow gaze to distant locations and around barriers. *Journal of Comparative Psychology, 119*, 145–154.

Bräuer, J., Call, J., & Tomasello, M. (2007). Chimpanzees really know what others can see in a competitive situation. *Animal Cognition, 10*, 439–448.

Braun, D. R., Plummer, T., Ferraro, J. V., Ditchfield, P., & Bishop, L. C. (2009). Raw material quality and Oldowan hominin toolstone preferences: Evidence from Kanjera South, Kenya. *Journal of Archaeological Science, 36*, 1605–1614.

Braun, K., & Stach, T. (2017). Structure and ultrastructure of eyes and brains of *Thalia democratica* (Thaliacea, Tunicata, Chordata). *Journal of Morphology, 278*, 1421–1437.

Braun, M. H., & Likowiak, K. (2011). Intermediate and long-term memory are different at the neuronal level in *Lymnaea stagnalis* (L.) *Neurobiology of Learning and Memory, 96*, 403–416.

Brazeau, M. D., & Ahlberg, P. E. (2006). Tetrapod-like middle ear architecture in a Devonian fish. *Nature, 439*, 318–321.

Brecht, K. F., Wagener, L., Ostojic, L., Clayton, N. S., & Nieder, A. (2017). Comparing the face inversion effect in crows and humans. *Journal of Comparative Physiology, A, 203*, 1017–1027.

Breedlove, S. M., Watson, N. V., & Rosenzweig, M. R. (2010). *Biological psychology*. Sunderland, MA: Sinauer Associates.

Breitmeyer, B. G. (1984). *Visual masking: An integrative approach*. Oxford: Clarendon Press.

Bremmer, F. (2005). Navigation in space – the role of the macaque ventral intraparietal area. *Journal of Physiology, 566*, 29–35.

Breuer, T., Ndoundou-Hockemba, M., & Fishlock, V. (2005). First observation of tool use in wild gorillas. *PLoS Biology, 3*, 2041–2043.

Broadfield, D. C., Holloway, R. L., Mowbray, K., Silvers, A., Yuan, M. S., & Márquez, S. (2001). Endocast of Sambungmacan 3 (Sm 3): A new *Homo erectus* from Indonesia. *The Anatomical Record, 262*, 369–379.

Broglio, C., Gómez, A., Durán, E., Ocaña, F. M., Jiménez-Moya, F., Rodríguez, F., & Salas, C. (2005). Hallmarks of a common forebrain vertebrate plan: Specialized pallial areas for spatial, temporal and emotional memory in actinopterygian fish. *Brain Research Bulletin, 66*, 277–281.

Brown, C. (2012). Tool use in fishes. *Fish and Fisheries, 13*, 105–115.

Brown, F. H. (1992). Methods of dating. In S. Jones, R. Martin, & D. Pilbeam (Eds.), *The Cambridge encyclopedia of human evolution* (pp. 179–186). Cambridge, UK: Cambridge University Press.

Brown, J. C., & Golston, C. (2006). Embedded structure and the evolution of phonology. *Interaction Studies, 7*, 17–41.

Brown, P., Sutikna, T., Morwood, M. J., Soejono, R. P., Jatmiko, Saptomo, E. W., & Due, R. A. (2004). A new small-bodied hominin from the Late Pleistocene of Flores, Indonesia. *Nature, 431*, 1055–1061.

Bruck, J. N., Allen, N. A., Brass, K. E., Horn, B. A., & Campbell, P. (2017). Species differences in egocentric navigation: The effect of burrowing ecology on a spatial cognitive trait in mice. *Animal Behaviour, 127*, 67–73.

Brugger, J., Feulner, G., & Petri, S. (2017). Baby, it's cold outside: Climate model simulations of the effects of the asteroid impact at the end of the Cretaceous. *Geophysical Research Letters, 44*, 419–427.

Brüne, M., & Brüne-Cohrs, U. (2006). Theory of mind – evolution, ontogeny, brain mechanisms and psychopathology. *Neuroscience and Biobehavioral Reviews, 30*, 437–455.

Bruner, E. (2004). Geometric morphometrics and paleoneurology: Brain shape evolution in the genus *Homo*. *Journal of Human Evolution, 47*, 279–303.

Bruner, E. (2010). Morphological differences in the parietal lobes within the human genus. *Current Anthropology, 51*, S77-S88.

Bruner, E., Preuss, T. M., Chen, X., & Rilling, J. K. (2017). Evidence for expansion of the precuneus in human evolution. *Brain Structure and Function, 222*, 1053–1060.

Brunet, M., Guy, F., Boisserie, J-R., Djimdoumalbaye, A., Lehmann, T., Lihoreau, F., . . . Zollikofer, C. (2004). "Toumaï", Miocène supérieur du Tchad, le nouveau doyen du rameau humain. *Comptes Rendus Palevol, 3*, 277–285.

Budd, G. E., & Jensen, S. (2004). The limitations of the fossil record and the dating of the origin of the Bilateria. In P. C. J. Donoghue & M. P. Smith (Eds.), *Telling the evolutionary time: Molecular clocks and the fossil record* (pp. 166–189). Boca Raton: CRC Press.

Bugnyar, T., Stöwe, M., & Heinrich, B. (2004). Ravens, *Corvus corax*, follow gaze direction of humans around obstacles. *Proceedings of the Royal Society of London, B., 271*, 1331–1336.

Burgihel, P., Lane, N. J., Fabio, G., Stefano, T., Zaniolo, G., Carnevali, M. D. C., & Manni, L. (2003). Novel, secondary sensory cell organ in ascidians: In search of the ancestor of the vertebrate lateral line. *The Journal of Comparative Neurology, 461,* 236–249.

Bussey, T. J., Saksida, L. M., & Murray, E. A. (2002). Perirhinal cortex resolves feature ambiguity in complex visual discriminations. *European Journal of Neuroscience, 15,* 365–374.

Butler, A. B. (2000). Chordate evolution and origin of craniates: An old brain in a new head. *The Anatomical Record Part B, The New Anatomist, 261,* 111–125.

Butterfield, N. J. (2003). Exceptional fossil preservation and the Cambrian explosion. *Integrative and Comparative Biology, 43,* 166–177.

Buxhoeveden, D. P., Switala, A. E., Litaker, M., Roy, E., & Casanova, M. F. (2001). Lateralization of minicolumns in human planum temporale is absent in nonhuman primate cortex. *Brain, Behavior, and Evolution, 57,* 349–358.

Byrne, R. W. (2004). The manual skills and cognition that lie behind hominid tool use. In A. E. Russon & D. R. Begun (Eds.), *The evolution of thought: Evolutionary origins of great ape intelligence* (pp. 31–33). Cambridge, UK: Cambridge University Press.

Cabeza, R., & Nyberg, L. (2000). Imaging cognition II: An empirical review of 275 PET and fMRI studies. *Journal of Cognitive Neuroscience, 12,* 1–47.

Caetano-Anollés, G., Nasir, A., Zhou, K., Castano-Anollés, D., Mittenthal, J. E., Sun, F-J., & Kim, K. M. (2014). Archaea: The first domain of diversified life. *Archaea, 2014,* article 590214.

Calcutt, S. E., Rubin, T. L., Pokorny, J. J., & de Waal, F. B. M. (2017). Discrimination of emotional facial expressions by tufted capuchin monkeys (*Sapajus apella*). *Journal of Comparative Psychology, 131,* 40–49.

Call, J., Bräuer, J., Kaminski, J., & Tomasello, M. (2003). Domestic dogs (*Canis familiaris*) are sensitive to the attentional state of humans. *Journal of Comparative Psychology, 117,* 257–263.

Calvin, W. H. (2002). Rediscovery and the cognitive aspects of toolmaking: Lessons from the handaxe. *Behavioral and Brain Sciences, 25,* 403–404.

Cameron, D. W., & Groves, C. P. (2004). *Bones, stones, and molecules: "Out of Africa" and human origins.* Amsterdam: Elsevier Academic Press.

Camprodon, J. A., Zohary, E., Brodbeck, V., & Pascual-Leone, A. (2010). Two phases of V1 activity for visual recognition of natural images. *Journal of Cognitive Neuroscience, 22,* 1262–1269.

Cañestro, C., Albalat, R., Irimia, M., & Garcia-Fernàndez, J. (2013). Impact of gene gains, losses and duplication modes on the origin and diversification of vertebrates. *Seminars in Cell & Developmental Biology, 24,* 83–94.

Cantalupo, C., Pilcher, D. L., & Hopkins, W. D. (2003). Are planum temporale and sylvian fissure asymmetries directly related?: A MRI study in great apes. *Neuropsychologia, 41,* 1975–1981.

Cantone, G., Xiao, J., McFarlane, N., & Levitt, J. B. (2005). Feedback connections to ferret striate cortex: Direct evidence for visuotopic convergence of feedback inputs. *The Journal of Comparative Neurology, 487,* 312–331.

Capitani, E., Laiacona, M., Mahon, B., & Caramazza, A. (2003). What are the facts of semantic category-specific deficits? A critical review of the clinical evidence. *Cognitive Neuropsychology, 20,* 213–261.

Caron, J. B., & Rudkin, D. (2009). *A burgess shale primer: History, geology, and research highlights.* Toronto: The Burgess Shale Consortium.

Cartmill, E. A., & Byrne, R. W. (2007). Orangutans modify their gestural signaling according to their audience's comprehension. *Current Biology, 17,* 1345–1348.

Cartmill, E. A., & Byrne, R. W. (2010). Semantics of primate gestures: Intentional meanings of orangutan gestures. *Animal Cognition, 13,* 793–804.

Cartmill, M. (1992). Non-human primates. In S. Jones, R. Martin, & D. Pilbeam (Eds.), *The Cambridge encyclopedia of human evolution* (pp. 24–32). Cambridge, UK: Cambridge University Press.

Carvalho, S., Biro, D., Cunha, E., Hockings, K., Richmond, B. G., & Matsuzawa, T. (2012). Chimpanzee carrying behaviour and the origins of human bipedality. *Current Biology, 22,* R180–R181.

Cashmore, L., Uomini, N., & Chapelain, A. (2008). The evolution of handedness in humans and great apes: A review and current issues. *Journal of Anthropological Sciences, 86,* 7–35.

Catania, K. C., & Henry, E. C. (2006). Touching on somatosensory specializations in mammals. *Current Opinion in Neurobiology, 16,* 467–473.

Cauchoix, M., Crouzet, S. M., Fize, D., & Serre, T. (2016). Fast ventral stream neural activity enables rapid visual categorization. *NeuroImage, 125,* 280–290.

Cavanagh, J. P. (1972). Relation between the immediate memory span and the memory search rate. *Psychological Review, 79,* 525–530.

Censky, E. J., Hodge, K., & Dudley, J. (1998). Over-water dispersal of lizards due to hurricanes. *Nature, 395,* 556.

Cerri, G., Cabinio, M., Blasi, V., Borroni, P., Iadanza, A., Fava, E., . . . Bello, L. (2015). The mirror neuron system and the strange case of Broca's area. *Human Brain Mapping, 36,* 1010–1027.

Chang, L., Fang, Q., Zhang, S., Poo, M., & Gong, N. (2015). Mirror-induced self-directed behaviors in rhesus monkeys after visual-somatosensory training. *Current Biology, 25,* 212–217.

Chapman, K. M., Weiss, D. J., & Rosenbaum, D. A. (2010). Evolutionary roots of motor planning: The end-state comfort effect in lemurs. *Journal of Comparative Psychology, 124,* 229–232.

Chatterjee, H. J., Ho, S. Y. W., Barnes, I., & Groves, C. (2009). Estimating the phylogeny and divergence times of primates using a supermatrix approach. *BMC Evolutionary Biology, 9,* 259.

Cheney, D. L., & Seyfarth, R. M. (2007). *Baboon metaphysics: The evolution of a social mind.* Chicago: University of Chicago Press.

Chester, S. G. B., Bloch, J. I., Boyer, D. M., & Clemens, W. A. (2015). Oldest known euarchontan tarsals and affinities of Paleocene *Purgatorius* to primates. *Proceedings of the National Academy of Sciences, 112,* 1487–1492.

Chirchir, H., Kivell, T. L., Ruff, C. B., Hublin, J-J., Carlson, K. J., Zipfel, B., & Richmond, B. G. (2015). Recent origin of low trabecular bone density in modern humans. *Proceedings of the National Academy of Sciences, 112,* 366–371.

Choleris, E., & Kavaliers, M. (1999). Social learning in animals: Sex differences and neurobiological analysis. *Pharmacology Biochemistry and Behavior, 64,* 767–776.

Chow, R. L., & Lang, R. A. (2001). Early eye development in vertebrates. *Annual Review of Cell and Developmental Biology, 17,* 255–296.

Clark, G. (1967). *The stone age hunters.* New York, NY: McGraw-Hill.

Clark, M., Ferrara, T., Jones, D., Marion, A., Rose, K., & Yaeger, L. (1990). Koko's Mac II: A preliminary report. In B. Laurel (Ed.), *The art of human-computer interface design* (pp. 95–102). Reading, MA: Addison Wesley.

Clary, D., & Kelly, D. M. (2016). Graded mirror self-recognition by Clark's nutcrackers. *Scientific Reports, 6,* article 36459.

Claxton, A. G., Hammond, A. S., Romano, J., Oleinik, E., & DeSilva, J. M. (2016). Virtual reconstruction of the *Australopithecus africanus* pelvis Sts 65 with implications for obstetrics and locomotion. *Journal of Human Evolution, 99,* 10–24.

Clayton, N. S., Bussey, T. J., & Dickinson, A. (2003). Can animals recall the past and plan for the future? *Nature Reviews Neuroscience, 4,* 685–691.

Coates, M. I., Jeffery, J. E., & Ruta, M. (2002). Fish to limbs: What the fossils say. *Evolution & Development, 4,* 390–401.

Cochet, H., & Byrne, R. W. (2013). Evolutionary origins of human handedness: Evaluating contrasting hypotheses. *Animal Cognition, 16,* 531–542.

Coleman, M. N., & Ross, C. F. (2004). Primate auditory diversity and its influence on hearing performance. *The Anatomical Record Part A, 281A,* 1123–1137.

Collier-Baker, E., Davis, J. M., Nielsen, M., & Suddendorf, T. (2006). Do chimpanzees (*Pan troglodytes*) understand single invisible displacement? *Animal Cognition, 9,* 55–61.

Collins, A. G., Lipps, J. H., & Valentine, J. W. (2000). Modern mucociliary creeping trails and the bodyplans of neoproterozoic trace-makers. *Paleobiology, 26,* 47–55.

Cook, R., Bird, G., Catmure, C., Press, C., & Heyes, C. (2014). Mirror neurons: From origin to function. *Behavioral and Brain Sciences, 37,* 177–192.

Corballis, M. C. (2003). From mouth to hand: Gesture, speech, and the evolution of right-handedness. *Behavioral and Brain Sciences, 26,* 199–260.

Coren, S., & Porac, C. (1977). Fifty centuries of right-handedness: The historical record. *Science, 198,* 631–632.

Cornford, J. M. (1986). Specialized resharpening techniques and evidence of handedness. In P. Callow & J. M. Cornford (Eds.), *La Cotte de St. Brelade 1961–1978* (pp. 337–351). Norwich, England: Geo Books.

Cortés-Ortiz, L., Bermingham, E., Rico, C., Rodríguez-Luna, E., Sampaio, I., & Ruiz-García, M. (2003). Molecular systematics and biogeography of the neotropical monkey genus, *Alouatta*. *Molecular Phylogenetics and Evolution, 26*, 64–81.

Coull, J. T., Frackowiak, R. S. J., & Frith, C. D. (1998). Monitoring for target objects: Activation of right frontal and parietal cortices with increasing time on task. *Neuropsychologia, 36*, 1325–1334.

Cowen, R. (2005). *History of life* (4th ed.). Oxford: Blackwell Publishing.

Cowey, A., & Alexander, I. (2012). Are hemianopic monkeys and a human hemianope aware of visual events in the blind field? *Experimental Brain Research, 219*, 47–57.

Coyne, J. (2009). *Why evolution is true*. New York, NY: Viking Press.

Cracraft, J., & Donoghue, M. J. (2004). Assembling the tree of life: Where we stand at the beginning of the 21st century. In J. Cracraft & M. J. Donohue (Eds.), *Assembling the tree of life* (pp. 553–561). Oxford: Oxford University Press.

Cracraft, J., Feinstein, J., García-Moreno, J., Barker, F. K., Stanley, S., Sorenson, M. D., . . . Mindell, D. P. (2004). Phylogenetic relationships among modern birds (neornithes): Toward an avian tree of life. In J. Cracraft & M. J. Donohue (Eds.), *Assembling the tree of life* (pp. 468–489). Oxford: Oxford University Press.

Cronin, T., Arshad, Q., & Seemungal, B. M. (2017). Vestibular deficits in neurodegenerative disorders: Balance, dizziness, and spatial disorientation. *Frontiers in Neurology, 8*, article 538.

Crook, R., & Basil, J. (2008). A biphasic memory curve in the chambered nautilus, *Nautilus pompilius* L. (Cephalopoda: Nautiloidea). *The Journal of Experimental Biology, 211*, 1992–1998.

Culham, J. C., Cavina-Pratesi, C., & Singhal, A. (2006). The role of parietal cortex in visuomotor control: What have we learned from neuroimaging? *Neuropsychologia, 44*, 2668–2684.

Cunningham, J. A., Vargas, K., Yin, Z., Bengtson, S., & Donohue, P. C. J. (2017). The Weng'an biota (Doushantuo Formation): An Ediacaran window on soft-bodied and multicellular microorganisms. *Journal of the Geological Society, 174*, 793–802.

Custance, D., & Bard, K. A. (1994). The comparative and developmental study of self-recognition and imitation: The importance of social factors. In S. T. Parker, R. W. Mitchell, & M. L. Boccia (Eds.), *Self-awareness in animals and humans* (pp. 207–226). Cambridge, UK: Cambridge University Press.

Cutini, S., Scatturin, P., Moro, S. B., & Zorzi, M. (2014). Are the neural correlates of subitizing and estimation dissociable? An fNIRS investigation. *NeuroImage, 85*, 391–399.

Cyran, K. A., & Kimmel, M. (2010). Alternatives to the Wright-Fisher model: The robustness of mitochondrial Eve dating. *Theoretical Population Biology, 78*, 165–172.

Dadda, M., Cantalupo, C., & Hopkins, W. D. (2006). Further evidence of an association between handedness and neuroanatomical asymmetries in the primary motor cortex of chimpanzees (*Pan troglodytes*). *Neuropsychologia, 44*, 2582–2586.

Daeschler, E. B., Shubin, N. H., & Jenkins, F. A. (2006). A Devonian tetrapod-like fish and the evolution of the tetrapod body plan. *Nature, 440*, 757–763.

Dagosto, M. (2002). The origins and diversification of anthropoid primates: Introduction. In W. C. Hartwig (Ed.), *The primate fossil record* (pp. 125–132). Cambridge, UK: Cambridge University Press.

Dale, G., & Arnell, K. M. (2013). Investigating the stability of and relationships among global/local processing measures. *Attention, Perception, & Psychophysics, 75*, 394–406.

Danchin, E. G. J., Gouret, P., & Pontarotti, P. (2006). Eleven ancestral gene families lost in mammals and vertebrates while otherwise universally conserved in animals. *BMC Evolutionary Biology, 6*, 5.

D'Août, K., Vereecke, E., Schoonaert, K., De Clercq, D., Van Elsacker, L., & Aerts, P. (2004). Locomotion in bonobos (*Pan paniscus*): Differences and similarities between bipedal and quadrupedal terrestrial walking, and a comparison with other locomotor modes. *Journal of Anatomy, 204*, 353–361.

Darwin, C. (1859). *On the origin of species by means of natural selection*. London: John Murray.

Dávid-Barrett, T., & Dunbar, R. I. M. (2016). Bipedality and hair loss in human evolution revisited: The impact of altitude and activity scheduling. *Journal of Human Evolution, 94,* 72–82.

Davidson, I., & McGrew, W. C. (2005). Stone tools and the uniqueness of human culture. *Journal of the Royal Anthropological Institute, 11,* 793–817.

Davies, W. I. L., Collin, S. P., & Hunt, D. M. (2012). Molecular ecology and adaptation of visual photopigments in craniates. *Molecular Ecology, 21,* 3121–3158.

Deacon, T. W. (1992b). The human brain. In S. Jones, R. Martin, & D. Pilbeam (Eds.), *The Cambridge encyclopedia of human evolution* (pp. 115–123). Cambridge, UK: Cambridge University Press.

Deamer, D., & Weber, A. L. (2010). Bioenergetics and life's origins. *Cold Spring Harbor Perspectives in Biology, 2,* article a004929.

Deaner, R. O., Isler, K., Burkart, J., & van Schaik, C. (2007). Overall brain size, and not encephalization quotient, best predicts cognitive ability across non-human primates. *Brain, Behavior and Evolution, 70,* 115–124.

de Beaune, S. A. (2004). The invention of technology: Prehistory and cognition. *Current Anthropology, 45,* 139–151.

de Boer, B. (2012). Loss of air sacs improved hominin speech abilities. *Journal of Human Evolution, 62,* 1–6.

Deecke, V. B. (2012). Tool-use in the brown bear (*Ursus arctos*). *Animal Cognition, 15,* 725–730.

Dehaene, S., & Cohen, L. (2007). Cultural recycling of cortical maps. *Neuron, 56,* 384–398.

de Heinzelin, J., Clark, J. D., White, T., Hart, W., Renne, P., WoldeGabriel, G., . . . Vrba, E. (1999). Environment and behavior of 2.5-million-year-old Bouri hominids. *Science, 284,* 625–629.

de Ibarra, N. H., Vorobyev, M., Brandt, R., & Giurfa, M. (2000). Detection of bright and dim colours by honeybees. *The Journal of Experimental Biology, 203,* 3289–3298.

de Ibarra, N. H., Vorobyev, M., & Menzel, R. (2014). Mechanisms, functions and ecology of colour vision in the honeybee. *Journal of Comparative Physiology, A, 200,* 411–433.

de la Rosa, S., Schillinger, F. L., Bülthoff, H. H., Schulz, J., & Uludag, K. (2016). fMRI adaptation between action observation and action execution reveals cortical areas with mirror neuron properties in human BA 44/45. *Frontiers in Human Neuroscience, 10,* article 78.

Delgado-Bonal, A., & Martín-Torres, J. (2016). Human vision is determined based on information theory. *Scientific Reports, 6,* article 36038.

De Lillo, C., Palumbo, M., Spinozzi, G., & Giustino, G. (2012). Effects of pattern redundancy and hierarchical grouping on global-local visual processing in monkeys (*Cebus apella*) and humans (*Homo sapiens*). *Behavioural Brain Research, 226,* 445–455.

de Lussanet, M. H. E., & Osse, J. W. M. (2010). An ancestral axial twist explains the contralateral forebrain and the optic chiasm in vertebrates. *Animal Biology, 62,* 193–216.

de Manuel, M., Kuhlwilm, M., Frandsen, P., Sousa, V. C., Desai, T., Prado-Martinez, J., . . . Marques-Bonet, T. (2016). Chimpanzee genomic diversity reveals ancient admixture with bonobos. *Science, 354,* 477–481.

Dembo, M., Matzke, N. J., Mooers, A. O., & Collard, M. (2015). Bayesian analysis of a morphological supermatrix sheds light on controversial fossil hominin relationships. *Proceedings of the Royal Society, B, 282,* article 20150943.

Dembo, M., Radovcic, D., Garvin, H. M., Laird, M. F., Schroeder, L., Scott, J. E., . . . Collard, M. (2016). The evolutionary relationships and age of *Homo naledi*: An assessment using dated Bayesian phylogenetic methods. *Journal of Human Evolution, 97,* 17–26.

De Moraes, B., Souto, A., & Schiel, N. (2018). Adaptability in stone tool use by wild capuchin monkeys. *American Journal of Primatology, 76,* 967–977.

Denes, G. (1999). Disorders of body awareness and body knowledge. In G. Denes & L. Pizzamiglio (Eds.), *Handbook of clinical and experimental neuropsychology* (pp. 497–506). Hove, UK: Psychology Press.

Dennett, D. C. (1981). What is it like to be me? *New Scientist, 91,* 806–809.

Denys, K., Vanduffel, W., Fize, D., Nelissen, K., Peuskens, H., Van Essen, D., & Orban, G. A. (2004). The processing of visual shape in the cerebral cortex of human and nonhuman primates: A functional magnetic resonance imaging study. *The Journal of Neuroscience, 24,* 2551–2565.

de Oliveira, F. B., Molina, E. C., & Marroig, G. (2009). Paleogeography of the South Atlantic: A route for primates and rodents into the new world? In P. A. Garber, A. Estrada, J. C. Bicca-Marques, E. W. Heymann, & K. B. Strier (Eds.), *South American primates, developments in primatology: Progress and prospects* (pp. 55–68). New York, NY: Springer.

d'Errico, F., Henshilwood, C., Vanhaeren, M., & van Niekerk, K. (2005). *Nassarius kraussianus* shell beads from Blombos Cave: Evidence for symbolic behaviour in the middle stone age. *Journal of Human Evolution, 48*, 3–24.

Derst, C., & Karschin, A. (1998). Evolutionary link between prokaryotic and eukaryotic K+ channels. *The Journal of Experimental Biology, 201*, 2791–2799.

Desimone, R., Albright, T. D., Gross, C. G., & Bruce, C. (1984). Stimulus-selective properties of inferior temporal neurons in the macaque. *The Journal of Neuroscience, 4*, 2051–2062.

de Sousa, A. A., Sherwood, C. C., Mohlberg, H., Amunts, K., Schleicher, A., MacLeod, C. E., Hof, P. R., Frahm, H., & Zilles, K. (2010). Hominoid visual brain structure volumes and the position of the lunate sulcus. *Journal of Human Evolution, 58*, 281–292.

De Valois, R. L., & De Valois, K. K. (1990). *Spatial vision*. New York, NY: Oxford University Press.

de Veer, M. W., Gallup, G. G., Theall, L. A., van den Bos, R., & Povinelli, D. J. (2003). An 8-year longitudinal study of mirror self-recognition in chimpanzees (*Pan troglodytes*). *Neuropsychologia, 41*, 229–234.

de Waal, F. B. M. (2003). Darwin's legacy and the study of primate visual communication. *Annals of the New York Academy of Sciences, 1000*, 7–31.

de Waal, F. B. M., Dindo, M., Freeman, C. A., & Hall, M. J. (2005). The monkey in the mirror: Hardly a stranger. *Proceedings of the National Academy of Sciences, 102*, 11140–11147.

Diez-Martin, F., Yustos, P. S., Domínguez-Rodrigo, M., Mabulla, A. Z. P., Bunn, H. T., Ashley, G. M., . . . Baquedano, E. (2010). New insights into hominin lithic activities at FLK North Bed I, Olduvai Gorge, Tanzania. *Quaternary Research, 74*, 376–387.

Diez-Martin, F., Yustos, P. S., Uribelarrea, D., Baquedano, E., Mark, D. F., Mabulla, A., . . . Domínguez-Rodrigo, M. (2015). The origin of the Acheulean: The 1.7 million-year-old site of FLK West, Olduvai Gorge (Tanzania). *Scientific Reports, 5*, article 17839.

Dirks, P. H. G. M., Roberts, E. M., Hilbert-Wolf, H., Kramers, J. D., Hawks, J., Dosseto, A., . . . Berger, L. R. (2017). The age of *Homo naledi* and associated sediments in the Rising Star Cave, South Africa. *eLife, 6*, article e24231.

Do, M. T. H., & Yau, K-W. (2010). Intrinsically photosensitive retinal ganglion cells. *Physiological Reviews, 90*, 1547–1581.

Doglioni, C., Pignatti, J., & Coleman, M. (2016). Why did life develop on the surface of the earth in the Cambrian? *Geoscience Frontiers, 7*, 865–873.

Domínguez-Rodrigo, M., Pickering, T. R., & Bunn, H. T. (2010). Configurational approach to identifying the earliest hominin butchers. *Proceedings of the National Academy of Sciences, 107*, 20929–20934.

Dominguez-Rodrigo, M., Serrallonga, J., Juan-Tresserras, J., Alcala, L., & Luque, L. (2001). Woodworking activities by early humans: A plant residue analysis on Acheulian stone tools from Peninj (Tanzania). *Journal of Human Evolution, 40*, 289–299.

Dominy, N. J. (2004). Fruits, fingers, and fermentation: The sensory cues available to foraging primates. *Integrative and Comparative Biology, 44*, 295–303.

Dominy, N. J., & Lucas, P. W. (2001). Ecological importance of trichromatic vision to primates. *Nature, 410*, 363–366.

dos Reis, M., Thawomwattana, Y., Angelis, K., Telford, M. J., Donoghue, P. C. J., & Yang, Z. (2015). Uncertainty in the timing of origin of animals and the limits of precision in molecular timescales. *Current Biology, 25*, 2929–2950.

Drapeau, M. S. M. (2015). Metacarpal torsion in apes, humans, and early *Australopithecus*: Implications for manipulative abilities. *PeerJ, 3*, article e1311.

Drayton, L. A., & Santos, L. R. (2017). Do rhesus macaques, *Macaca mulatta*, understand what others know when gaze following? *Animal Behaviour, 134*, 193–199.

Drummond, A. J., Ho, S. Y. W., Phillips, M. J., & Rambaut, A. (2006). Relaxed phylogenetics and dating with confidence. *PLoS Biology, 4*, 699–710.

Duncan, J. S., & Fritzsch, B. (2012). Evolution of sound and balance perception: Innovations that aggregate single hair cells into the ear and transform a gravistatic sensor into the Organ of Corti. *The Anatomical Record, 295*, 1760–1774.

Dunsworth, H., & Walker, A. (2002). Early genus *Homo*. In W. C. Hartwig (Ed.), *The primate fossil record* (pp. 419–435). Cambridge, UK: Cambridge University Press.

Eccles, J. C. (1966). Conscious experience and memory. In J. E. Eccles (Ed.), *Brain and conscious experience*. New York, NY: Springer-Verlag.

Eichenbaum, H., & Fortin, N. J. (2005). Bridging the gap between brain and behavior: Cognitive and neural mechanisms of episodic memory. *Journal of the Experimental Analysis of Behavior, 84*, 619–629.

Ekstrom, A. D., Kahana, M. J., Caplan, J. B., Fields, T. A., Isham, E. A., Newman, E. L., & Fried, I. (2003). Cellular networks underlying human spatial navigation. *Nature, 425*, 184–187.

Elton, S. (2008). The environmental context of human evolutionary history in Eurasia and Africa. *Journal of Anatomy, 212*, 377–393.

Emery, N. J. (2006). Cognitive ornithology: The evolution of avian intelligence. *Philosophical Transactions of the Royal Society, B, 361*, 23–43.

Emonet, E-G., Andossa, L., Mackaye, H. T., & Brunet, M. (2014). Subocclusal dental morphology of *Sahelanthropus tchadensis* and the evolution of teeth in hominins. *American Journal of Physical Anthropology, 153*, 116–123.

Erwin, D. H., Laflamme, M., Tweedt, S. M., Sperling, E. A., Pisani, D., & Peterson, K. J. (2011). The Cambrian conundrum: Early divergence and later ecological success in the early history of animals. *Science, 334*, 1091–1097.

Estalrrich, A., & Rosas, A. (2013). Handedness in Neandertals from the El Sidrón (Asturias, Spain): Evidence from instrumental striations with ontogenetic inferences. *PLoS One, 8*, article e62797.

Fagot, J., & Deruelle, C. (1997). Processing of global and local visual information and hemispheric specialization in humans (*Homo sapiens*) and baboons (*Papio papio*). *Journal of Experimental Psychology: Human Perception and Performance, 23*, 429–442.

Fagot, J., & Tomonaga, M. (1999). Global and local processing in humans (*Homo sapiens*) and chimpanzees (*Pan troglodytes*): Use of a visual search task with compound stimuli. *Journal of Comparative Psychology, 113*, 3–12.

Falk, D. (1983). Cerebral cortices of East African early hominids. *Science, 221*, 1072–1074.

Falk, D. (2014). Interpreting sulci on hominin endocasts: Old hypotheses and new findings. *Frontiers in Human Neuroscience, 8*, article 134.

Falk, D., Hildebolt, C., Smith, K., Morwood, M. J., Sutikna, T., Brown, P., . . . Prior, F. (2005). The brain of LB1. *Homo floresiensis. Science, 308*, 242–245.

Falk, D., Hildebolt, C., Smith, K., Morwood, M. J., Sutikna, T., Jatmiko., . . . Prior, F. (2006). Response to comment on "The Brain of LB1, *Homo floresiensis*". *Science, 312*, 999c.

Falk, D., Hildebolt, C., Smith, K., Morwood, M. J., Sutikna, T., Jatmiko., . . . Prior, F. (2007). Brain shape in human microcephalics and *Homo floresiensis. Proceedings of the National Academy of Sciences, 104*, 2513–2518.

Falk, D., Hildebolt, C., Smith, K., Morwood, M. J., Sutikna, T., Jatmiko., . . . Prior, F. (2009). LB1's virutal endocast, microcephaly, and hominin brain evolution. *Journal of Human Evolution, 57*, 601.

Faurie, C., & Raymond, M. (2004). Handedness frequency over more than ten thousand years. *Proceedings of the Royal Society of London, B, 271*, S43-S45.

Faurie, C., & Raymond, M. (2005). Handedness, homicide and negative frequency-dependent selection. *Proceedings of the Royal Society of London, B, 272*, 25–28.

FBI. (2018). Retrieved May 1, 2018, from https://ucr.fbi.gov/crime-in-the-u.s/2015/crime-in-the-u.s.-2015/tables/expanded_homicide_data_table_8_murder_victims_by_weapon_2011-2015.xls

Feder, J. L., Opp, S. B., Wlazlo, B., Reynolds, K., Go, W., & Spisak, S. (1994). Host fidelity is an effective premating barrier between sympatric races of the apple maggot fly. *Proceedings of the National Academy of Sciences, 91*, 7990–7994.

Fedonkin, M. A. (2003). The origin of the Metazoa in the light of the Proterozoic fossil record. *Paleontological Research, 7*, 9–41.

Fernald, R. D. (2000). Evolution of eyes. *Current Opinion in Neurobiology, 10*, 444–450.

Ferrier, D. E. K. (2016). The origin of the hox/parahox genes, the ghost locus hypothesis and the complexity of the first animal. *Briefings in Functional Genomics, 15*, 333–341.

Ferris, J. P. (2005). Mineral catalysis and prebiotic synthesis: Montmorillonite-catalyzed formation of RNA. *Elements, 1*, 145–149.

Ferris, J. P., Joshi, P. C., Wang, K-J., Miyakawa, S., & Huang, W. (2004). Catalysis in prebiotic chemistry: Application to the synthesis of RNA oligomers. *Advances in Space Research, 33*, 100–105.

Field, G. D., & Sampath, A. P. (2017). Behavioural and physiological limits to vision in mammals. *Philosophical Transactions of the Royal Society, B, 372*, article 20160072.

Field, Y., Boyle, E. A., Telis, N., Gao, Z., Gaulton, K. J., Golan, D., . . . Pritchard, J. K. (2016). Direction of human adaptation during the past 2000 years. *Science, 354*, 760–764.

Fincham, J. M., Carter, C. S., van Veen, V., Stenger, V. A., & Anderson, J. R. (2002). Neural mechanisms of planning: A computational analysis using event-related fMRI. *Proceedings of the National Academy of Sciences, 99*, 3346–3351.

Finnerty, J. R. (2005). Did internal transport, rather than directed locomotion, favor the evolution of bilateral symmetry in animals? *BioEssays, 27*, 1174–1180.

Fischer, M. S., Schilling, N., Schmidt, M., Haarhaus, D., & Witte, H. (2002). Basic limb kinematics of small therian mammals. *The Journal of Experimental Biology, 205*, 1315–1338.

Fleagle, J. G. (1992). Primate locomotion and posture. In S. Jones, R. Martin, & D. Pilbeam (Eds.), *The Cambridge encyclopedia of human evolution* (pp. 75–79). Cambridge, UK: Cambridge University Press.

Foley, N. M., Springer, M. S., & Teeling, E. C. (2016). Mammal madness: Is the mammal tree of life not yet resolved? *Philosophical Transactions of the Royal Society, B, 371*, article 20150140.

Follmann, H., & Brownson, C. (2009). Darwin's warm little pond revisited: From molecules to the origin of life. *Naturwissenschaften, 96*, 1265–1292.

Forbes, A. A., Powell, T. H. Q., Stelinski, L. L., Smith, J. J., & Feder, J. L. (2009). Sequential sympatric speciation across trophic levels. *Science, 323*, 776–779.

Foundas, A. L., Leonard, C. M., Gilmore, R., Fennell, E., & Heilman, K. M. (1994). Planum temporale asymmetry and language dominance. *Neuropsychologia, 32*, 1225–1231.

Fouts, R. S., Fouts, D. H., & Van Cantfort, T. E. (1989). The infant Loulis learns signs from cross-fostered chimpanzees. In R. A. Gardner, B. T. Gardner, & T. E. Van Cantfort (Eds.), *Teaching sign language to chimpanzees* (pp. 280–292). Albany: State University of New York Press.

Fouts, R. S., Jensvold, M. L. A., & Fouts, D. H. (2002). Chimpanzee signing: Darwinian realities and Cartesian delusions. In M. Beckoff, C. Allen, & G. M. Burghardt (Eds.), *The cognitive animal: Empirical and theoretical perspectives on animal cognition* (pp. 285–291). Cambridge, MA: MIT Press.

Fouts, R. S., & Waters, G. (2003). Unbalanced human apes and syntax. *Behavioral and Brain Sciences, 26*, 221–222.

Fox, C. L., & Frayer, D. W. (1997). Non-dietary marks in the anterior dentition of the Krapina Neanderthals. *International Journal of Osteoarchaeology, 7*, 133–149.

Fox, E. A., Sitompul, A. F., & Van Schaik, C. P. (1999). Intelligent tool use in wild Sumatran orangutans. In S. T. Parker, R. W. Mitchell, & H. L. Miles (Eds.), *The mentalities of gorillas and orangutans: Comparative perspectives* (pp. 99–116). Cambridge, UK: Cambridge University Press.

Fragaszy, D., Izar, P., Visalberghi, E., Ottoni, E. B., & de Oliviera, M. G. (2004). Wild capuchin monkeys (*Cebus libidinosus*) use anvils and stone pounding tools. *American Journal of Primatology, 64*, 359–366.

Franchi, M., & Gallori, E. (2005). A surface-mediated origin of the RNA world: Biogenic activities of clay-adsorbed RNA molecules. *Gene, 346*, 205–214.

Francks, C., DeLisi, L. E., Shaw, S. H., Fisher, S. E., Richardson, A. J., Stein, J. F., & Monaco, A. P. (2003). Parent-of-origin effects on handedness and schizophrenia susceptibility on chromosome 2p12-q11. *Human Molecular Genetics, 12*, 3225–3230.

Frayer, D. W., Clarke, R. J., Fiore, I., Blumenschine, R. J., Pérez-Pérez, A., Martinez, L. M.,. . . . Bondioli, L. (2016). OH-65: The earliest evidence for right-handedness in the fossil record. *Journal of Human Evolution, 100*, 65–72.

Frey, S. H., Vinton, D., Norlund, R., & Grafton, S. T. (2005). Cortical topography of human anterio intraparietal cortex active during visually guided grasping. *Cognitive Brain Research, 23*, 397–405.

Friederici, A. D. (2017). Evolution of the neural language network. *Psychonomic Bulletin & Review, 24*, 41–47.

Fritzsch, B. (1987). Inner ear of the coelacanth fish *Latimeria* has tetrapod affinities. *Nature, 327*, 153–154.

Fritzsch, B., Beisel, K. W., Jones, K., Fariñas, I., Maklad, A., Lee, J., & Reichardt, L. F. (2002). Development and evolution of inner ear sensory epithelia and their innervation. *Journal of Neurobiology, 53*, 143–156.

Fritzsch, B., Pan, N., Jahan, I., Duncan, J. S., Kopecky, B. J., Elliott, K. L., . . . Yang, T. (2013). Evolution and development of the tetrapod auditory system: An organ of corti-centric perspective. *Evolution & Development, 15*, 63–79.

Fritzsch, B., Pauley, S., & Beisel, K. W. (2006). Cells, molecules and morphogenesis: The making of the vertebrate ear. *Brain Research, 1091*, 151–171.

Fyhn, M., Hafting, T., Witter, M. P., Moser, E. I., & Moser, M-B. (2008). Grid cells in mice. *Hippocampus, 18*, 1230–1238.

Galán, A. B., & Domínguez-Rodrigo, M. (2014). Testing the efficiency of simple flakes, retouched flakes and small handaxes during butchery. *Archaeometry, 56*, 1054–1074.

Gallistel, C. R., & Gelman, R. (1992). Preverbal and verbal counting and computation. *Cognition, 44*, 43–74.

Gallup, G. G. (1968). Mirror-image stimulation. *Psychological Bulletin, 70*, 782–793.

Gallup, G. G. (1970). Chimpanzees: Self-recognition. *Science, 167*, 86–87.

Gallup, G. G. (1977). Self-recognition in primates. *American Psychologist, 32*, 329–338.

Gallup, G. G. (1982). Self-awareness and the emergence of mind in primates. *American Journal of Primatology, 2*, 237–248.

Gallup, G. G. (1998). Can animals empathize? Yes. *Scientific American Presents, 9*, 66–71.

Gallup, G. G., Anderson, J. R., & Shillito, D. J. (2002). The mirror test. In M. Bekoff, C. Allen, & G. M. Burghardt (Eds.), *The cognitive animal: Empirical and theoretical perspectives on animal cognition* (pp. 325–334). Cambridge, MA: MIT Press.

Gamberini, M., Bò, G. D., Breveglieri, R, Briganti, S., Passarelli, L., Fattori, P., & Galletti, C. (2018). Sensory properties of the caudal aspect of the macaque's superior parietal lobule. *Brain Structure and Function, 223*, 1863–1879.

Gamberini, M., Galletti, C., Bosco, A., Breveglieri, R., & Fattori, P. (2011). Is the medial posterior parietal area V6a a single functional area? *The Journal of Neuroscience, 31*, 5145–5157.

Gannon, P. J., Kheck, N., & Hof, P. R. (2008). Leftward interhemispheric asymmetry of macaque monkey temporal lobe language area homolog is evident at the cytoarchitectural, but not gross anatomic level. *Brain Research, 1199*, 62–73.

Garde, M. M., & Cowey, A. (2000). "Deaf hearing": Unacknowledged detection of auditory stimuli in a patient with cerebral deafness. *Cortex, 2000, 36*, 71–80.

Gardner, R. A., & Gardner, B. T. (1998). *The structure of learning: From sign stimuli to sign language*. Mahwah, NJ: Lawrence Erlbaum Associates.

Garm, A., Oskarsson, M., & Nilsson, D-E. (2011). Box jellyfish use terrestrial visual cues for navigation. *Current Biology, 21*, 798–803.

Gasc, J-P. (2001). Comparative aspects of gait, scaling and mechanics in mammals. *Comparative Biochemistry and Physiology Part A, 131*, 121–133.

Gatesy, S. M. (1990). Caudefemoral musculature and the evolution of therapod locomotion. *Paleobiology, 16*, 170–186.

Gebo, D. L., Dagosto, M., Beard, K. C., Qi, T., & Wang, J. (2000). The oldest known anthropoid postcranial fossils and the early evolution of higher primates. *Nature, 404*, 276–278.

Gebo, D. L., MacLatchy, L., Kityo, A., Deino, A., Kingston, J., & Pilbeam, D. (1997). A hominoid genus from the early Miocene of Uganda. *Science, 276*, 401–404.

Gehring, W. J. (2005). New perspectives on eye development and the evolution of eyes and photoreceptors. *Journal of Heredity, 96*, 171–184.

Georgieva, S., Peeters, R., Kolster, H., Todd, J. T., & Orban, G. A. (2009). The processing of three-dimensional shape from disparity in the human brain. *The Journal of Neuroscience, 29,* 727–742.

Gerbault, P., Liebert, A., Itan, Y., Powell, A., Currat, M., Burger, J., . . . Thomas, M. G. (2011). Evolution of lactase persistence: An example of human niche construction. *Philosophical Transactions of the Royal Society, B, 366,* 863–877.

Ghazanfar, A. A., & Hauser, M. D. (2001). The auditory behaviour of primates: A neuroethological perspective. *Current Opinion in Neurobiology, 11,* 712–720.

Gibbons, A. (1997). Y chromosome shows that Adam was an African. *Science, 278,* 804–805.

Gibbons, A. (2009). A new kind of ancestor: *Ardipithecus* unveiled. *Science, 326,* 36–40.

Gibson, E. K., McKay, D. S., Thomas-Keptra, K., & Romanek, C. S. (2000). The case for relic life on Mars. In D. H. Levy (Ed.), *The Scientific American book of the cosmos* (pp. 283–291). New York, NY: St. Martin's Press.

Gierlinski, G. D., Niedzwiedzki, G., Lockley, M. G., Athanassiou, A., Fassoulas, C., Dubicka, Z., . . . Ahlberg, P. E. (2017). Possible hominin footprints from the late Miocene (c. 5.7 Ma) of Crete? *Proceedings of the Geologists' Association, 128,* 697–710.

Gilad, Y., Bustamante, C. D., Lancet, D., & Pääbo, S. (2003). Natural selection on the olfactory receptor gene family in humans and chimpanzees. *American Journal of Human Genetics, 73,* 489–501.

Gilad, Y., Man, O., & Glusman, G. (2005). A comparison of the human and chimpanzee olfactory receptor gene repertoires. *Genome Research, 15,* 224–230.

Gilad, Y., Wiebe, V., Przeworski, M., Lancet, D., & Pääbo, S. (2004). Loss of olfactory receptor genes coincides with the acquisition of full trichromatic vision in primates. *PLoS Biology, 2,* 120–125.

Gilbert, S. J., Spengler, S., Simons, J. S., Steele, J. D., Lawrie, S. M., Frith, C. D., & Burgess, P. W. (2006). Functional specialization within rostral prefrontal cortex (area 10): A meta-analysis. *Journal of Cognitive Neuroscience, 18,* 932–948.

Gilligan, I. (2010). The prehistoric development of clothing: Archaeological implications of a thermal model. *Journal of Archaeological Method and Theory, 17,* 15–80.

Gilliland, A. R. (1948). The rate of forgetting. *Journal of Educational Psychology, 39,* 19–26.

Glezer, V. D. (1995). *Vision and mind: Modeling mental functions.* Mahwah, NJ: Lawrence Erlbaum Associates.

Glover, S. (2004). Separate visual representations in the planning and control of action. *Behavioral and Brain Sciences, 27,* 3–78.

Glueck, M., Crane, K., Anderson, S., Rutnik, A., & Khan, A. (2009). Multiscale 3D reference visualization. *Proceedings of I3D 2009: The 2009 ACM SIGGRAPH Symposium on Interactive 3D Graphics and Games,* 225–232.

Go, Y. (2006). Lineage-specific expansions and contractions of the bitter taste receptor gene repertoire in vertebrates. *Molecular Biology and Evolution, 23,* 964–972.

Go, Y., & Niimura, Y. (2008). Similar numbers but different repertoires of olfactory receptor genes in humans and chimpanzees. *Molecular Biology and Evolution, 25,* 1897–1907.

Gommery, D., & Senut, B. (2006). La phalange distale du pouce d'*Orrorin tugenensis* (Miocène supérieur du Kenya). *Geobios, 39,* 372–384.

Goodchild, A. K., Ghosh, K. K., & Martin, P. R. (1996). Comparison of photoreceptor spatial density and ganglion cell morphology in the retina of human, macaque monkey, cat, and the marmoset *Callithrix jacchus. The Journal of Comparative Neurology, 366,* 55–75.

Gough, P. M., Connally, E. L., Howell, P., Ward, D., Chesters, J., & Watkins, K. E. (2018). Planum temporale asymmetry in people who stutter. *Journal of Fluency Disorders, 55,* 94–105.

Goulas A., Bastiani M., Bezgin G., Uylings H. B. M., Roebroeck A., et al. (2014). Comparative analysis of the macroscale structural connectivity in the macaque and human brain. *PLoS Computational Biology, 10*(3): e1003529.

Gowlett, J. A. J. (1992a). Early human mental abilities. In S. Jones, R. Martin, & D. Pilbeam (Eds.), *The Cambridge encyclopedia of human evolution* (pp. 341–345). Cambridge, UK: Cambridge University Press.

Gowlett, J. A. J. (1992b). Tools – The Paleolithic record. In S. Jones, R. Martin, & D. Pilbeam (Eds.), *The Cambridge encyclopedia of human evolution* (pp. 350–360). Cambridge, UK: Cambridge University Press.

Gowlett, J. A. J. (2016). The discovery of fire by humans: A long and convoluted process. *Philosophical Transactions of the Royal Society, B, 371*, article 20150164.

Granatosky, M. C., Tripp, C. H., Fabre, A-C., & Schmitt, D. (2016). Patterns of quadrupedal locomotion in a vertical clinging and leaping primate (*Propithecus coquereli*) with implications for understanding the functional demands of primate quadrupedal locomotion. *American Journal of Physical Anthropology, 160*, 644–652.

Granit, R. (1977). *The purposive brain.* Cambridge, MA: MIT Press.

Graphodatsky, A. S., Trifonov, V. A., & Stanyon, R. (2011). The genome diversity and karyotype evolution of mammals. *Molecular Cytogenetics, 4*, article 22.

Green, R. E., Krause, J., Briggs, A. W., Maricic, T., Stenzel, U., Kircher, M., . . . Pääbo, S. (2010). A draft sequence of the Neandertal genome. *Science, 328*, 710–722.

Green, R. E., Krause, J., Ptak, S. E., Briggs, A. W., Ronan, M. T., Simons, J. F., . . . Pääbo, S. (2006). Analysis of one million base pairs of Neanderthal DNA. *Nature, 444*, 330–336.

Green, R. E., Malaspinas, A-S., Krause, J., Briggs, A. W., Johnson, P. L. F., Uhler, C., . . . Pääbo, S. (2008). A complete Neandertal mitochondrial genome sequence determined by high-throughput sequencing. *Cell, 134*, 416–426.

Greenfield, P. M., & Savage-Rumbaugh, E. S. (1991). Imitation, grammatical development, and the invention of protogrammar by an ape. In N. A. Krasnegor, D. M. Rumbaugh, R. L. Schiefelbusch, & M. Studdert-Kennedy (Eds.), *Biological and behavioral determinants of language development* (pp. 235–258). Hillsdale, NJ: Lawrence Erlbaum Associates.

Greenwood, J. A., Szinte, M., Sayim, B., & Cavanagh, P. (2017). Variations in crowding, saccadic precision, and spatial localization reveal the shared topology of spatial vision. *Proceedings of the National Academy of Sciences, 114*, E3573–E3582.

Grefkes, C., & Fink, G. R. (2005). The functional organization of the intraparietal sulcus in humans and monkeys. *Journal of Anatomy, 207*, 3–17.

Grefkes, C., Weiss, P. H., Zilles, K., & Fink, G. R. (2002). Crossmodal processing of object features in human anterior intraparietal cortex: An fMRI study implies equivalencies between humans and monkeys. *Neuron, 35*, 173–184.

Grimm, R. E., & Marchi, S. (2018). Direct thermal effects of the Hadrean bombardment did not limit early subsurface habitability. *Earth and Planetary Science Letters, 485*, 1–8.

Gross, C. G. (2002). Genealogy of the "grandmother cell". *The Neuroscientist, 8*, 512–518.

Haaland, K. Y., Harrington, D. L., & Knight, R. T. (2000). Neural representations of skilled movement. *Brain, 123*, 2306–2313.

Haile-Selassie, Y., Suwa, G., & White, T. D. (2004). Late Miocene teeth from Middle Awash, Ethiopia, and early hominid dental evolution. *Science, 303*, 1503–1505.

Hallos, J. (2005). "15 minutes of fame": Exploring the temporal dimension of Middle Pleistocene lithic technology. *Journal of Human Evolution, 49*, 155–179.

Halsband, U., & Lange, R. K. (2006). Motor learning in man: A review of functional and clinical studies. *Journal of Physiology – Paris, 99*, 414–424.

Halsey, L. G., & White, C. R. (2012). Comparative energetics of mammalian locomotion: Humans are not different. *Journal of Human Evolution, 63*, 718–722.

Hampton, R. R., & Schwartz, B. L. (2004). Episodic memory in nonhumans: What, and where, is when? *Current Opinion in Neurobiology, 14*, 192–197.

Hänelt, I., Tholema, N., Kröning, N., Vor der Brüggen, M., & Wunnicke, D. (2011). KtrB, a member of the superfamily of K+ transporters. *European Journal of Cell Biology, 90*, 696–704.

Hansma, H. G. (2010). Possible origin of life between mica sheets. *Journal of Theoretical Biology, 266*, 175–188.

Hansma, H. G. (2014). The power of crowding for the origins of life. *Origins of Life and Evolution of Biospheres, 44*, 307–311.

Harcourt-Smith, W. E. H., & Aiello, L. C. (2004). Fossils, feet and the evolution of human bipedal locomotion. *Journal of Anatomy, 204*, 403–416.

Harmand, S., Lewis, J. E., Feibel, C. S., Lepre, C. J., Prat, S., Lenoble, A., . . . Roche, H. (2015). 3.3-million-year-old stone tools from Lomekwi 3, West Turkana, Kenya. *Nature, 521*, 310–316.

Harrington, A. R., Silcox, M. T., Yapunich, G. S., Boyer, D. M., & Bloch, J. I. (2016). First virtual endocasts of adapiform primates. *Journal of Human Evolution, 99*, 52–78.

Harrison, T. (2002). Late Oligocene to middle Miocene catarrhines from Afro-Arabia. In W. C. Hartwig (Ed.), *The primate fossil record* (pp. 311–338). Cambridge, UK: Cambridge University Press.

Hart, B. L., Hart, L. A., & Pinter-Wollman, N. (2008). Large brains and cognition: Where do elephants fit in? *Neuroscience and Biobehavioral Reviews, 32*, 86–98.

Hatala, K. G., Demes, B., & Richmond, B. G. (2016). Laetoli footprints reveal bipedal gait biomechanics different from those of modern humans and chimpanzees. *Proceedings of the Royal Society, B, 283*, article 20160235.

Havrilak, J. A., Faltine-Gonzalez, D., Wen, Y., Fodera, D., Simpson, A. C., Magie, C. R., & Layden, M. J. (2017). Characterization of *NvLWamide-like* neurons reveals stereotypy in *Nematostella* nerve net development. *Developmental Biology, 431*, 336–346.

Hawks, J. (2009). *Ardipithecus FAQ*. Retrieved February 15, 2019 from http://johnhawks.net/weblog/fossils/ardipithecus/ardipithecus-faq-2009.html

Hayden, B. (2015). Insights into early lithic technologies from ethnography. *Philosophical Transactions of the Royal Society, B, 370*, article 20140356.

He, L., Zuo, Z., Chen, L., & Humphrey, G. (2013). Effects of number magnitude and notation at 7T: Separating the neural response to small and large, symbolic and nonsymbolic number. *Cerebral Cortex, 24*, 2199–2209.

Hedges, S. B., Marin, J., Suleski, M., Paymer, M., & Kumar, S. (2015). Tree of life reveals clocklike speciation and diversification. *Molecular Biology and Evolution, 32*, 835–845.

Heesy, C. P. (2004). On the relationship between orbit orientation and binocular visual field overlap in mammals. *The Anatomical Record Part A, 281A*, 1104–1110.

Heesy, C. P. (2008). Ecomorphology of orbit orientation and the adaptive significance of binocular vision in primates and other mammals. *Brain, Behavior, and Evolution, 71*, 54–67.

Heffner, R. S. (2004). Primate hearing from a mammalian perspective. *The Anatomical Record Part A, 281A*, 1111–1122.

Heilbroner, P. L., & Holloway, R. L. (1988). Anatomical brain asymmetries in new world and old world monkeys: Stages of temporal lobe development in primate evolution. *American Journal of Physical Anthropology, 76*, 39–48.

Heilman, K. M. (2005). *Creativity and the brain*. New York, NY: Psychology Press.

Henke, J. M., & Bassler, B. L. (2004). Bacterial social engagements. *Trends in Cell Biology, 14*, 648–656.

Heritage, S. (2014). Modeling olfactory bulb evolution through primate phylogeny. *PLoS One, 9*, article e113904.

Hernández-Pérez, R., Cuaya, L. V., Rojas-Hortelano, E., Reyes-Aguilar, A., Concha, L., & De Lafuente, V. (2017). Tactile object categories can be decoded from the parietal and lateral-occipital cortices. *Neuroscience, 352*, 226–235.

Herrera, J. P., & Dávalos, L. M. (2016). Phylogeny and divergence times of lemurs inferred with recent and ancient fossils in the tree. *Systematic Biology, 65*, 772–791.

Hickman, C. P., Roberts, L. S., & Larson, A. (1997). *Integrated principles of zoology*. Boston: McGraw-Hill.

Hobaiter, C., & Byrne, R. W. (2010). Able-bodied chimpanzees imitate a motor procedure used by a disabled individual to overcome handicap. *PLoS One, 5*, e11959.

Hobaiter, C., & Byrne, R. W. (2014). The meanings of chimpanzee gestures. *Current Biology, 24*, 1596–1600.

Hobaiter, C., Poisot, T., Zuberbühler, K., Hoppitt, W., & Gruber, T. (2014). Social network analysis shows direct evidence for social transmission of tool use in wild chimpanzees. *PLoS Biology, 12*, article e1001960.

Hodges, B. H. (2017). Carrying, caring, and conversing: Constraints on the emergence of cooperation, conformity, and language. *Interaction Studies, 18*, 26–54.

Hoffecker, J. F. (2005). Innovation and technological knowledge in the Upper Paleolithic of northern Eurasia. *Evolutionary Anthropology, 14,* 186–198.

Hoffman, J. N., Montag, A. G., & Dominy, N. J. (2004). Meissner corpuscles and somatosensory acuity: The prehensile appendages of primates and elephants. *Anatomical Record Part A, 281A,* 1138–1147.

Hohn-Schulte, B., Preuschoft, H., Witzel, U., & Distler-Hoffmann, C. (2013). Biomechanics and functional preconditions for terrestrial lifestyle in basal tetrapods, with special consideration of *Tiktaalik roseae. Historical Biology, 25,* 167–181.

Hoke, K. L., & Fernald, R. D. (1997). Rod photoreceptor neurogenesis. *Progress in Retinal and Eye Research, 16,* 31–49.

Holdstock, J. S., Hocking, J., Notley, P., Devlin, J. T., & Price, C. J. (2009). Integrating visual and tactile information in the perirhinal cortex. *Cerebral Cortex, 19,* 2993–3000.

Holland, N. D., & Chen, J. (2001). Origin and early evolution of the vertebrates: New insights from advances in molecular biology, anatomy, and palaeontology. *BioEssays, 23,* 142–151.

Holliday, T. W., Hutchinson, V. T., Morrow, M. M. B., & Livesay, G. A. (2010). Geometric morphometric analyses of hominid proximal femora: Taxonomic and phylogenetic considerations. *HOMO – Journal of Comparative Human Biology, 61,* 3–15.

Hong, S. W., & Tong, F. (2017). Neural representation of form-contingent color filling-in in the early visual cortex. *Journal of Vision, 17,* 1–10.

Hopkins, W. D. (2006). Comparative and familial analysis of handedness in great apes. *Psychological Bulletin, 132,* 538–559.

Hopkins, W. D., & Cantalupo, C. (2004). Handedness in chimpanzees (*Pan troglodytes*) is associated with asymmetries of the primary motor cortex but not with homologous language areas. *Behavioral Neuroscience, 118,* 1176–1183.

Hopkins, W. D., & Cantero, M. (2003). From hand to mouth in the evolution of language: The influence of vocal behavior on lateralized hand use in manual gestures by chimpanzees (*Pan troglodytes*). *Developmental Science, 6,* 55–61.

Hopkins, W. D., Hopkins, A. M., Misiura, M., Latash, E. M., Mareno, M. C., Schapiro, S. J., & Phillips, K. A. (2016). Sex differences in the relationship between planum temporale asymmetry and corpus callosum morphology in chimpanzees (*Pan troglodytes*): A combined MRI and DTI analysis. *Neuropsychologia, 93(Pt. B),* 325–334.

Hopkins, W. D., Meguerditchian, A., Coulon, O., Misiura, M., Pope, S., Mareno, M. C., & Schapiro, S. J. (2017). Motor skill for tool-use is associated with asymmetries in Broca's area and the motor hand area of the precentral gyrus in chimpanzees (*Pan troglodytes*). *Behavioural Brain Research, 318,* 71–81.

Hopkins, W. D., Phillips, K. A., Bania, A., Calcutt, S. E., Gardner, M., Russell, J., . . . Schapiro, S. J. (2011). Hand preferences for coordinated bimanual actions in 777 great apes: Implications for the evolution of handedness in hominins. *Journal of Human Evolution, 60,* 605–611.

Hopkins, W. D., Pilcher, D. L., & MacGregor, L. (2000). Sylvian fissure asymmetries in nonhuman primates revisited: A comparative MRI study. *Brain, Behavior, and Evolution, 56,* 293–299.

Hopkins, W. D., Russell, J. L., & Cantalupo, C. (2007). Neuroanatomical correlates of handedness for tool use in chimpanzees (*Pan troglodytes*). *Psychological Science, 18,* 971–977.

Hopkins, W. D., Russell, J. L., Cantalupo, C., Freeman, H., & Schapiro, S. J. (2005). Factors influencing the prevalence and handedness for throwing in captive chimpanzees (*Pan troglodytes*). *Journal of Comparative Psychology, 119,* 363–370.

Hopkins, W. D., Russell, J. L., Freeman, H., Buehler, N., Reynolds, E., & Schapiro, S. J. (2005). The distribution and development of handedness for manual gestures in captive chimpanzees (*Pan troglodytes*). *Psychological Science, 16,* 487–493.

Hopkins, W. D., Russell, J. L., Hostetter, A., Pilcher, D., & Dahl, J. F. (2005). Grip preference, dermatoglyphics, and hand use in captive chimpanzees (*Pan troglodytes*). *American Journal of Physical Anthropology, 128,* 57–62.

Hopkins, W. D., Schaeffer, J., Russell, J. L., Bogart, S. L., Meguerditchian, A., & Coulon, O. (2015). A comparative assessment of handedness and its potential neuroanatomical correlates in chimpanzees (*Pan troglodytes*) and bonobos (*Pan paniscus*). *Behaviour, 152,* 461–492.

Hopkins, W. D., & Washburn, D. A. (2002). Matching visual stimuli on the basis of global and local features by chimpanzees (*Pan troglodytes*) and rhesus monkeys (*Macaca mulatta*). *Animal Cognition, 5*, 27–31.

Hopkins, W. D., Wesley, M. J., Russell, J. L., & Schapiro, S. J. (2006). Parental and perinatal factors influencing the development of handedness in captive chimpanzees. *Developmental Psychobiology, 48*, 428–435.

Hosfield, R. (2016). Walking in a winter wonderland? Strategies for early and middle Pleistocene survival in midlatitude Europe. *Current Anthropology, 57*, 653–682.

Houle, A. (1999). The origin of platyrrhines: An evaluation of the Antarctic scenario and the floating island model. *American Journal of Physical Anthropology, 109*, 541–559.

Hublin, J-J., Ben-Ncer, A., Bailey, S. E., Freidline, S. E., Neubauer, S., Skinner, M. M., . . . Gunz, P. (2017). New fossils from Jebel Irhoud, Morocco and the pan-African origin of *Homo sapiens*. *Nature, 546*, 289–292.

Hutchinson, J., Bates, K. T., Molnar, J., Allen, V., & Makovicky, P. J. (2011). A computational analysis of limb and body dimensions in tyrannosaurus rex with implications for locomotion, ontogeny, and growth. *PLoS ONE, 6*(10): e26037.

Hughes, A. L., da Silva, J., & Friedman, R. (2001). Ancient genome duplications did not structure the human Hox-bearing chromosomes. *Genome Research, 11*, 771–780.

Humphreys, G. L., & Forde, E. M. E. (2001). Hierarchies, similarity, and interactivity in object recognition: "Category-specific" neuropsychological deficits. *Behavioral and Brain Sciences, 24*, 453–476.

Humphreys, G. L., & Riddoch, M. J. (2003). A case series analysis of "category-specific" deficits of living things: The HIT account. *Cognitive Neuropsychology, 20*, 263–306.

Hung, C-C., Yen, C. C., Ciuchta, J. L., Papoti, D., Bock, N. A., Leopold, D. A., & Silva, A. C. (2015). Functional mapping of face-selective regions in the extrastriate visual cortex of the marmoset. *The Journal of Neuroscience, 35*, 1160–1172.

Hung, C. P., Kreiman, G., Poggio, T., & DiCarlo, J. J. (2005). Fast readout of object identity from macaque inferior temporal cortex. *Science, 310*, 863–866.

Hunley, K. L., Cabana, G. S., & Long, J. C. (2016). The apportionment of human diversity revisited. *American Journal of Physical Anthropology, 160*, 561–569.

Hunt, K. D. (1994). The evolution of human bipedality: Ecology and functional morphology. *Journal of Human Evolution, 26*, 183–202.

Hunt, K. D. (2016). Why are there apes? Evidence for the co-evolution of ape and monkey ecomorphology. *Journal of Anatomy, 228*, 630–685.

Hupé, J-M., James, A. C., Girard, P., Lomber, S. G., Payne, B. R., & Bullier, J. (2001). Feedback connections act on the early part of the responses in monkey visual cortex. *Journal of Neurophysiology, 85*, 134–145.

Hutchinson, C. V., Prados, J., & Davidson, C. (2015). Persistent conditioned place preference to cocaine and withdrawal hypo-locomotion to mephedrone in the flatworm planaria. *Neuroscience Letters, 593*, 19–23.

Huxley, T. H. (1870). Further evidence of the affinity between the dinosaurian reptiles and birds. *Quarterly Journal of the Geological Society*. Retrieved July 24, 2006 from http://aleph0.clarku.edu/huxley/SM3/Dino-boid.html

Hyatt, C. W. (1998). Responses of gibbons (*Hylobates lar*) to their mirror images. *American Journal of Primatology, 45*, 307–311.

Iachini, T., Ruggiero, G., Conson, M., & Trojano, L. (2009). Lateralization of egocentric and allocentric spatial processing after parietal brain lesions. *Brain and Cognition, 69*, 514–520.

Ibrahim, N., & Kutschera, U. (2013). The ornithologist Alfred Russel Wallace and the controversy surrounding the dinosaurian origin of birds. *Theory in Biosciences, 132*, 267–275.

Ichim, I., Kieser, J., & Swain, M. (2007). Tongue contractions during speech may have led to the development of the bony geometry of the chin following the evolution of human language: A mechanobiological hypothesis for the development of the human chin. *Medical Hypotheses, 69*, 20–24.

Ida, Y., & Bryden, M. P. (1996). A comparison of hand preference in Japan and Canada. *Canadian Journal of Experimental Psychology, 50*, 234–239.

Iidaka, T. (2014). Role of the fusiform gyrus and superior temporal sulcus in face perception and recognition: An empirical review. *Japanese Psychological Research, 56*, 33–45.

Indriati, E., Swisher, C. C., Lepre, C., Quinn, R. L., Suriyanto, R. A., Hascaryo, A. T., . . . Antón, S. C. (2011). The age of the 20 meter Solo River Terrace, Java, Indonesia and the survival of *Homo erectus* in Asia. *PLoS One, 6*, e21562.

Ingman, M., Kaessmann, H., Pääbo, S., & Gyllensten, U. (2000). Mitochondrial genome variation and the origin of modern humans. *Nature, 408*, 708–713.

Irisarri, I., & Meyer, A. (2016). The identification of the closest living relative(s) of tetrapods: Phylogenomic lessons for resolving short ancient internodes. *Systematic Biology, 65*, 1057–1075.

Irish, J. D., Guatelli-Steinberg, D., Legge, S. S., de Ruiter, D. J., & Berger, L. R. (2013). Dental morphology and the phylogenetic "place" of *Australopithecus sediba. Science, 340*, article 1233062.

Ishikawa, T., & Montello, D. R. (2006). Spatial knowledge acquisition from direct experience in the environment: Individual differences in the development of metric knowledge and the integration of separately learned places. *Cognitive Psychology, 52*, 93–129.

Isik, L., Meyers, E. M., Leibo, J. Z., & Poggio, T. (2014). The dynamics of invariant object recognition in the human visual system. *Journal of Neurophysiology, 111*, 91–102.

Islas, S., Velasco, A. M., Becerra, A., Delaye, L., & Lazcano, A. (2003). Hyperthermophily and the origin and earliest evolution of life. *International Microbiology, 6*, 87–94.

Iwaniuk, A. N., & Nelson, J. E. (2003). Developmental differences are correlated with relative brain size in birds: A comparative analysis. *Canadian Journal of Zoology, 81*, 1913–1928.

Jaakkola, K. (2014). Do animals understand invisible displacement? A critical review. *Journal of Comparative Psychology, 128*, 225–239.

Jablonski, N. G., & Chaplin, G. (2017). The colours of humanity: The evolution of pigmentation in the human lineage. *Philosophical Transactions of the Royal Society, B, 372*, article 20160349.

Jacobs, J., Weidermann, C. T., Miller, J. F., Solway, A., Burke, J. F., Wei, X-X., . . . Kahana, M. J. (2013). Direct recordings of grid-like neuronal activity in human spatial navigation. *Nature Neuroscience, 16*, 1188–1191.

James, W. (1892). *Psychology*. New York, NY: Henry Holt.

Janmaat, K. R. L., Ban, S. D., & Boesch, C. (2013). Chimpanzees use long-term spatial memory to monitor large fruit trees and remember feeding experiences across seasons. *Animal Behaviour, 86*, 1183–1205.

Jax, S. A., Buxbaum, L. J., & Moll, A. D. (2006). Deficits in movement planning and intrinsic coordinate control in ideomotor apraxia. *Journal of Cognitive Neuroscience, 18*, 2063–2076.

Jerison, H. J. (1973). *Evolution of the brain and intelligence*. New York, NY: Academic Press.

Ji, Q., Luo, Z-X., Yuan, C-X., Wible, J. R., Zhang, J-P., & Georgi, J. A. (2002). The earliest known eutherian mammal. *Nature, 416*, 816–822.

Jiggins, C. D., & Bridle, J. R. (2004). Speciation in the apple maggot fly: A blend of vintages? *Trends in Ecology and Evolution, 19*, 111–114.

Johanson, D. (2017). The paleoanthropology of Hadar, Ethiopia. *Comptes Rendus Palevol, 16*, 140–154.

Johnson, G. (1995, June 6). Chimp talk debate: Is it really language? *New York Times*, pp. C1, C10.

Johnson, M. C., & Wuensch, K. L. (1994). An investigation of habituation in the jellyfish *Aurelia aurita. Behavioral and Neural Biology, 61*, 54–59.

Johnson, R. R., & Burkhalter, A. (1996). Microcircuitry of forward and feedback connections within rat visual cortex. *The Journal of Comparative Neurology, 368*, 383–398.

Jones, S., Martin, R., & Pilbeam, D. (1992). Glossary. In *The Cambridge encyclopedia of human evolution* (pp. 458–472). Cambridge, UK: Cambridge University Press.

Joordens, J. C. A., d'Errico, F., Wesselingh, F. P., Munro, S., de Vos, J., Wallinga, J., . . . Roebroeks, W. (2015). *Homo erectus* at Trinil on Java used shells for tool production and engraving. *Nature, 518*, 228–231.

Josse, G., Mazoyer, B., Crivello, F., & Tzourio-Mazoyer, N. (2003). Left planum temporale: An anatomical marker of left hemispheric specialization for language comprehension. *Cognitive Brain Research, 18*, 1–14.

Kaas, J. H. (2004a). Evolution of somatosensory and motor cortex in primates. *The Anatomical Record Part A, 281A*, 1148–1156.

Kaas, J. H. (2004b). The evolution of the visual system in primates. In L. M. Chalupa & J. S. Werner (Eds.), *The visual neurosciences* (Vol. 2, pp. 1563–1572). Cambridge, MA: MIT Press.

Kaas, J. H. (2012). The evolution of neocortex in primates. *Progress in Brain Research, 195*, 91–102.

Kaas, J. H. (2013). The evolution of brains from early mammals to humans. *Wiley Interdisciplinary Reviews: Cognitive Science, 4*, 33–45.

Kahlenberg, S. M., & Wrangham, R. W. (2011). Sex differences in chimpanzees' use of sticks as play objects resemble those of children. *Current Biology, 20*, R1067–R1068.

Kaltenbach, S. L., Yu, J-K., & Holland, N. D. (2009). The origin and migration of the earliest-developing sensory neurons in the peripheral nervous system of amphioxus. *Evolution & Development, 11*, 142–151.

Kaminski, J., Riedel, J., Call, J., & Tomasello, M. (2005). Domestic goats, *Capra hircus*, follow gaze direction and use social cues in an object choice task. *Animal Behaviour, 69*, 11–18.

Kano, F., & Call, J. (2014). Cross-species variation in gaze following and conspecific preference among great apes, human infants and adults. *Animal Behaviour, 91*, 137–150.

Kanwisher, N., & Yovel, G. (2006). The fusiform face area: A cortical region specialized for the perception of faces. *Philosophical Transactions of the Royal Society B, 361*, 2109–2128.

Kapoor, B. G., & Khanna, B. (2004). *Ichthyology handbook*. New York, NY: Springer-Verlag.

Kappelman, J. (1996). The evolution of body mass and relative brain size in fossil hominids. *Journal of Human Evolution, 30*, 243–276.

Kasting, J. F., & Catling, D. (2003). Evolution of a habitable planet. *Annual Review of Astronomy and Astrophysics, 41*, 429–463.

Kavaklioglu, T., Ajmal, M., Hameed, A., & Francks, C. (2016). Whole exome sequencing for handedness in a large and highly consanguineous family. *Neuropsychologia, 93*, 342–349.

Kay, R. F., Campbell, V. M., Rossie, J. B., Colbert, M. W., & Rowe, T. B. (2004). Olfactory fossa of *Tremacebus harringtoni* (Platyrrhini, Early Miocene, Sacanana, Argentina): Implications for activity pattern. *The Anatomical Record Part A, 281A*, 1157–1172.

Kemp, A. D., & Kirk, E. C. (2014). Eye size and visual acuity influence vestibular anatomy in mammals. *The Anatomical Record, 297*, 781–790.

Kemp, T. S. (2005). *The origin and evolution of mammals*. Oxford: Oxford University Press.

Kherdjemil, Y., Lalonde, R. L., Sheth, R., Dumouchel, A., de Martino, G., Pineault, K. M., . . . Kmita, M. (2016). Evolution of *Hoxa11* regulation in vertebrates is linked to the pentadactyl state. *Nature, 539*, 89–92.

Kielan-Jaworowska, Z., & Hurum, J. H. (2006). Limb posture in early mammals: Sprawling or parasagittal. *Acta Paleontologica Polonica, 51*, 393–406.

Killian, N. J., Jutras, M. J., & Buffalo, E. A. (2012). A map of visual space in the primate entorhinal cortex. *Nature, 491*, 761–764.

Kim, J-W., Yang, H-J., Oel, A. P., Brooks, M. J., Jia, L., Plachetzki, D. C., . . . Swaroop, A. (2016). Recruitment of rod photoreceptors from short-wavelength-sensitive cones during the evolution of nocturnal vision in mammals. *Developmental Cell, 37*, 520–532.

Kircher, T., Nagles, A., Kirner-Veselinovic, A., & Krach, S. (2011). Neural correlates of rhyming vs. lexical and semantic fluency. *Brain Research, 1391*, 71–80.

Kirino, M., Parnes, J., Hansen, A., Kiyohara, S., & Finger, T. E. (2013). Evolutionary origins of taste buds: Phylogenetic analysis of purinergic neurotransmission in epithelial chemosensors. *Open Biology, 3*, article 130015.

Kirk, E. C. (2006). Visual influences on primate encephalization. *Journal of Human Evolution, 51*, 76–90.

Kivell, T. L., Deane, A. S., Tocheri, M. W., Orr, C. M., Schmid, P., Hawks, J., . . . Churchill, S. E. (2015). The hand of *Homo naledi*. *Nature Communications, 6*, article 8431.

Kivell, T. L., & Schmitt, D. (2009). Independent evolution of knuckle-walking in African apes shows that humans did not evolve from a knuckle-walking ancestor. *Proceedings of the National Academy of Sciences, 106*, 14241–14246.

Klatzky, R. L. (1980). *Human memory: Structures and processes*. San Francisco: W. H. Freeman and Co.

Köhler, W. (1927). *The mentality of apes*. New York, NY: Harcourt, Brace & Co.

Kolb, B., & Whishaw, I. Q. (2003). *Fundamentals of human neuropsychology*. New York, NY: W. H. Freeman and Co.

Kolb, B., & Whishaw, I. Q. (2009). *Fundamentals of human neuropsychology*. New York, NY: Worth Publishers.

Kooyman, B. P. (2000). *Understanding stone tools and archaeological sites*. Calgary, Canada: University of Calgary Press.

Koudouna, E., Winkler, M., Mikula, E., Juhasz, T., Brown, D. J., & Jester, J. V. (2018). Evolution of the vertebrate corneal stroma. *Progress in Retinal and Eye Research, 64*, 65–76.

Kovács, G., Sáry, G., Köteles, K., Chadaide, Z., Tompa, T., Vogels, R., & Benedek, G. (2003). Effects of surface cues on macaque inferior temporal cortical responses. *Cerebral Cortex, 13*, 178–188.

Kozmik, Z. (2008). The role of Pax genes in eye evolution. *Brain Research Bulletin, 75*, 335–339.

Krause, J., Lalueza-Fox, C., Enard, W., Green, R. E., Burbano, H. A., Hublin, J-J., . . . Pääbo, S. (2007). The derived FOXP2 variant of modern humans was shared with Neanderthals. *Current Biology, 17*, 1908–1912.

Krauzlis, R. J., Bogadhi, A. R., Herman, J. P., & Bollimunta, A. (2018). Selective attention without a neocortex. *Cortex, 102*, 161–175.

Kröger, R. H. H., Gustafsson, O. S. E., & Tuminaite, I. (2014). Suspension and optical properties of the crystalline lens in the eyes of basal vertebrates. *Journal of Morphology, 275*, 613–622.

Kuczaj, S., Tranel, K., Trone, M., & Hill, H. (2001). Are animals capable of deception or empathy? Implications for animal consciousness and animal welfare. *Animal Welfare, 10*, S161-S173.

Kühl, H. S., Kalan, A. K., Arandjelovic, M., Aubert, F., D'Auvergne, L., Goedmakers, A., . . . Boesch, C. (2016). Chimpanzee accumulative stone throwing. *Scientific Reports, 6*, article 22219.

Kuhlmeier, V. A., Boysen, S. T., & Mukobi, K. L. (1999). Scale-model comprehension by chimpanzees (*Pan troglodytes*). *Journal of Comparative Psychology, 113*, 396–402.

Kumar, V., Croxson, P. L., & Simonyan, K. (2016). Structural organization of the laryngeal motor cortical network and its implication for evolution of speech production. *The Journal of Neuroscience, 36*, 4170–4181.

Kvavadze, E., Bar-Yosef, O., Belfer-Cohen, A., Boaretto, E., Jakeli, N., Matskevich, Z., & Meshveliani, T. (2009). 30,000-year-old wild flax fibers. *Science, 325*, 1359.

Lacalli, T. C. (1994). Apical organs, epithelial domains, and the origin of the chordate central nervous system. *American Zoologist, 34*, 533–541.

Lacalli, T. C. (1996). Frontal eye circuitry, rostral sensory pathways and brain organization in amphioxus larvae: Evidence from 3D reconstructions. *Philosophical Transactions of the Royal Society of London, B, 351*, 243–263.

Lacalli, T. C. (2004). Sensory systems in amphioxus: A window on the ancestral chordate. *Brain, Behavior, and Evolution, 64*, 148–162.

Lacalli, T. C. (2018). Amphioxus, motion detection, and the evolutionary origin of the vertebrate retinotectal map. *EvoDevo, 9*, article 6.

Lacalli, T. C., & Kelly, S. J. (2003). Ventral neurons in the anterior nerve cord of amphioxus larvae. I. An inventory of cell types and synaptic patterns. *Journal of Morphology, 257*, 190–211.

Lacey, S., Tal, N., Amedi, A., & Sathian, K. (2009). A putative model of multisensory object representation. *Brain Topography, 21*, 269–274.

Lafer-Sousa, R., Conway, B. R., & Kanwisher, N. G. (2016). Color-biased regions of the ventral visual pathway lie between face- and place-selective regions in humans, as in macaques. *The Journal of Neuroscience, 36*, 1682–1697.

Lamb, T. D., Collin, S. P., & Pugh, E. N. (2007). Evolution of the vertebrate eye: Opsins, photoreceptors, retina and eye cup. *Nature Reviews Neuroscience, 8*, 960–975.

Lambertz, M., Grommes, K., Kohlsdorf, T., & Perry, S. F. (2015). Lungs of the first amniotes: Why simple if they can be complex? *Biology Letters, 11*, article 20130848.

Lamme, V. A. F. (2003). Why visual attention and awareness are different. *Trends in Cognitive Sciences*, 7, 12–18.

Lamme, V. A. F., Rodriguez-Rodriguez, V., & Spekreijse, H. (1999). Separate processing dynamics for texture elements, boundaries and surfaces in primary visual cortex of the macaque monkey. *Cerebral Cortex*, 9, 406–413.

Land, M. F., & Nilsson, D-E. (2012). *Animal eyes*. Oxford: Oxford University Press.

Landin-Romero, R., Tan, R., Hodges, J. R., & Kumfor, F. (2016). An update on semantic dementia: Genetics, imaging, and pathology. *Alzheimer's Research & Therapy*, 8, article 52.

Lane, N., & Martin, W. (2010). The energetics of genome complexity. *Nature*, 467, 929–934.

Larney, E., & Larson, S. G. (2004). Compliant walking in primates: Elbow and knee yield in primates compared to other mammals. *American Journal of Physical Anthropology*, 125, 42–50.

Larson, S. G., & Stern, J. T. (2006). Maintenance of above-branch balance during primate arboreal quadrupedalism: Coordinated use of forearm rotators and tail motion. *American Journal of Physical Anthropology*, 129, 71–81.

Laurin, M., Girondot, M., & de Ricqlès, A. (2000). Early tetrapod evolution. *TREE*, 15, 118–123.

Laval, S. H., Dann, J. C., Butler, R. J., Loftus, J., Rue, J., Leask, S. J., . . . Crow, T. J. (1998). Evidence for linkage to psychosis and cerebral asymmetry (relative hand skill) on the X chromosome. *American Journal of Medical Genetics (Neuropsychiatric Genetics)*, 81, 420–427.

Lee, M. S. Y., Reeder, T. W., Slowinski, J. B., & Lawson, R. (2004). Resolving reptile relationships: Molecular and morphological markers. In J. Cracraft & M. J. Donohue (Eds.), *Assembling the tree of life* (pp. 451–467). Oxford: Oxford University Press.

Lefebvre, L., Marino, L., Sol, D., Lemieux-Lefebvre, S., & Arshad, S. (2006). Large brains and lengthened life history periods in Odontocetes. *Brain, Behavior, and Evolution*, 68, 218–228.

Lefebvre, L., Nicolakakis, N., & Boire, D. (2002). Tools and brains in birds. *Behaviour*, 139, 939–973.

Leh, S. E., Mullen, K. T., & Ptito, A. (2006). Absence of S-cone input in human blindsight following hemispherectomy. *European Journal of Neuroscience*, 24, 2954–2960.

Leh, S. E., Ptito, A., Schönwiesner, M., Chakravarty, M. M., & Mullen, K. T. (2010). Blindsight mediated by an S-cone-independent colllicular pathway: An fMRI study in hemispherectomized patients. *Journal of Cognitive Neuroscience*, 22, 670–682.

Leonard, W. R., & Robertson, M. L. (1997). Rethinking the energetics of bipedality. *Current Anthropology*, 38, 304–309.

Lepre, C. J., Roche, H., Kent, D. V., Harmand, S., Quinn, R. L., Brugal, J-P., . . . Feibel, C. S. (2011). An earlier origin for the Acheulian. *Nature*, 477, 82–85.

Leslie, A. M. (1987). Pretense and representation: The origins of "theory of mind". *Psychological Review*, 94, 412–426.

Leung, H-C., Gore, J. C., & Goldman-Rakic, P. S. (2005). Differential anterior prefrontal activation during the recognition stage of a spatial working memory task. *Cerebral Cortex*, 15, 1742–1749.

Levin, H. L. (2010). *The earth through time*. Hoboken, NJ: John Wiley & Sons.

Lewin, R., & Foley, R. A. (2004). *Principles of human evolution*. Malden, MA: Blackwell Publishing.

Lewis, A., Call, J., & Bemtsen, D. (2017). Non-goal-directed recall of specific events in apes after long delays. *Proceedings of the Royal Society, B*, 284, article 20170518.

Lewis, J. E., & Harmand, S. (2016). An earlier origin for stone tool making: Implications for cognitive evolution and the transition to *Homo*. *Philosophical Transactions of the Royal Society, B*, 371, article 20150233.

Leys, S. P., Cronin, T. W., Degnan, B. M., & Marshall, J. N. (2002). Spectral sensitivity in a sponge larva. *Journal of Comparative Physiology A – Neuroethology Sensory Neural and Behavioral Physiology*, 188, 199–202.

Li, D., & Zhang, J. (2013). Diet shapes the evolution of the vertebrate bitter taste receptor gene repertoire. *Molecular Biology and Evolution*, 31, 303–309.

Li, G., Liu, X., Xing, C., Zhang, H., Shimeld, S. M., & Wang, Y. (2017). Cerberus-Nodal-Lefty-Pitx signaling cascade controls left-right asymmetry in amphioxus. *Proceedings of the National Academy of Sciences, 114*, 3684–3689.

Libby, T., Moore, T. Y., Chang-Siu, E., Li, D., Cohen, D. J., Jusufi, A., & Full, R. J. (2012). Tail-assisted pitch control in lizards, robots and dinosaurs. *Nature, 481*, 181–186.

Liebenthal, E., Sabri, M., Beardsley, S. A., Mangalathu-Arumana, J., & Desai, A. (2013). Neural dynamics of phonological processing in the dorsal auditory stream. *The Journal of Neuroscience, 33*, 15414–15424.

Lieberman, P. (2006). *Toward an evolutionary biology of language.* Cambridge, MA: Belknap Press of Harvard University Press.

Lieberman, P. (2007). The evolution of human speech: Its anatomical and neural bases. *Current Anthropology, 48*, 39–66.

Lieberman, P. (2012). Vocal tract anatomy and the neural bases of talking. *Journal of Phonetics, 40*, 608–622.

Liebeskind, B. J., Hillis, D. M., Zakon, H. H., & Hofman, H. A. (2016). Complex homology and the evolution of nervous systems. *Trends in Ecology & Evolution, 31*, 127–135.

Light, J. E., & Reed, D. L. (2009). Multigene analysis of phylogenetic relationships and divergence times of primate sucking lice (Phthiraptera: Anoplura). *Molecular Phylogenetics and Evolution, 50*, 376–390.

Lincoln, T. A., & Joyce, G. F. (2009). Self-sustained replication of an RNA enzyme. *Science, 323*, 1229–1232.

Lind, J., Enquist, M., & Ghirlanda, S. (2015). Animal memory: A review of delayed matching-to-sample data. *Behavioural Processes, 117*, 52–58.

Lindshield, S. M., & Rodrigues, M. A. (2009). Tool use in wild spider monkeys (*Ateles geoffroyi*). *Primates, 50*, 269–272.

Liu, Y., Su, Y., Xu, G., & Pei, M. (2018). When do you know what you know? The emergence of memory monitoring. *Journal of Experimental Child Psychology, 166*, 34–48.

Logan, G. A., Hayes, J. M., Hieshima, G. B., & Summons, R. E. (1995). Terminal Proterozoic reorganization of biogeochemical cycles. *Nature, 376*, 53–56.

Long, G. M., & McCarthy, P. R. (1982). Rod persistence on a partial-report task with scotopic and photopic backgrounds. *American Journal of Psychology, 95*, 309–322.

Lonsdorf, E. V., & Hopkins, W. D. (2005). Wild chimpanzees show population-level handedness for tool use. *Proceedings of the National Academy of Sciences, 102*, 12634–12638.

Lordkipanidze, D., de León, M. S. P., Margvelashvili, A., Rak, Y., Rightmire, G. P., Vekua, A., & Zollikofer, C. P. E. (2013). A complete skull from Dmanisi, Georgia, and the evolutionary biology of early *Homo. Science, 342*, 326–331.

Lovejoy, C. O., Latimer, B., Suwa, G., Asfaw, B., & White, T. D. (2009). Combining prehension and propulsion: The foot of *Ardipithecus ramidus. Science, 326*, 72e1–72e8.

Lovejoy, C. O., Meindl, R. S., Ohman, J. C., Heiple, K. G., & White, T. D. (2002). The Maka femur and its bearing on the antiquity of human walking: Applying contemporary concepts of morphogenesis to the human fossil record. *American Journal of Physical Anthropology, 119*, 97–133.

Lovejoy, C. O., Simpson, S. W., White, T. D., Asfaw, B., & Suwa, G. (2009). Careful climbing in the Miocene: The forelimbs of *Ardipithecus ramidus* and humans are primitive. *Science, 326*, 70–70e6.

Lovejoy, C. O., Suwa, G., Spurlock, L., Asfaw, B., & White, T. D. (2009). The pelvis and femur of *Ardipithecus ramidus*: The emergence of upright walking. *Science, 326*, 71.

Lozano, M., Mosquera, M., de Castro, J. M. B., Arsuaga, J. L., & Carbonell, E. (2009). Right handedness of *Homo heidelbergensis* from Sima de los Huesos (Atapuerca, Spain) 500,000 years ago. *Evolution and Human Behavior, 30*, 369–376.

Lü, Z-L., Williamson, S. J., & Kaufman, L. (1992). Human auditory primary and association cortex have differing lifetimes for activation traces. *Brain Research, 572*, 236–241.

Lucas, S. F. (2015). Thinopus and a critical review of Devonian tetrapod footprints. *Ichnos, 22*, 136–154.

Lundberg, Y. W., Xu, Y., Thiessen, K. D., & Kramer, K. L. (2015). Mechanisms of otoconia and otolith development. *Developmental Dynamics, 244*, 239–253.

Lunine, J. I. (2005). *Astrobiology: A multidisciplinary approach*. San Francisco: Addison Wesley.

Luo, Z-X. (2001). A new mammaliaform from the early Jurassic and evolution of mammalian characteristics. *Science, 292*, 1535–1540.

Luo, Z-X., Yuan, C-X., Meng, Q.-J., & Ji, Q. (2011). A Jurassic eutherian mammal and divergence of marsupials and placentals. *Nature, 476*, 442–445.

Lycett, S. J., & von Cramon-Taubadel, N. (2008). Acheulean variability and hominin dispersals: A model-bound approach. *Journal of Archaeological Science, 35*, 553–562.

Lyn, H. (2007). Mental representation of symbols as revealed by vocabulary errors in two bonobos (*Pan paniscus*). *Animal Cognition, 10*, 461–475.

Lyn, H., Greenfield, P. M., & Savage-Rumbaugh, E. S. (2011). Semiotic combinations in *Pan*: A comparison of communication in a chimpanzee and two bonobos. *First Language, 31*, 300–325.

Lyn, H., & Savage-Rumbaugh, S. (2013). The use of emotion symbols in language-using apes. In S. Watanabe & S. Kuczaj (Eds.), *Emotions of animals and humans: Comparative perspectives* (pp. 113–127). Tokyo: Springer.

Lyras, G. A., Dermitzakis, M. D., Van der Geer, A. A. E., Van der Geer, S. B., & De Vos, J. (2009). The origin of *Homo floresiensis* and its relation to evolutionary processes under isolation. *Anthropological Science, 117*, 33–43.

Machemer, H. (2001). The swimming cell and its world: Structures and mechanisms of orientation in protists. *European Journal of Protistology, 37*, 3–14.

Macintyre, B. (2007, February 16). A large vocabulary isn't just useful – it can be sexy too. *The Times*. Retrieved February 15, 2019 from http://entertainment.timesonline.co.uk/tol/ arts_and_entertainment/books/article2317122.ece

MacLarnon, A., & Hewitt, G. P. (1999). The evolution of human speech: The role of enhanced breathing control. *American Journal of Physical Anthropology, 109*, 341–363.

MacLarnon, A., & Hewitt, G. P. (2004). Increased breathing control: Another factor in the evolution of human language. *Evolutionary Anthropology, 13*, 181–197.

MacLatchy, L. (2004). The oldest ape. *Evolutionary Anthropology, 13*, 90–103.

Mafee, M. F., Pruzansky, S., Corrales, M. M., Phatak, M. G., Valvassori, G. E., Dobben, G. D., & Capek, V. (1986). CT in the evaluation of the orbit and the bony interorbital distance. *American Journal of Neuroradiology, 7*, 265–269.

Maisey, J. G. (1996). *Discovering fossil fishes*. New York, NY: Henry Holt and Co.

Malinzak, M. D., Kay, R. F., & Hullar, T. E. (2012). Locomotor head movements and semicircular canal morphology in primates. *Proceedings of the National Academy of Sciences, 109*, 17914–17919.

Mallary, M. (2004). *Our improbable universe*. New York, NY: Thunder's Mouth Press.

Mallatt, J., & Chen, J-Y. (2003). Fossil sister group of craniates: Predicted and found. *Journal of Morphology, 258*, 1–31.

Mandler, G., & Shebo, B. J. (1982). Subitizing: An analysis of its component processes. *Journal of Experimental Psychology: General, 111*, 1–22.

Manley, G. A. (2000). Cochlear mechanisms from a phylogenetic viewpoint. *Proceedings of the National Academy of Sciences, 97*, 11736–11743.

Manley, G. A., & Clack, J. A. (2004). An outline of the evolution of vertebrate hearing organs. In G. A. Manley, A. N. Popper, & R. R. Fay (Eds.), *Evolution of the vertebrate auditory system* (pp. 1–26). New York, NY: Springer-Verlag.

Mann, J., & Patterson, E. M. (2013). Tool use by aquatic animals. *Philosophical Transactions of the Royal Society, B, 368*, article 20120424.

Mansy, S. S., Schrum, J. P., Krishnamurthy, M., Tobé, S., Treco, D. A., & Szostak, J. W. (2008). Template-directed synthesis of a genetic polymer in a model protocell. *Nature, 454*, 122–126.

Marchant, L. F., & McGrew, W. C. (2007). Ant fishing by wild chimpanzees is not lateralised. *Primates, 48*, 22–26.

Marean, C. W., Bar-Matthews, M., Bernatchez, J., Fisher, E., Goldberg, P., Herries, A. I. R., . . . Williams, H. M. (2007). Early human use of marine resources and pigment in South Africa during the Middle Pleistocene. *Nature, 449*, 905–909.

Marguet, E., & Forterre, P. (1998). Protection of DNA by salts against thermodegration at temperatures typical for hyperthermophiles. *Extremophiles, 2*, 115–122.

Marie, D., Roth, M., Lacoste, R., Nazarin, B., Bertello, A., Anton, J-L., . . . Meguerditchian, A. (2018). Left brain asymmetry of the planum temporale in a nonhominid primate: Redefining the origin of brain specialization for language. *Cerebral Cortex*, *28*, 1808–1815.

Markey, M. J., & Marshall, C. R. (2007). Terrestrial-style feeding in a very early aquatic tetrapod is supported by evidence from experimental analysis of suture morphology. *Proceedings of the National Academy of Sciences*, *104*, 7134–7138.

Marshall, J. W. B., Baker, H. F., & Ridley, R. M. (2002). Contralesional neglect in monkeys with small unilateral parietal cortical ablations. *Behavioural Brain Research*, *136*, 257–265.

Martin, R. D. (1981). Relative brain size and basal metabolic rate in terrestrial vertebrates. *Nature*, *293*, 57–60.

Martin, R. D. (1992). Walking on two legs. In S. Jones, R. Martin, & D. Pilbeam (Eds.), *The Cambridge encyclopedia of human evolution* (p. 78). Cambridge, UK: Cambridge University Press.

Martínez, I., Arsuaga, J. L., Quam, R., Carretero, J. M., Gracia, A., & Rodríguez, L. (2008). Human hyoid bones from the middle Pleistocene site of the Sima de los Huesos (Sierra de Atapuerca, Spain). *Journal of Human Evolution*, *54*, 118–124.

Martínez, I., Rosa, M., Quam, R., Jarabo, P., Lorenzo, C., Bonmatí, A., . . . Arsuaga, J. L. (2013). Communicative capacities in Middle Pleistocene humans from the Sierra de Atapuerca in Spain. *Quaternary International*, *295*, 94–101.

Martin-Ordas, G., Haun, D., Colmenares, F., & Call, J. (2010). Keeping track of time: Evidence for episodic-like memory in great apes. *Animal Cognition*, *13*, 331–340.

Marzke, M. W. (1997). Precision grips, hand morphology, and tools. *American Journal of Physical Anthropology*, *102*, 91–110.

Marzke, M. W., & Marzke, R. F. (2000). Evolution of the human hand: Approaches to acquiring, analysing and interpreting the anatomical evidence. *Journal of Anatomy*, *197*, 121–140.

Masao, F. T., Ichumbaki, E. B., Cherin, M., Barili, A., Boschian, G., Iurino, D. A., . . . Manzi, G. (2016). New footprints from Laetoli (Tanzania) provide evidence for marked body size variation in early hominins. *eLife*, *5*, article 19568.

Masterton, B., Heffner, J., & Ravizza, R. (1969). The evolution of human hearing. *The Journal of the Acoustical Society of America*, *45*, 966–985.

McBrearty, S., & Jablonski, N. G. (2005). First fossil chimpanzee. *Nature*, *437*, 105–108.

McCarthy, G., Puce, A., Belger, A., & Allison, T. (1999). Electrophysiological studies of human face perception. II: Response properties of face-specific potentials generated in occipitotemporal cortex. *Cerebral Cortex*, *9*, 431–444.

McCormick, C., Ciaramelli, E., De Luca, F., & Maguire, E. A. (2018). Comparing and contrasting the cognitive effects of hippocampal and ventromedial prefrontal cortex damage: A review of human lesion studies. *Neuroscience*, *374*, 295–318.

McCourt, M. E., Garlinghouse, M., & Reuter-Lorenz, P. A. (2005). Unilateral visual cueing and asymmetric line geometry share a common attentional origin in the modulation of pseudoneglect. *Cortex*, *41*, 499–511.

McGrew, W. C. (1992). *Chimpanzee material culture: Implications for human evolution*. Cambridge, UK: Cambridge University Press.

McHenry, H. M., & Coffing, K. (2000). *Australopithecus* to *Homo*: Transformations in body and mind. *Annual Review of Anthropology*, *29*, 125–146.

McKeever, W. F. (2000). A new family handedness sample with findings consistent with X-linked transmission. *British Journal of Psychology*, *91*, 21–39.

McMahon, R. (2004). Genes and language. *Community Genetics*, *7*, 2–13.

McManus, I. C., & Bryden, M. P. (1992). The genetics of handedness, cerebral dominance and lateralization. In I. Rapin & S. J. Segalowitz (Eds.), *Handbook of neuropsychology* (Vol. 6, pp. 115–144). Amsterdam: Elsevier Academic Press.

McNulty, K. P., Begun, D. R., Kelley, J., Mnthi, F. K., & Mbua, E. N. (2015). A systematic revision of *Proconsul* with the description of a new genus of early Miocene hominoid. *Journal of Human Evolution*, *84*, 42–61.

McPherron, S. P., Alemseged, Z., Marean, C. W., Wynn, J. G., Reed, D., Geraads, D., . . . Béarat, H. A. (2010). Evidence for stone-tool-assisted consumption of animal tissues before 3.39 million years ago at Dikika, Ethiopia. *Nature*, *466*, 857–860.

Medland, S. E., Duffy, D. L., Wright, M. J., Geffen, G. M., & Martin, N. G. (2006). Handedness in twins: Joint analysis of data from 35 samples. *Twin Research and Human Genetics*, *9*, 46–53.

Meisenzahl, E. M., Zetsche, T., Preuss, U., Frodl, T., Leinsinger, G., & Möller, H-J. (2002). Does the definition of borders of the planum temporale influence the results in schizophrenia? *American Journal of Psychiatry*, *159*, 1198–1200.

Melin, A. D., Chiou, K. L., Walco, E. R., Bergstrom, M. L., Kawamura, S., & Fedigan, L. M. (2017). Trichromacy increases fruit intake rates of wild capuchins (*Cebus capucinus imitator*). *Proceedings of the National Academy of Sciences*, *114*, 10402–10407.

Melin, A. D., Khetpal, V., Matsushita, Y., Zhou, K., Campos, F. A., Welker, B., & Kawamura, S. (2017). Howler monkey foraging ecology suggests convergent evolution of routine trichromacy as an adaptation for folivory. *Ecology and Evolution*, *7*, 1421–1434.

Mendez, F. L., Poznik, G. D., Castellano, S., & Bustamante, C. D. (2016). The divergence of neandertal and modern human Y chromosomes. *The American Journal of Human Genetics*, *98*, 728–734.

Menzel, C. R. (1999). Unprompted recall and reporting of hidden objects by a chimpanzee (*Pan troglodytes*) after extended delays. *Journal of Comparative Psychology*, *113*, 426–434.

Mercader, J., Panger, M., & Boesch, C. (2002). Excavation of a chimpanzee stone tool site in the African rainforest. *Science*, *296*, 1452–1455.

Merker, B. (2007). Consciousness without a cerebral cortex: A challenge for neuroscience and medicine. *Behavioral and Brain Sciences*, *30*, 63–134.

Meunier, H., Fizet, J., & Vauclair, J. (2013). Tonkean macaques communicate with their right hand. *Brain & Language*, *126*, 181–187.

Mgeladze, A., Lordkipanidze, D., Moncel, M-H., Despriee, J., Chagelishvili, R., Nioradze, M., & Nioradze, G. (2011). Hominin occupations in the Dmanisi site, Georgia, Southern Causcasus: Raw materials and technical behaviours of Europe's first hominins. *Journal of Human Evolution*, *60*, 571–596.

Miceli, G. (1999). Grammatical deficits in aphasia. In G. Denes & L. Pizzamiglio (Eds.), *Handbook of clinical and experimental neuropsychology* (pp. 245–272). Hove, UK: Psychology Press.

Miles, H. L. W. (1990). The cognitive foundations for reference in a signing orangutan. In S. T. Parker & K. R. Gibson (Eds.), *"Language" and intelligence in monkeys and apes* (pp. 511–539). Cambridge, UK: Cambridge University Press.

Miljkovic-Licina, M., Gauchat, D., & Galliot, B. (2004). Neuronal evolution: Analysis of regulatory genes in a first-evolved nervous system, the hydra nervous system. *BioSystems*, *76*, 75–87.

Miller, G. A. (1956). The magical number seven, plus or minus two: Some limits on our capacity for processing information. *The Psychological Review*, *63*, 81–97.

Miller, G. A. (1962). *Psychology: The science of mental life*. New York, NY: Harper & Row.

Moffat, S. D., Elkins, W., & Resnick, S. M. (2006). Age differences in the neural systems supporting human allocentric spatial navigation. *Neurobiology of Aging*, *27*, 965–972.

Montgomery, S. H. (2013). Primate brains, the 'island rule' and the evolution of *Homo floresiensis*. *Journal of Human Evolution*, *65*, 750–760.

Monti, M. M., & Osherson, D. N. (2012). Logic, language and the brain. *Brain Research*, *1428*, 33–42.

Moore, B. R. (2004). The evolution of learning. *Biological Reviews*, *79*, 301–335.

Moore, T., Rodman, H. R., Repp, A. B., & Gross, C. G. (1995). Localization of visual stimuli after striate cortex damage in monkeys: Parallels with human blindsight. *Proceedings of the National Academy of Sciences*, *92*, 8215–8218.

Morelle, R. (2006). *Arctic fossils mark move to land*. Retrieved February 15, 2019 from http://news.bbc.co.uk/2/hi/science/nature/4879672.stm

Morino, L. (2011). Left-hand preference for a complex manual task in a population of wild siamangs (*Symphalangus syndactylus*). *International Journal of Primatology*, *32*, 793–800.

Morino, L., Uchikoshi, M., Bercovitch, F., & Hopkins, W. D. (2017). Tube task hand preference in captive hylobatids. *Primates*, *58*, 403–412.

Morris, S. C., & Caron, J. B. (2014). A primitive fish from the Cambrian of North America. *Nature*, *512*, 419–422.

Morrison, R., & Reiss, D. (2018). Precocious development of self-awareness in dolphins. *PLoS One*, *13*, article e0189813.

Mort, D. J., Malhotra, P., Mannan, S. K., Rorden, C., Pambakian, A., Kennard, C., & Husain, M. (2003). The anatomy of visual neglect. *Brain, 126,* 1986–1997.

Morwood, M. J., Brown, P., Jamitko, Sutikna, T., Saptomo, E. W., Westaway, K. E., . . . Djubian-tono, T. (2005). Further evidence for small-bodied hominins from the Late Pleistocene of Flores, Indonesia. *Nature, 437,* 1012–1017.

Morwood, M. J., & Jungers, W. L. (2009). Conclusions: Implications of the Liang Bua excavations for hominin evolution and biogeography. *Journal of Human Evolution, 57,* 640–648.

Moulton, V., Gardner, P. P., Pointon, R. F., Creamer, L. K., Jameson, G. B., & Pennyl, D. (2000). RNA folding argues against a hot-start origin of life. *Journal of Molecular Evolution, 51,* 416–421.

Mühlau, M., Hermsdörfer, J., Goldenberg, G., Wohlschläger, A. M., Castrop, F., Stahl, R., . . . Boecker, H. (2005). Left inferior parietal dominance in gesture imitation: An fMRI study. *Neuropsychologia, 43,* 1086–1098.

Mukhopadhyay, R. (2014). Quantum mechanics, objective reality, and the problem of consciousness. *Journal of Consciousness Studies, 21,* 57–80.

Murofushi, K. (1997). Numerical matching behavior by a chimpanzee (*Pan troglodytes*): Subitizing and analogue magnitude estimation. *Japanese Psychological Research, 39,* 140–153.

Musgrave, S., Morgan, D., Lonsdorf, E., Mundry, R., & Sanz, C. (2016). Tool transfers are a form of teaching among chimpanzees. *Scientific Reports, 6,* article 34783.

Nakamura, R. K., & Mishkin, M. (1980). Blindness in monkeys following non-visual cortical lesions. *Brain Research, 188,* 572–577.

Naqshbandi, M., Feeney, M. C., McKenzie, T. L. B., & Roberts, W. A. (2007). Testing for episodic-like memory in rats in the absence of time of day cues: Replication of Babb and Crystal. *Behavioural Processes, 74,* 217–225.

Nation, P. (1993). Using dictionaries to estimate vocabulary size: Essential, but rarely followed, procedures. *Language Today, 10,* 27–40.

Navon, D. (1977). Forest before trees: The precedence of global features in visual perception. *Cognitive Psychology, 9,* 353–383.

Nee, D. E., & Brown, J. W. (2013). Dissociable frontal-striatal and frontal-parietal networks involved in updating hierarchical contexts in working memory. *Cerebral Cortex, 23,* 2146–2158.

Neggers, S. F. W., Van der Lubbe, R. H. J., Ramsey, N. F., & Postma, A. (2006). Interactions between ego- and allocentric neuronal representations of space. *NeuroImage, 31,* 320–331.

Neiworth, J. J., Whillock, K. M., Kim, S. H., Greenberg, J. R., Jones, K. B., Patel, A. R., . . . Kudura, A. G. (2014). Gestalt principle use in college students, children with autism, toddlers (*Homo sapiens*), and cotton top tamarins (*Saguinus oedipus*). *Journal of Comparative Psychology, 128,* 188–198.

Nelson, E. L., O'Karma, J. M., Ruperti, F. S., & Novak, M. A. (2009). Laterality in semi-free-ranging black and white ruffed lemurs (*Varecia variegata variegata*): Head-tilt correlates with hand use during feeding. *American Journal of Primatology, 71,* 1032–1040.

Nelson, T. O. (1996). Consciousness and metacognition. *American Psychologist, 51,* 102–116.

Nettle, D. (1999). Towards a future history of macrofamily research. In C. Renfrew & D. Nettle (Eds.), *Nostratic: Examining a linguistic macrofamily* (pp. 403–419). Cambridge, UK: The McDonald Institute for Archaeological Research.

Neufuss, J., Humle, T., Cremaschi, A., & Kivell, T. L. (2017). Nut-cracking behaviour in wild-born, rehabilitated bonobos (*Pan paniscus*): A comprehensive study of hand-preference, hand grips and efficiency. *American Journal of Primatology, 79,* article e22589.

Ni, X., Wang, Y., Hu, Y., & Li, C. (2004). A euprimate skull from the early Eocene of China. *Nature, 427,* 65–68.

Nieder, A., Freedman, D. J., & Miller, E. K. (2002). Representation of the quantity of visual items in the primate prefrontal cortex. *Science, 297,* 1708–1711.

Niedzwiedzki, G., Szrek, P., Narkiewicz, K., Narkiewicz, M., & Ahlberg, P. E. (2010). Tetrapod trackways from the early Middle Devonian period of Poland. *Nature, 463,* 43–48.

Nieuwenhuys, R. (2002). Deuterostome brains: Synopsis and commentary. *Brain Research Bulletin, 57,* 257–270.

Niimura, Y. (2012). Olfactory receptor multigene family in vertebrates: From the viewpoint of evolutionary genomics. *Current Genomics, 13,* 103–114.

Nilsson, D-E. (2013). Eye evolution and its functional basis. *Visual Neuroscience, 30,* 5–20.

Noonan, J. P., Coop, G., Kudaravalli, S., Smith, D., Krause, J., Alessi, J., . . . Rubin, E. M. (2006). Sequencing and analysis of Neanderthal genomic DNA. *Science, 314,* 1113–1118.

Norman, E. (2017). Metacognition and mindfulness: The role of fringe consciousness. *Mindfulness, 8,* 95–100.

Normand, E., & Boesch, C. (2009). Sophisticated Euclidean maps in forest chimpanzees. *Animal Behaviour, 77,* 1195–1201.

Norton, O. R. (1998). *Rocks from space* (2nd ed.). Missoula, MT: Mountain Press Publishing Company.

Novacek, M. J. (1999). 100 million years of land vertebrate evolution: The cretaceous-early tertiary transition. *Annals of the Missouri Botanical Garden, 86,* 230–258.

Nunez, V., Shapley, R. M., & Gordon, J. (2018). Cortical double-opponent cells in color perception: Perceptual scalling and chromatic visual evoked potentials. *i-Perception, Jan-Feb 2018,* 1–16.

Nyakatura, J. A., Allen, V. R., Lauströer, J., Andikfar, A., Danczak, M., Ullrich, H-J., . . . Fischer, M. S. (2015). A three-dimensional skeletal reconstruction of the stem amniote *Orobates pabsti* (Diadectidae): Analyses of body mass, centre of mass position, and joint mobility. *PLoS One, 10,* article e0137284.

Oftedal, O. T. (2002). The mammary gland and its origin during synapsid evolution. *Journal of Mammary Gland Biology and Neoplasia, 7,* 225–252.

Ogawa, T., Shoji, E., Suematsu, N. J., Nishimori, H., Izumi, S., Awazu, A., & Iima, M. (2016). The flux of *Euglena gracilis* cells depends on the gradient of light intensity. *PLoS One, 11,* article e0168114.

Öğmen, H., Breitmeyer, B. G., Todd, S., & Mardon, L. (2006). Target recovery in metacontrast: The effect of contrast. *Vision Research, 46,* 4726–4734.

Oh, Y-M., & Koh, E-J. (2009). Language lateralization in patients with temporal lobe epilepsy: A comparison between volumetric analysis and the Wada test. *Journal of the Korean Neurosurgical Society, 45,* 329–335.

Olkowicz, S., Kocourek, M., Lucan, R. K., Portes, M., Fitch, W. T., Herculano-Houzel, S., & Nemec, P. (2016). Birds have primate-like numbers of neurons in the forebrain. *Proceedings of the National Academy of Sciences, 113,* 7255–7260.

Olson, I. R., Gatenby, J. C., Leung, H-C., Skudlarski, P., & Gore, J. C. (2003). Neuronal representation of occluded objects in the human brain. *Neuropsychologia, 42,* 95–104.

Orban, G. A., Claeys, K., Nelissen, K., Smans, R., Sunaert, S., Todd, J. T., . . . Vanduffel, W. (2006). Mapping the parietal cortex of human and non-human primates. *Neuropsychologia, 44,* 2647–2667.

Orban, G. A., Van Essen, D., & Vanduffel, W. (2004). Comparative mapping of higher visual areas in monkeys and humans. *Trends in Cognitive Sciences, 8,* 315–324.

Organ, C. L., Schweitzer, M. H., Zheng, W., Freimark, L. M., Cantley, L. C., & Asara, J. M. (2008). Molecular phylogenetics of mastodon and *Tyrannosaurus rex. Science, 320,* 499.

Orlando, L., Darlu, P., Toussaint, M., Bonjean, D., Otte, M., & Hanni, C. (2006). Revisiting Neandertal diversity with a 100,000 year old mtDNA sequence. *Current Biology, 16,* R400–R401.

Ornstein, R. E. (1972). *The psychology of consciousness.* New York, NY: Viking Press.

Orwant, R. (2005). First convincing chimp fossil discovered. *New Scientist, 187,* 7.

Papademetriou, E., Sheu, C-F., & Michel, G. F. (2005). A meta-analysis of primate hand preferences, particularly for reaching. *Journal of Comparative Psychology, 119,* 33–48.

Papaleontiou-Louca, E. (2008). *Metacognition and theory of mind.* Newcastle, UK: Cambridge Scholars Publishing.

Parker, E. T., Cleaves, H. J., Dworkin, J. P., Glavin, D. P., Callahan, M., Aubrey, A., . . . Bada, J. L. (2011). Primordial synthesis of amines and amino acids in a 1958 Miller H_2S-rich spark discharge experiment. *Proceedings of the National Academy of Sciences, 108,* 5526–5531.

Parker, S. T. (2004). The cognitive complexity of social organization and socialization in wild baboons and chimpanzees: Guided exploration, socializing interactions, and event representation.

In A. E. Russon & D. R. Begun (Eds.), *The evolution of thought: Evolutionary origins of great ape intelligence* (pp. 45–60). Cambridge, UK: Cambridge University Press.

Parker, S. T., Kerr, M., Markowitz, H., & Gould, J. (1999). A survey of tool use in zoo gorillas. In S. T. Parker, R. W. Mitchell, & H. L. Miles (Eds.), *The mentalities of gorillas and orangutans* (pp. 188–193). Cambridge, UK: Cambridge University Press.

Parr, L. A. (2011). The evolution of face processing in primates. *Philosophical Transactions of the Royal Society, B, 366,* 1764–1777.

Pascual-Anaya, J., D'Aniello, S., Kuratani, S., & Garcia-Fernàndez, J. (2013). Evolution of *Hox* gene clusters in deuterostomes. *BMC Developmental Biology, 13,* 26.

Pascual-Leone, A., & Walsh, V. (2001). Fast backprojections from the motion to the primary visual area necessary for visual awareness. *Science, 292,* 510–512.

Pasupathy, A., & Connor, C. E. (2001). Shape representation in area V4: Position-specific tuning for boundary conformation. *Journal of Neurophysiology, 86,* 2505–2519.

Pasupathy, A., & Connor, C. E. (2002). Population coding of shape in area V4. *Nature Neuroscience, 5,* 1332–1338.

Paterson, J. R., García-Bellido, D. C., Lee, M. S. Y., Brock, G. A., Jago, J. B., & Edgecombe, G. D. (2011). Acute vision in the giant Cambrian predator *Anomalocaris* and the origin of compound eyes. *Nature, 480,* 237–240.

Paulson, A. (2007). In Iowa, English 101 for bonobos. *Christian Science Monitor.* Retrieved February 16, 2008, from www.usatoday.com/tech/science/discoveries/2007-05-10-bonobo-studies_N.htm

Payne, B. R., Lomber, S. G., MacNeil, M. A., & Cornwell, P. (1996). Evidence for greater sight in blindsight following damage of primary visual cortex early in life. *Neuropsychologia, 34,* 741–774.

Pearce, B. K. D., & Pudritz, R. E. (2015). Seeding the pregenetic earth: Meteoric abundances of nucleobases and potential reaction pathways. *The Astrophysical Journal, 807,* article 85.

Pearce, E., & Dunbar, R. (2012). Latitudinal variation in light levels drives human visual system size. *Biological Letters, 8,* 90–93.

Peeters, R., Simone, L., Nelissen, K., Fabbri-Destro, M., Vanduffel, W., Rizzolatti, G., & Orban, G. A. (2009). The representation of tool use in human and monkeys: Common and uniquely human features. *The Journal of Neuroscience, 29,* 11523–11539.

Peirce, J. W., Leigh, A. E., daCosta, A. P. C., & Kendrick, K. M. (2001). Human face recognition in sheep: Lack of configurational coding and right hemisphere advantage. *Behavioural Processes, 55,* 13–26.

Peirce, J. W., Leigh, A. E., & Kendrick, K. M. (2000). Configurational coding, familiarity and the right hemisphere advantage for face recognition in sheep. *Neuropsychologia, 38,* 475–483.

Peng, K., Steele, S. C., Becerra, L., & Borsook, D. (2018). Brodmann area 10: Collating, integrating and high level processing of nociception and pain. *Progress in Neurobiology, 161,* 1–22.

Perelman, P., Johnson, W. E., Roos, C., Seuánez, H. N., Horvath, J. E., Moreira, M. A. M., . . . Pecon-Slattery, J. (2011). A molecular phylogeny of living primates. *PLoS Genetics, 7,* e1001342.

Perry, S. E., Barrett, B. J., & Godoy, I. (2017). Older, sociable capuchins (*Cebus capucinus*) invent more social behaviors, but younger monkeys innovate more in other contexts. *Proceedings of the National Academy of Sciences, 114,* 7806–7813.

Peterson, K. J., & Butterfield, N. J. (2005). Origin of the eumetazoa: Testing ecological predictions of molecular clocks against the Proterozoic fossil record. *Proceedings of the National Academy of Sciences, 102,* 9547–9552.

Petrides, M. (2005). Lateral prefrontal cortex: Architectonic and functional organization. *Philosophical Transactions of the Royal Society of London, B, 360,* 781–795.

Petrides, M., & Pandya, D. N. (1999). Dorsolateral prefrontal cortex: Comparative cytoarchitectonic analysis in the human and the macaque brain and corticocortical connections patterns. *European Journal of Neuroscience, 11,* 1011–1036.

Petrides, M., & Pandya, D. N. (2009). Distinct parietal and temporal pathways to the homologues of Broca's area in the monkey. *PLoS Biology, 7,* e1000170.

Petryshyn, V. A., Bottjer, D. J., Chen, J-Y., & Gao, F. (2013). Petrographic analysis of new specimens of the putative microfossil *Vernanimalcula guizhouena* (Doushantuo Formation, South China). *Precambrian Research, 225*, 58–66.

Peuskens, H., Sunaert, S., Dupont, P., Van Hecke, P., & Orban, G. A. (2001). Human brain regions involved in heading estimation. *The Journal of Neuroscience, 21*, 2451–2461.

Phillipson, L. (1997). Edge modification as an indicator of function and handedness of Acheulian handaxes from Kariandusi, Kenya. *Lithic Technology, 22*, 171–183.

Pichaud, F., & Desplan, C. (2002). Pax genes and eye organogenesis. *Current Opinion in Genetics & Development, 12*, 430–434.

Pierrot-Deseilligny, C., Ploner, C. J., Müri, R. M., Gaymard, B., & Rivaud-Péchoux, S. (2002). Effects of cortical lesions on saccadic eye movements in humans. *Annals of the New York Academy of Sciences, 956*, 216–229.

Pilbeam, D. (1972). *The ascent of man: An introduction to human evolution.* New York, NY: Palgrave Macmillan.

Piñeiro, G., Ferigolo, J., Meneghel, M., & Laurin, M. (2012). The oldest known amniotic embryos suggest viviparity in mesosaurs. *Historical Biology, 24*, 620–630.

Pisani, D., Mohun, S. M., Harris, S. R., McInerney, J. O., & Wilkinson, M. (2006). Molecular evidence for dim-light vision in the last common ancestor of the vertebrates. *Current Biology, 16*, R318–R319.

Plachetzki, D. C., Degnan, B. M., & Oakley, T. H. (2007). The origins of novel protein interactions during opsin evolution. *PLoS One, 2*, e1054.

Plotnik, J. M., de Waal, F. B. M., & Reiss, D. (2006). Self-recognition in an Asian elephant. *Proceedings of the National Academy of Sciences, 103*, 17053–17057.

Pobiner, B. L. (1999). The use of stone tools to determine handedness in hominids. *Current Anthropology, 40*, 90–92.

Pollick, A. S., & de Waal, F. B. M. (2007). Ape gestures and language evolution. *Proceedings of the National Academy of Sciences, 104*, 8184–8189.

Poremba, A., Malloy, M., Saunders, R. C., Carson, R. E., Herscovitch, P., & Mishkin, M. (2004). Species-specific calls evoke asymmetric activity in the monkey's temporal poles. *Nature, 427*, 448–451.

Porter, M. L., & Crandall, K. A. (2003). Lost along the way: The significance of evolution in reverse. *Trends in Ecology and Evolution, 18*, 541–547.

Posada, S., & Colell, M. (2007). Another gorilla (*Gorilla gorilla gorilla*) recognizes himself in a mirror. *American Journal of Primatology, 69*, 576–583.

Posner, M. I. (1978). *Chronometric explorations of mind.* Hillsdale, NJ: Lawrence Erlbaum Associates.

Posner, M. I. (1992). Attention as a cognitive and neural system. *Current Directions in Psychological Science, 1*, 11–14.

Posner, M. I., Walker, J. A., Friedrich, F. J., & Rafal, R. D. (1984). Effects of parietal injury on covert orienting of attention. *The Journal of Neuroscience, 4*, 1863–1874.

Poss, S. R., Kuhar, C., Stoinski, T. S., & Hopkins, W. D. (2006). Differential use of attentional and visual communicative signaling by orangutans (*Pongo pygmaeus*) and gorillas (*Gorilla gorilla*) in response to the attentional status of a human. *American Journal of Primatology, 68*, 978–992.

Potts, R. (1992). The hominid way of life. In S. Jones, R. Martin, & D. Pilbeam (Eds.), *The Cambridge encyclopedia of human evolution* (pp. 325–334). Cambridge, UK: Cambridge University Press.

Pouydebat, E., Coppens, Y., & Gorce, P. (2006). Évolution de la préhension chez les primates humains et non humains: la précision et l'utilisation d'outils revisitées. *L'anthropologie, 110*, 687–697.

Povinelli, D. J. (1989). Failure to find self-recognition in Asian elephants (*Elephas maximus*) in contrast to their use of mirror cues to discover hidden food. *Journal of Comparative Psychology, 103*, 122–131.

Povinelli, D. J., & Bering, J. M. (2003). The mentality of apes revisited. *Current Directions in Psychological Science, 11*, 115–119.

Povinelli, D. J., Reaux, J. E., Theall, L. A., & Giambrone, S. (2000). *Folk physics for apes*. Oxford: Oxford University Press.

Povinelli, D. J., Rulf, A. B., Landau, K. R., & Bierschwale, D. T. (1993). Self-recognition in chimpanzees (*Pan troglodytes*): Distribution, ontogeny, and patterns of emergence. *Journal of Comparative Psychology, 107*, 347–372.

Poznik, G. D., Henn, B. M., Yee, M-C., Sliwerska, E., Euskirchen, G. M., Lin, A. A., . . . Bustamante, C. D. (2013). Sequencing Y chromosomes resolves discrepancy in time to common ancestor of males versus females. *Science, 341*, 562–565.

Pozzi, L., Hodgson, J. A., Burrell, A. S., Sterner, K. N., Rauum, R. L., & Distotell, T. R. (2014). Primate phylogenetic relationships and divergence dates inferred from complete mitochondrial genomes. *Molecular Phylogenetics and Evolution, 75*, 165–183.

Premack, D., & Woodruff, G. (1978). Does the chimpanzee have a theory of mind? *The Behavioral and Brain Sciences, 4*, 515–526.

Prieur, J., Barbu, S., & Blois-Heulin, C. (2018). Human laterality for manipulation and gestural communication related to 60 everyday activities: Impact of multiple individual-related factors. *Cortex, 99*, 118–134.

Prieur, J., Pika, S., Barbu, S., & Blois-Heulin, C. (2016a). A multifactorial investigation of captive chimpanzees' intraspecific gestural laterality. *Animal Behaviour, 116*, 31–43.

Prieur, J., Pika, S., Barbu, S., & Blois-Heulin, C. (2016b). Gorillas are right-handed for their most frequent intraspecific gestures. *Animal Behaviour, 118*, 165–170.

Prinzmetal, W. (1981). Principles of feature integration in visual perception. *Perception & Psychophysics, 30*, 330–340.

Prior, H., Schwarz, A., & Güntürkün, O. (2008). Mirror-induced behavior in the magpie (*Pica pica*): Evidence of self-recognition. *PLoS Biology, 6*, 1642–1650.

Proffitt, T., Luncz, V. L., Malaivijitnond, S., Gumert, M., Svensson, M. S., & Haslam, M. (2018). Analysis of wild macaque stone tools used to crack oil palm nuts. *Royal Society Open Science, 5*, article 171904.

Proops, L., Rayner, J., Taylor, A. M., & McComb, K. (2013). The responses of young domestic horses to human-given cues. *PLoS One, 8*, article e67000.

Prothero, D. R. (2007). *Evolution: What the fossils say and why it matters*. New York, NY: Columbia University Press.

Pruetz, J. D., Bertolani, P., Ontl, K. B., Lindshield, S., Shelley, M., & Wessling, E. G. (2015). New evidence on the tool-assisted hunting exhibited by chimpanzees (*Pan troglodytes verus*) in a savannah habitat at Fongoli, Sénégal. *Royal Society Open Science, 2*, article 140507.

Prüfer, K., Munch, K., Hellmann, I., Akagi, K., Miller, J. R., Walenz, B.,. . . . Pääbo, S. (2012). The bonobo genome compared with the chimpanzee and human genomes. *Nature, 486*, 527–531.

Ptak, R., & Schnider, A. (2011). The attention network of the human brain: Relating structural damage associated with spatial neglect to functional imaging correlates of spatial attention. *Neuropsychologia, 49*, 3063–3070.

Pugh, K. R., Mencl, W. E., Shaywitz, B. A., Shaywitz, S. E., Fulbright, R. K., Constable, R. T., . . . Gore, J. C. (2000). The angular gyrus in developmental dyslexia: Task-specific differences in functional connectivity within posterior cortex. *Psychological Science, 11*, 51–56.

Quraishi, S., Heider, B., & Siegel, R. M. (2007). Attentional modulation of receptive field structure in area 7a of the behaving monkey. *Cerebral Cortex, 17*, 1841–1857.

Raaum, R. L., Sterner, K. N., Noviello, C. M., Stewart, C-B., & Distotell, T. R. (2005). Catarrhine primate divergence dates estimated from complete mitochondrial genomes: Concordance with fossil and nuclear DNA evidence. *Journal of Human Evolution, 48*, 237–257.

Rafal, R. D. (2006). Oculomotor functions of the parietal lobe: Effects of chronic lesions in humans. *Cortex, 42*, 730–739.

Rajala, A. Z., Reininger, K. R., Lancaster, K. M., & Populin, L. C. (2010). Rhesus monkeys (*Macaca mulatta*) do recognize themselves in the mirror: Implications for the evolution of self-recognition. *PLoS One, 5*, e12865.

Ramsier, M. A., Cunningham, A. J., Finneran, J. J., & Dominy, N. J. (2012). Social drive and evolution of primate hearing. *Philosophical Transactions of the Royal Society, B, 367*, 1860–1868.

Rantala, M. J. (2007). Evolution of nakedness in *Homo sapiens*. *Journal of Zoology, 273*, 1–7.

Rasmussen, D. T. (2002a). Early catarrhines of the African Eocene and Oligocene. In W. C. Hartwig (Ed.), *The primate fossil record* (pp. 203–220). Cambridge, UK: Cambridge University Press.

Rasmussen, D. T. (2002b). The origin of primates. In W. C. Hartwig (Ed.), *The primate fossil record* (pp. 5–9). Cambridge, UK: Cambridge University Press.

Raup, D. M. (1991). *Extinction: Bad genes or bad luck?* New York, NY: W. W. Norton & Company.

Raymann, K., Brochier-Armanet, C., & Gribaldo, S. (2015). The two-domain tree of life is linked to a new root for the Archaea. *Proceedings of the National Academy of Sciences, 112*, 6670–6675.

Raymond, M., & Pontier, D. (2004). Is there geographic variation in human handedness? *Laterality, 9*, 35–51.

Raymond, M., Pontier, D., Dufour, A-B., & Møller, A. P. (1996). Frequency-dependent maintenance of left handedness in humans. *Proceedings of the Royal Society of London, B, 263*, 1627–1633.

Reader, J. (1981). *Missing links: The hunt for earliest man*. Boston: Little, Brown and Company.

Reddy, P. C., Unni, M. K., Gungi, A., Agarwal, P., & Galande, S. (2015). Evolution of Hox-like genes in Cnidaria: Study of Hydra Hox repertoire reveals tailor-made Hox-code for Cnidarians. *Mechanisms of Development, 138*, 87–96.

Rees, G., & Frith, C. (2007). Methodologies for identifying the neural correlates of consciousness. In M. Velmans & S. Schneider (Eds.), *The Blackwell companion to consciousness* (pp. 553–566). Malden, MA: Blackwell Publishing.

Regaiolli, B., Spiezio, C., & Hopkins, W. D. (2016). Hand preference on unimanual and bimanual tasks in strepsirrhines: The case of the ring-tailed lemur (*Lemur catta*). *American Journal of Primatology, 78*, 851–860.

Reiss, D., & Marino, L. (2001). Mirror self-recognition in the bottlenose dolphin: A case of cognitive convergence. *Proceedings of the National Academy of Sciences, 98*, 5937–5942.

Rhodes, G., Byatt, G., Michie, P. T., & Puce, A. (2004). Is the fusiform face area specialized for faces, individuation, or expert individuation? *Journal of Cognitive Neuroscience, 16*, 189–203.

Richmond, B. G., Aiello, L. C., & Wood, B. A. (2002). Early hominin limb proportions. *Journal of Human Evolution, 43*, 529–548.

Richmond, B. G., & Jungers, W. L. (2008). *Orrorin tugenensis* femoral morphology and the evolution of hominin bipedalism. *Science, 319*, 1662–1665.

Rightmire, G. P. (2004). Brain size and encephalization in early to mid-Pleistocene *Homo*. *American Journal of Physical Anthropology, 124*, 109–123.

Rilling, J. K. (2006). Human and nonhuman primate brains: Are they allometrically scaled versions of the same design? *Evolutionary Anthropology, 15*, 65–77.

Rilling, J. K. (2014). Comparative primate neurobiology and the evolution of the brain language systems. *Current Opinion in Neurobiology, 28*, 10–14.

Rilling, J. K., Glasser, M. F., Preuss, T. M., Ma, X., Zhao, T., Hu, X., & Behrens, T. E. J. (2008). The evolution of the arcuate fasciculus revealed with comparative DTI. *Nature Neuroscience, 11*, 426–428.

Rivas, E. (2005). Recent use of signs by chimpanzees (*Pan troglodytes*) in interactions with humans. *Journal of Comparative Psychology, 119*, 404–417.

Rivera, A. S., Ozturk, N., Fahey, B., Plachetzki, D. C., Degnan, B. M., Sancar, A., & Oakley, T. H. (2012). Blue-light-receptive cryptochrome is expressed in a sponge eye lacking neurons and opsin. *The Journal of Experimental Biology, 215*, 1278–1286.

Rizzolatti, G., Fogassi, L., & Gallese, V. (2002). Motor and cognitive functions of the ventral premotor cortex. *Current Opinion in Neurobiology, 12*, 149–154.

Roberts, A. I., Roberts, S. G. B., & Vick, S-J. (2014). The repertoire and intentionality of gestural communication in wild chimpanzees. *Animal Cognition, 17*, 317–336.

Roca, M., Torralva, T., Gleichgerrcht, E., Woolgar, A., Thompson, R., Duncan, J., & Manes, F. (2011). The role of area 10 (BA10) in human multitasking and in social cognition: A lesion study. *Neuropsychologia, 49*, 3525–3531.

Rockland, K. S., & Knutson, T. (2000). Feedback connections from area MT of the squirrel monkey to areas V1 and V2. *The Journal of Comparative Neurology, 425*, 345–368.

Rodriguez, J., Jones, T. H., Sierwald, P., Marek, P. E., Shear, W. A., Brewer, M. S., . . . Bond, J. E. (2018). Step-wise evolution of complex chemical defenses in millipedes: A phylogenomic approach. *Scientific Reports, 8*, article 3209.

Roebroeks, W., & Villa, P. (2011). On the earliest evidence for habitual use of fire in Europe. *Proceedings of the National Academy of Sciences, 108*, 5209–5214.

Rogers, A. R., Iltis, D., & Wooding, S. (2004). Genetic variation at the MC1R locus and the time since loss of human body hair. *Current Anthropology, 45*, 105–108.

Rogers, T. T., Hocking, J., Noppeney, U., Michelli, A., Gorno-Tempini, M. L., Patterson, K., & Price, C. J. (2006). Anterior temporal cortex and semantic memory: Reconciling findings from neuropsychology and functional imaging. *Cognitive, Affective, & Behavioral Neuroscience, 6*, 201–213.

Roma, P. G., Silberberg, A., Huntsberry, M. E., Christensen, C. J., Ruggiero, A. M., & Suomi, S. J. (2007). Mark tests for mirror self-recognition in capuchin monkeys (*Cebus apella*) trained to touch marks. *American Journal of Primatology, 69*, 989–1000.

Rosa, M. G. P., & Krubitzer, M. A. (1999). The evolution of visual cortex: Where is V2? *Trends in Neurosciences, 22*, 242–248.

Rosa, M. G. P., & Manger, P. R. (2005). Clarifying homologies in the mammalian cerebral cortex: The case of the third visual area (V3). *Clinical and Experimental Pharmacology and Physiology, 32*, 327–339.

Rosenbaum, R. S., Gilboa, A., & Moscovitch, M. (2014). Case studies continue to illuminate the cognitive neuroscience of memory. *Annals of the New York Academy of Sciences, 1316*, 105–133.

Rosenberger, A. L., & Hartwig, W. C. (2002). New world monkeys. In *Encyclopedia of life sciences* (Vol. 13, pp. 164–167). London: Nature Publishing Group.

Rosenzweig, M. R., Bennett, E. L., Colombo, P. J., Lee, D. W., & Serrano, P. A. (1993). Short-term, intermediate-term, and long-term memories. *Behavioural Brain Research, 57*, 193–198.

Rosenzweig, M. R., Breedlove, S. M., & Watson, N. V. (2005). *Biological psychology*. Sunderland, MA: Sinauer Associates.

Ross, C. F. (2000). Into the light: The origin of Anthropoidea. *Annual Review of Anthropology, 29*, 147–194.

Ross, C. F., & Kirk, E. C. (2007). Evolution of eye size and shape in primates. *Journal of Human Evolution, 52*, 294–313.

Roth, T. C., Brodin, A., Smulders, T. V., LaDage, L. D., & Pravosudov, V. V. (2010). Is bigger always better? A critical appraisal of the use of volumetric analysis in the study of the hippocampus. *Philosophical Transactions of the Royal Society, B, 365*, 915–931.

Rowe, T. B. (2004). Chordate phylogeny and development. In J. Cracraft & M. J. Donohue (Eds.), *Assembling the tree of life* (pp. 384–409). Oxford: Oxford University Press.

Rowe, T. B., Macrini, T. E., & Luo, Z-X. (2011). Fossil evidence on origin of the mammalian brain. *Science, 332*, 955–957.

Rubin, D. C., Hinton, S., & Wenzel, A. (1999). The precise time course of retention. *Journal of Experimental Psychology: Learning, Memory, and Cognition, 25*, 1161–1176.

Ruff, C. B., & Higgins, R. (2013). Femoral neck structure and function in early hominins. *American Journal of Physical Anthropology, 150*, 512–525.

Ruff, C. B., Holt, B., Niskanen, M., Sladek, V., Berner, M., Garofalo, E., . . . Whittey, E. (2015). Gradual decline in mobility with the adoption of food production in Europe. *Proceedings of the National Academy of Sciences, 112*, 7147–7152.

Ruhlen, M. (1994a). *On the origins of languages*. Stanford: Stanford University Press.

Ruhlen, M. (1994b). *The origin of language*. New York, NY: John Wiley & Sons.

Ruiz, A., Gómez, J. C., Roeder, J. J., & Byrne, R. W. (2009). Gaze following and gaze priming in lemurs. *Animal Cognition, 12*, 427–434.

Russon, A. E. (2004). Great ape cognitive systems. In A. E. Russon & D. R. Begun (Eds.), *The evolution of thought: Evolutionary origins of great ape intelligence* (pp. 76–100). Cambridge, UK: Cambridge University Press.

Rutschmann, F. (2006). Molecular dating of phylogenetic trees: A brief review of current methods that estimate divergence times. *Diversity and Distributions, 12*, 35–48.

Rutz, C., Klump, B. C., Komarczyk, L., Leighton, R., Kramer, J., Wischnewski, S., . . . Masuda, B. M. (2016). Discovery of species-wide tool use in the Hawaiian crow. *Nature, 537*, 403–407.

Rutz, C., Sugasawa, S., van der Wal, J. E. M., Klump, B. C., & St. Clair, J. J. H. (2016). Tool bending in New Caledonian crows. *Royal Society Open Science*, 3, article 160439.

Sakamoto, M., Benton, M. J., & Venditti, C. (2016). Dinosaurs in decline tens of millions of years before their final extinction. *Proceedings of the National Academy of Sciences*, 113, 5036–5040.

Sakitt, B., & Long, G. M. (1979). Spare the rod and spoil the icon. *Journal of Experimental Psychology: Human Perception and Performance*, 5, 19–30.

Saladino, R., Carota, E., Botta, G., Kapralov, M., Timoshenko, G. N., Rozanov, A. Y., . . . Di Mauro, E. (2015). Meteorite-catalyzed syntheses of nucleosides and of other prebiotic compounds from formamide under proton irradiation. *Proceedings of the National Academy of Sciences*, 112(21), E2746–E2755.

San Antonio, J. D., Schweitzer, M. H., Jensen, S. T., Kalluri, R., Buckley, M., & Orgel, J. P. R. O. (2011). Dinosaur peptides suggest mechanisms of protein survival. *PLoS One*, 6, article e20381.

Sandel, A. A., MacLean, E. L., & Hare, B. (2011). Evidence from four lemur species that ring-tailed lemur social cognition converges with that of haplorhine primates. *Animal Behaviour*, 81, 925–931.

Santini, L., Rojas, D., & Donati, G. (2015). Evolving through day and night: Origin and diversification of activity pattern in modern primates. *Behavioral Ecology*, 26, 789–796.

Sanz, C. M., & Morgan, D. B. (2007). Chimpanzee tool technology in the Goualougo Triangle, Republic of Congo. *Journal of Human Evolution*, 52, 420–433.

Sanz, C. M., Morgan, D. B., & Hopkins, W. D. (2016). Lateralization and performance asymmetries in the termite fishing of wild chimpanzees in the Goualougo Triangle, Republic of Congo. *American Journal of Primatology*, 78, 1190–1200.

Sato, N., Sakata, H., Tanaka, Y. L., & Taira, M. (2006). Navigation-associated medial parietal neurons in monkeys. *Proceedings of the National Academy of Sciences*, 103, 17001–17006.

Savage-Rumbaugh, E. S. (1992). Language training of apes. In S. Jones, R. Martin, & D. Pilbeam (Eds.), *The Cambridge encyclopedia of human evolution* (pp. 138–141). Cambridge, UK: Cambridge University Press.

Savage-Rumbaugh, E. S., McDonald, K., Sevcik, R. A., Hopkins, W. D., & Rubert, E. (1986). Spontaneous symbol acquisition and communicative use by pygmy chimpanzees (*Pan paniscus*). *Journal of Experimental Psychology: General*, 115, 211–235.

Sawamura, H., Georgieva, S., Vogels, R., Vanduffel, W., & Orban, G. A. (2005). Using functional magnetic resonance imaging to assess adaptation and size invariance of shape processing by humans and monkeys. *The Journal of Neuroscience*, 25, 4294–4306.

Saxena, A., Towers, M., & Cooper, K. L. (2017). The origins, scaling and loss of tetrapod digits. *Philosophical Transactions of the Royal Society, B*, 372, article 20150482.

Schall, U., Johnston, P., Lagopoulos, J., Jüptner, M., Jentzen, W., Thienel, R., . . . Ward, P. B. (2003). Functional brain maps of tower of London performance: A positron emission tomography and functional magnetic resonance imaging study. *NeuroImage*, 20, 1154–1161.

Scharlau, I. (2004). Evidence for split foci of attention in a priming paradigm. *Perception & Psychophysics*, 66, 988–1002.

Schilling, T. (2003). Making jaws. *Heredity*, 90, 3–5.

Schinazi, V. R., Nardi, D., Newcombe, N. S., Shipley, T. F., & Epstein, R. A. (2013). Hippocampal size predicts rapid learning of a cognitive map in humans. *Hippocampus*, 23, 515–528.

Schlosser, G. (2017). From so simple a beginning – what amphioxus can teach us about placode evolution. *International Journal of Developmental Biology*, 61, 633–648.

Schmid, P., Churchill, S. E., Nalla, S., Weissen, E., Carlson, K. J., de Ruiter, D. J., & Berger, L. R. (2013). Mosaic morphology in the thorax of *Australopithecus sediba*. *Science*, 340, article 1234598.

Schmitt, D. (2003). Insights into the evolution of human bipedalism from experimental studies of humans and other primates. *The Journal of Experimental Biology*, 206, 1437–1448.

Schmitt, D., Cartmill, M., Griffin, T. M., Hanna, J. B., & Lemelin, P. (2006). Adaptive value of ambling gaits in primates and other mammals. *The Journal of Experimental Biology*, 209, 2042–2049.

Schmitt, D., & Lemelin, P. (2002). Origins of primate locomotion: Gait mechanics of the wooly opossum. *American Journal of Physical Anthropology*, 118, 231–238.

Schmitz, L., & Motani, R. (2011). Nocturnality in dinosaurs inferred from scleral ring and orbit morphology. *Science, 332*, 705–708.

Schmolesky, M. (2018). The primary visual cortex. *Webvision*, Moran Eye Center. Retrieved August 20, 2018, from https://webvision.med.utah.edu/book/part-ix-psychophysics-of-vision/the-primary-visual-cortex

Schneider, D., Schneider, L., Claussen, C-F., & Kolchev, C. (2001). Cortical representation of the vestibular system as evidenced by brain electrical activity mapping of vestibular late evoked potentials. *Ear, Nose, & Throat Journal, 80*, 251–265.

Schoene, B., Samperton, K. M., Eddy, M. P., Keller, G., Adatte, T., Bowring, S. A., . . . Gertsch, B. (2015). U-Pb geochronology of the Deccan traps and relation to the end-Cretaceous mass extinction. *Science, 347*, 182–184.

Schrago, C. G., & Russo, C. A. M. (2003). Timing the origin of New World monkeys. *Molecular Biology and Evolution, 20*, 1620–1625.

Schrago, C. G., & Voloch, C. M. (2013). The precision of the hominid timescale estimated by relaxed clock methods. *Journal of Evolutionary Biology, 26*, 746–755.

Schroeter, E. R., DeHart, C. J., Cleland, T. P., Zheng, W., Thomas, P. M., Kelleher, N. L., . . . Schweitzer, M. H. (2017). Expansion for the *Brachlyophosaurus canadensis* collagen I sequence and additional evidence of the preservation of cretaceous protein. *Journal of Proteome Research, 16*, 920–932.

Schulman, A. H., & Kaplowitz, C. (1977). Mirror-image response during the first two years of life. *Developmental Psychobiology, 10*, 133–142.

Schulte, P., Alegret, L., Arenillas, I., Arz, J. A., Barton, P. J., Bown, P. R., . . . Willumsen, P. S. (2010). The Chicxulub asteroid impact and mass extinction at the Cretaceous-Paleogene boundary. *Science, 327*, 1214–1218.

Schulze-Makuch, D., Fairen, A. G., & Davila, A. F. (2008). The case for life on Mars. *International Journal of Astrobiology, 7*, 117–141.

Schwab, I. R. (2018). The evolution of eyes: Major steps. The Keeler lecture 2017: Centenary of Keeler Ltd. *Eye, 32*, 302–313.

Schwartz, B. L., Colon, M. R., Sanchez, I. C., Rodriguez, I. A., & Evans, S. (2002). Single-trial learning of "what" and "who" information in a gorilla (*Gorilla gorilla gorilla*): Implications for episodic memory. *Animal Cognition, 5*, 85–90.

Schwartz, B. L., Hoffman, M. L., & Evans, S. (2005). Episodic-like memory in a gorilla: A review and new findings. *Learning and Motivation, 36*, 226–244.

Schwartz, J. H., & Tattersall, I. (2000). The human chin revisited: What is it and who has it? *Journal of Human Evolution, 38*, 367–409.

Schweitzer, M. H., Zheng, W., Organ, C. L., Avci, R., Suo, Z., Freimark, L. M., . . . Asara, J. M. (2009). Biomolecular characterization and protein sequences of the Campanian hadrosaur *B. canadensis*. *Science, 324*, 626–631.

Scozzari, R., Massaia, A., Trombetta, B., Bellusci, G., Myres, N. M., Novelletto, A., & Cruciani, F. (2014). An unbiased resource of novel SNP markers provides a new chronology for the human Y chromosome and reveals a deep phylogenetic structure in Africa. *Genome Research, 24*, 535–544.

Seghier, M. L., Lazeyras, F., Pegna, A. J., Annoni, J-M., Zimine, I., Mayer, E., . . . Khateb, A. (2004). Variability of fMRI activation during a phonological and semantic language task in healthy subjects. *Human Brain Mapping, 23*, 140–155.

Seipel, K., & Schmid, V. (2005). Evolution of striated muscle: Jellyfish and origin of triploblasty. *Developmental Biology, 282*, 14–26.

Selby, M. S., Simpson, S. W., & Lovejoy, C. O. (2016). The functional anatomy of the carpometacarpal complex in anthropoids and its implications for the evolution of the hominoid hand. *Anatomical Record, 299*, 583–600.

Seligman, M. E. P. (1970). On the generality of the laws of learning. *Psychological Review, 77*, 406–418.

Sellers, W. I., Cain, G. M., Wang, W., & Crompton, R. H. (2005). Stride lengths, speed and energy costs in walking of *Australopithecus afarensis*: Using evolutionary robotics to predict locomotion of early human ancestors. *Journal of the Royal Society Interface, 2*, 431–441.

Semaw, S. (2000). The world's oldest stone artefacts from Gona, Ethiopia: Their implications for understanding stone technology and patterns of human evolution between 2.6–1.5 million years ago. *Journal of Archaeological Science, 27*, 1197–1214.

Semendeferi, K., Armstrong, E., Schleicher, A., Zilles, K., & Van Hoesen, G. W. (2001). Prefrontal cortex in humans and apes: A comparative study of area 10. *American Journal of Physical Anthropology, 114*, 224–241.

Semendeferi, K., & Damasio, H. (2000). The brain and its main anatomical subdivisions in living hominoids using magnetic resonance imaging. *Journal of Human Evolution, 38*, 317–332.

Semenov, S. A. (1970). *Prehistoric technology*. Bath, UK: Adams & Dart.

Senut, B. (2006). Bipédie et climat. *Comptes Rendus Palevol, 5*, 89–98.

Seymour, K. J., Williams, M. A., & Rich, A. N. (2016). The representation of color across the human visual cortex: Distinguishing chromatic signals contributing to object form versus surface color. *Cerebral Cortex, 26*, 1997–2005.

Shelton, A. L., & Gabrieli, J. D. E. (2002). Neural correlates of encoding space from route and survey perspectives. *The Journal of Neuroscience, 22*, 2711–2727.

Shepherd, S. V., & Platt, M. L. (2008). Spontaneous social orienting and gaze following in ring-tailed lemurs (*Lemur catta*). *Animal Cognition, 11*, 13–20.

Shields, W. E., Smith, J. D., Guttmannova, K., & Washburn, D. A. (2005). Confidence judgments by humans and rhesus monkeys. *The Journal of General Psychology, 132*, 165–186.

Shoshani, J., Groves, C. P., Simons, E. L., & Gunnell, G. F. (1996). Primate phylogeny: Morphological vs molecular results. *Molecular Phylogenetics and Evolution, 5*, 102–154.

Shoshani, J., Kupsky, W. J., & Marchant, G. H. (2006). Elephant brain. Part 1: Gross morphology, functions, comparative anatomy, and evolution. *Brain Research Bulletin, 70*, 124–157.

Shreve, J. (2015, October). Mystery man: A trove of fossils found deep in a South African cave adds a baffling new branch to the human family tree. *National Geographic*, 30–57.

Shu, D-G., Luo, H-L., Morris, S. C., Zhang, X-L., Hu, S-X., Chen, L., . . . Chen, L-Z. (1999). Lower Cambrian vertebrates from South China. *Nature, 402*, 42–46.

Shubin, N. H. (2008). *Your inner fish*. New York, NY: Pantheon Books.

Shubin, N. H., Daeschler, E. B., & Coates, M. I. (2004). The early evolution of the tetrapod humerus. *Science, 304*, 90–93.

Sidor, C. A., & Hopson, J. A. (1998). Ghost lineages and "mammalness": Assessing the temporal pattern of character acquisition in the synapsida. *Paleobiology, 24*, 254–273.

Sihvonen, A. J., Ripollés, P., Leo, V., Rodríguez-Fornells, A., Soinila, S., & Särkämö, T. (2016). Neural basis of acquired amusia and its recovery after stroke. *The Journal of Neuroscience, 36*, 8872–8881.

Silcox, M. T., Benham, A. E., & Bloch, J. I. (2010). Endocasts of *Microsyops* (Microsyopidae, Primates) and the evolution of the brain in primitive primates. *Journal of Human Evolution, 58*, 505–521.

Silcox, M. T., Dalmyn, C. K., & Bloch, J. I. (2009). Virtual endocast of *Ignacius graybullianus* (Paromomyidae, Primates) and brain evolution in early primates. *Proceedings of the National Academy of Sciences, 106*, 10987–10992.

Simion, P., Philippe, H., Baurain, D., Jager, M., Richter, D. J., DiFranco, A., . . . Manuel, M. (2017). A large and consistent phylogenomic dataset supports sponges as the sister group to all other animals. *Current Biology, 27*, 958–967.

Simons, E. L. (1992). The fossil history of primates. In S. Jones, R. Martin, & D. Pilbeam (Eds.), *The Cambridge encyclopedia of human evolution* (pp. 199–208). Cambridge, UK: Cambridge University Press.

Simons, E. L., Seiffert, E. R., Ryan, T. M., & Attia, Y. (2007). A remarkable female cranium of the early Oligocene anthropoid *Aegyptopithecus zeuxis* (Catarrhini, Propliopithecidae). *Proceedings of the National Academy of Sciences, 104*, 8731–8736.

Simonti, C. N., Vernot, B., Bastarache, L., Bottinger, E., Carrell, D. S., Chisholm, R. L., . . . Capra, J. A. (2016). The phenotypic legacy of admixture between modern humans and Neandertals. *Science, 351*, 737–741.

Simonyan, K. (2014). The laryngeal motor cortex: Its organization and connectivity. *Current Opinion in Neurobiology, 28*, 15–21.

Simpson, S. W., Latimer, B., & Lovejoy, C. O. (2018). Why do knuckle-walking African apes knuckle-walk? *The Anatomical Record*, *301*, 496–514.

Simpson, S. W., Quade, J., Levin, N. E., Butler, R., Dupont-Nivet, G., Everett, M., & Semaw, S. (2008). A female *Homo erectus* pelvis from Gona, Ethiopia. *Science*, *322*, 1089–1092.

Skottun, B. C., & Skoyles, J. R. (2010). Backward masking as a test of magnocellular sensitivity. *Neuro-ophthalmology*, *34*, 342–346.

Slatkin, M., & Racimo, F. (2016). Ancient DNA and human history. *Proceedings of the National Academy of Sciences*, *113*, 6380–6387.

Smith, T. D., Rossie, J. B., & Bhatnagar, K. P. (2007). Evolution of the nose and nasal skeleton in primates. *Evolutionary Anthropology*, *16*, 132–146.

Smithson, T. R., Carroll, R. L., Panchen, A. L., & Andrews, S. M. (1994). *Westlothiana lizziae* from the Viséan of East Kirkton, West Lothian, Scotland, and the amniote stem. *Transactions of the Royal Society of Edinburgh: Earth Sciences*, *84*, 383–412.

Soler, M., Pérez-Contreras, T., & Peralta-Sánchez, J. M. (2014). Mirror-mark tests performed on jackdaws reveal potential methodological problems in the use of stickers in avian mark-test studies. *PLoS One*, *9*, article e86193.

Soligo, C., & Martin, R. D. (2006). Adaptive origins of primates revisited. *Journal of Human Evolution*, *50*, 414–430.

Souto, A., Bione, C. B. C., Bastos, M., Bezerra, B. M., Fragaszy, D., & Schiel, N. (2011). Critically endangered blonde capuchins fish for termites and use new techniques to accomplish the task. *Biology Letters*, *7*, 532–535.

Spelke, E. S. (1982). Perceptual knowledge of objects in infancy. In J. Mehler, E. C. T. Walker, & M. Garrett (Eds.), *Perspectives on mental representation* (pp. 409–430). Hillsdale, NJ: Lawrence Erlbaum Associates.

Sperling, G. (1963). A model for visual memory tasks. *Human Factors*, *5*, 19–31.

Spillmann, L. (2014). Receptive fields of visual neurons: The early years. *Perception*, *43*, 1145–1176.

Spinozzi, G., De Lillo, C., & Salvi, V. (2006). Local advantage in the visual processing of hierarchical stimuli following manipulations of stimulus size and element numerosity in monkeys (*Cebus apella*). *Behavioural Brain Research*, *166*, 45–54.

Spoor, F., Leakey, M. G., Gathogo, P. N., Brown, F. H., Antón, S. C., McDougall, I., . . . Leakey, L. N. (2007). Implications of new early *Homo* fossils from Ileret, east of lake Turkana, Kenya. *Nature*, *448*, 688–691.

Springer, M. S., Emerling, C. A., Meredith, R. W., Janecka, J. E., Eizirik, E., & Murphy, W. J. (2017). Waking the undead: Implications of a soft explosive model for the timing of placental mammal diversification. *Molecular Phylogenetics and Evolution*, *106*, 86–102.

Squires, T. M. (2004). Optimizing the vertebrate vestibular semicircular canal: Could we balance any better? *Physical Review Letters*, *93*, 198106, 1–4.

Stanford, C. B. (1996). The hunting ecology of wild chimpanzees: Implications for the evolutionary ecology of Pliocene hominids. *American Anthropologist*, *98*, 96–113.

Stanford, C. B. (2003). *Upright: The evolutionary key to becoming human*. Boston: Houghton Mifflin Co.

Stanford, C. B. (2006). Arboreal bipedalism in wild chimpanzees: Implications for the evolution of hominid posture and locomotion. *American Journal of Physical Anthropology*, *129*, 225–231.

Steele, A., Goddard, D. T., Stapleton, D., Toporski, J. K. W., Peters, V., Bassinger, V., . . . McKay, D. S. (2000). Investigations into an unknown organism on the martian meteorite Allan Hills 84001. *Meteoritics & Planetary Science*, *35*, 237–241.

Steenhuis, R. E., & Bryden, M. P. (1989). Different dimensions of hand preference that relate to skilled and unskilled activities. *Cortex*, *25*, 289–304.

Steinmetz, P. R. H., Arman, A., Kraus, J. E. M., & Technau, U. (2017). Gut-like ectodermal tissue in a sea anemone challenges germ layer homology. *Nature Ecology & Evolution*, *1*, 1535–1542.

Stepniewska, I., Preuss, T. M., & Kaas, J. H. (2006). Ipsilateral cortical connections of dorsal and ventral premotor areas in new world owl monkeys. *The Journal of Comparative Neurology*, *495*, 691–708.

Steudel-Numbers, K. L. (2001). Role of locomotor economy in the origin of bipedal posture and gait. *American Journal of Physical Anthropology, 116*, 171–173.

Stevens, N. J., Seiffert, E. R., O'Connor, P. M., Roberts, E. M., Schmitz, M. D., Krause, C., . . . Temu, J. (2013). Paleontological evidence for an Oligocene divergence between old world monkeys and apes. *Nature, 497*, 611–614.

Steventon, B., Mayor, R., & Streit, A. (2016). Directional cell movements downstream of Gbx2 and Otx2 control the assembly of sensory placodes. *Biology Open, 5*, 1620–1624.

Stone, A. C., Battistuzzi, F. U., Kubatko, L. S., Perry, G. H., Trudeau, E., Lin, H., & Kumar, S. (2010). More reliable estimates of divergence times in *Pan* using complete mtDNA sequences and accounting for population structure. *Philosophical Transactions of the Royal Society, B, 365*, 3277–3288.

Stout, D. (2002). Skill and cognition in stone tool production: An ethnographic case study from Irian Jaya. *Current Anthropology, 43*, 693–715.

Stout, D., Toth, N., Schick, K., & Chaminade, T. (2008). Neural correlates of early stone age toolmaking: Technology, language and cognition in human evolution. *Philosophical Transactions of the Royal Society, B, 363*, 1939–1949.

Strait, D. S., & Grine, F. E. (2004). Inferring hominoid and early hominid phylogeny using craniodental characters: The role of fossil taxa. *Journal of Human Evolution, 47*, 399–452.

Straube, B., & Chatterjee, A. (2010). Space and time in perceptual causality. *Frontiers in Human Neuroscience, 4*, article 28.

Straus, L. G. (2001). Africa and Iberia in the Pleistocene. *Quaternary International, 75*, 91–102.

Stringer, C., & McKie, R. (1996). *African exodus: The origins of modern humanity*. New York, NY: Henry Holt and Company.

Suárez, R., Gobius, I., & Richards, L. J. (2014). Evolution and development of interhemispheric connections in the vertebrate forebrain. *Frontiers in Human Neuroscience, 8*, article 497.

Suddendorf, T., & Whiten, A. (2001). Mental evolution and development: Evidence for secondary representation in children, great apes, and other animals. *Psychological Bulletin, 127*, 629–650.

Sugahara, F., Murakami, Y., Pascual-Anaya, J., & Kuratani, S. (2017). Reconstructing the ancestral vertebrate brain. *Development, Growth, & Differentiation, 59*, 163–174.

Sun, L., Kawano-Yamashita, E., Nagata, T., Tsukamoto, H., Furutani, Y., Koyanagi, M., & Terakita, A. (2014). Distribution of mammalian-like melanopsin in cyclostome retinas exhibiting a different extent of visual functions. *PLoS One, 9*, article e108209.

Sunaert, S., Van Hecke, P., Marchal, G., & Orban, G. A. (1999). Motion-responsive regions of the human brain. *Experimental Brain Research, 127*, 355–370.

Surridge, A. K., Osorio, D., & Mundy, N. I. (2003). Evolution and selection of trichromatic vision in primates. *Trends in Ecology and Evolution, 18*, 198–205.

Sussman, R. W., Rasmussen, D. T., & Raven, P. H. (2013). Rethinking primate origins again. *American Journal of Primatology, 75*, 95–106.

Sustaita, D., Pouydebat, E., Manzano, A., Abdala, V., Hertel, F., & Herrel, A. (2013). Getting a grip on tetrapod grasping: Form, function, and evolution. *Biological Reviews, 88*, 380–405.

Sutikna, T., Tocheri, M. W., Morwood, M. J., Saptomo, E. W., Jatmiko, Awe, R. D., . . . Roberts, R. G. (2016). Revised stratigraphy and chronology for *Homo floresiensis* at Liang Bua in Indonesia. *Nature, 532*, 366–369.

Suwa, G., Asfaw, B., Kono, R. T., Kubo, D., Lovejoy, C. O., & White, T. D. (2009). The *Ardipithecus ramidus* skull and its implications for hominid origins. *Science, 326*, 68.

Suzuki, W., Banno, T., Miyakawa, N., Abe, H., Goda, N., & Ichinohe, N. (2015). Mirror neurons in a new world monkey, common marmoset. *Frontiers in Neuroscience, 9*, article 459.

Swartz, K. B., Sarauw, D., & Evans, S. (1999). Comparative aspects of mirror self-recognition in great apes. In S. T. Parker, R. W. Mitchell, & H. L. Miles (Eds.), *The mentalities of gorillas and orangutans in comparative perspective* (pp. 283–294). Cambridge, UK: Cambridge University Press.

Tabiow, E., & Forrester, G. S. (2013). Structured bimanual actions and hand transfers reveal population-level right-handedness in captive gorillas. *Animal Behaviour, 86*, 1049–1057.

Taglialatela, J. P., Cantalupo, C., & Hopkins, W. D. (2006). Gesture handedness predicts asymmetry in the chimpanzee inferior frontal gyrus. *NeuroReport, 17*, 923–927.

Taglialatela, J. P., Russell, J. L., Schaeffer, J. A., & Hopkins, W. D. (2008). Communicative signaling activates 'Broca's' homolog in chimpanzees. *Current Biology, 18*, 1–6.

Takemoto, H., Kawamoto, Y., & Furuichi, T. (2015). How did bonobos come to range south of the Congo River? Reconsideration of the divergence of *Pan paniscus* from other *Pan* populations. *Evolutionary Anthropology, 24*, 170–184.

Takezaki, N., & Nishihara, H. (2017). Support for lungfish as the closest relative of tetrapods by using slowly evolving ray-finned fish as the outgroup. *Genome Biology and Evolution, 9*, 93–101.

Tanaka, K. (1997). Mechanisms of visual object recognition: Monkey and human studies. *Current Opinion in Neurobiology, 7*, 523–529.

Tang, S., Lee, T. S., Li, M., Zhang, Y., Xu, Y., Liu, F., . . . Jiang, H. (2018). Complex pattern selectivity in macaque primary visual cortex revealed by large-scale two-photon imaging. *Current Biology, 28*, 38–48.

Taubert, J., Van Belle, G., Vanduffel, W., Rossion, B., & Vogels, R. (2015). Neural correlate of the Thatcher face illusion in a monkey face-selective patch. *The Journal of Neuroscience, 35*, 9872–9878.

Taylor, J. G., & Fragopanagos, M. (2007). Resolving some confusions over attention and consciousness. *Neural Networks, 20*, 993–1003.

Teffer, K., Buxhoeveden, D. P., Stimpson, C. D., Fobbs, A. J., Schapiro, S. J., Baze, W. B., . . . Semendeferi, K. (2013). Developmental changes in the spatial organization of neurons in the neocortex of humans and common chimpanzees. *The Journal of Comparative Neurology, 521*, 4249–4259.

Terberger, T. (2006). From the first humans to the Mesolithic hunters in the Northern German lowlands – current results and trends. In K. M. Hansen & K. B. Pedersen (Eds.), *Across the Western Baltic* (pp. 111–184). Vordingborg, Denmark: Sydsjællands Museum.

Terrace, H. S., Petitto, L. A., Sanders, R. J., & Bever, T. G. (1979). Can an ape create a sentence? *Science, 206*, 891–902.

Thiergart, T., Landan, G., & Martin, W. F. (2014). Concatenated alignments and the case of the disappearing tree. *BMC Evolutionary Biology, 14*, 28–49.

Thompson, J. C., McPherron, S. P., Bobe, R., Reed, D., Barr, W. A., Wynn, J. G., . . . Alemseged, Z. (2015). Taphonomy of fossils from the hominin-bearing deposits at Dikika, Ethiopia. *Journal of Human Evolution, 86*, 112–135.

Thompson, R. L., & Boatright-Horowitz, S. L. (1994). The question of mirror-mediated self-recognition in apes and monkeys: Some new results and reservations. In S. T. Parker, R. W. Mitchell, & M. L. Boccia (Eds.), *Self-awareness in animals and humans* (pp. 330–349). Cambridge, UK: Cambridge University Press.

Thorpe, S. J., & Fabre-Thorpe, M. (2001). Seeking categories in the brain. *Science, 291*, 260–263.

Titchener, E. B. (1909). *A text-book of psychology*. New York, NY: Palgrave Macmillan.

Tobias, P. V. (1995). The brain of the first hominids. In J-P. Changeux & J. Chavaillon (Eds.), *Origins of the human brain* (pp. 61–81). New York, NY: Oxford University Press.

Tobias, P. V., & Campbell, B. (1981). The emergence of man in Africa and beyond. *Philosophical Transactions of the Royal Society of London, B, 292*, 43–56.

Tocheri, M. W., Marzke, M. W., Liu, D., Bae, M., Jones, G. P., Williams, R. C., & Razdan, A. (2003). Functional capabilities of modern and fossil hominid hands: Three-dimensional analysis of trapezia. *American Journal of Physical Anthropology, 122*, 101–112.

Tocheri, M. W., Orr, C. M., Larson, S. G., Sutikna, T., Jatmiko, Saptomo, E. W., . . . Jungers, W. L. (2007). The primitive wrist of *Homo floresiensis* and its implications for hominin evolution. *Science, 317*, 1743–1745.

Tomasello, M., & Call, J. (1997). *Primate cognition*. New York, NY: Oxford University Press.

Tomasello, M., Call, J., & Hare, B. (1998). Five primate species follow the visual gaze of conspecifics. *Animal Behaviour, 55*, 1063–1069.

Tomasello, M., Call, J., & Hare, B. (2003). Chimpanzees understand psychological states – the question is which ones and to what extent. *Trends in Cognitive Sciences, 7*, 153–156.

Tomasello, M., Hare, B., & Agnetta, B. (1999). Chimpanzees, *Pan troglodytes*, follow gaze direction geometrically. *Animal Behaviour, 58*, 769–777.

Topolski, R., & Inhoff, A. W. (1995). Loss of vision during the retinal stabilization of letters. *Psychological Research*, *58*, 155–162.

Toth, N. (1985). Archaeological evidence for preferential right-handedness in the lower and middle Pleistocene, and its possible implications. *Journal of Human Evolution*, *14*, 607–614.

Toups, M. A., Kitchen, A., Light, J. E., & Reed, D. L. (2010). Origin of clothing lice indicates early clothing use by anatomically modern humans in Africa. *Molecular Biology and Evolution*, *28*, 29–32.

Trainor, P. A., Melton, K. R., & Manzanares, M. (2003). Origins and plasticity of neural crest cells and their roles in jaw and craniofacial evolution. *International Journal of Developmental Biology*, *47*, 541–553.

Tranel, D., Damasio, H., & Damasio, A. R. (1997). A neural basis for the retrieval of conceptual knowledge. *Neuropsychologia*, *35*, 1319–1327.

Trevethan, C. T., Sahraie, A., & Weiskrantz, L. (2007). Form discrimination in a case of blindsight. *Neuropsychologia*, *45*, 2092–2103.

Trevors, J. T. (2003). Early assembly of cellular life. *Progress in Biophysics and Molecular Biology*, *81*, 201–217.

Trick, L. M., & Pylyshyn, Z. W. (1993). What enumeration studies can show us about spatial attention: Evidence for limited capacity preattentive processing. *Journal of Experimental Psychology: Human Perception and Performance*, *19*, 331–351.

Trinkaus, E. (1992). Evolution of human manipulation. In S. Jones, R. Martin, & D. Pilbeam (Eds.), *The Cambridge encyclopedia of human evolution* (pp. 346–349). Cambridge, UK: Cambridge University Press.

Truppa, V., De Simone, D. A., & De Lillo, C. (2016). Short-term memory effects on visual global/local processing in tufted capuchin monkeys (*Sapajus spp.*). *Journal of Comparative Psychology*, *130*, 162–173.

Tsao, D. Y., Vanduffel, W., Sasaki, Y., Fize, D., Knutsen, T. A., Mandeville, J. B., . . . Tootell, R. B. H. (2003). Stereopsis activates V3A and caudal intraparietal areas in macaques and humans. *Neuron*, *39*, 555–568.

Tschudin, A., Call, J., Dunbar, R. I. M., Harris, G., & van der Elst, C. (2001). Comprehension of signs by dolphins (*Tursiops truncatus*). *Journal of Comparative Psychology*, *115*, 100–105.

Tzourio-Mazoyer, N., Crivello, F., & Mazoyer, B. (2018). Is the planum temporale surface area a marker of hemispheric or regional language lateralization? *Brain Structure and Function*, *223*, 1217–1228.

Ujhelyi, M. (1998). Long-call structure in apes as a possible precursor for language. In J. R. Hurford, M. Studdert-Kennedy, & C. Knight (Eds.), *Approaches to the evolution of language* (pp. 177–189). Cambridge, UK: Cambridge University Press.

Ujhelyi, M., Merker, B., Buk, P., & Geissmann, T. (2000). Observations on the behavior of gibbons (*Hylobates leucogenys*, *H. gabriellae*, and *H. lar*) in the presence of mirrors. *Journal of Comparative Psychology*, *114*, 253–262.

Unterrainer, J. M., Rahm, B., Kaller, C. P., Ruff, C. C., Spreer, J., Krause, B. J., . . . Halsband, U. (2004). When planning fails: Individual differences and error-related brain activity in problem solving. *Cerebral Cortex*, *14*, 1390–1397.

Uomini, N. T. (2006). *In the Knapper's hands: Testing markers of laterality in hominin lithic production, with reference to the common substrate of language and handedness* (Thesis submitted for the degree of Doctor of Philosophy), University of Southampton.

Valentine, J. W., Jablonski, D., & Erwin, D. H. (1999). Fossils, molecules and embryos: New perspectives on the Cambrian explosion. *Development*, *126*, 851–859.

van den Bergh, G. D., Kaifu, Y., Kurniawan, I., Kono, R. T., Brumm, A., Setiyabudi, E., . . . Morwood, M. J. (2016). *Homo floresiensis*-like fossils from the early Middle Pleistocene of Flores. *Nature*, *534*, 245–248.

van den Heuvel, O. A., Groenewegen, H. J., Barkhof, F., Lazeron, R. H. C., van Dyck, R., & Veltman, D. J. (2003). Frontostriatal system in planning complexity: A parametric functional magnetic resonance version of Tower of London task. *NeuroImage*, *18*, 367–374.

van Duijn, M. (2017). Phylogenetic origins of biological cognition: Convergent patterns in the early evolution of learning. *Interface Focus*, 7, article 20160158.

Van Lawick-Goodall, J. (1971). *In the shadow of man*. New York, NY: Dell.

van Schaik, C. P., Ancrenaz, M., Borgen, G., Galdikas, B., Knott, C. D., Singleton, I., . . . Merrill, M. (2003). Orangutan cultures and the evolution of material culture. *Science*, *299*, 102–105.

van Tuinen, M., & Hadly, E. A. (2004). Error in estimation of rate and time inferred from the early amniote fossil record and avian molecular clocks. *Journal of Molecular Evolution*, *59*, 267–276.

Vargas, J. P., Bingman, V. P., Portavella, M., & López, J. C. (2006). Telencephalon and geometric space in goldfish. *European Journal of Neuroscience*, *24*, 2870–2878.

Vargha-Khadem, F., Watkins, K. E., Price, C. J., Ashburner, J., Alcock, K. J., Connelly, A., . . . Passingham, R. E. (1998). Neural basis of an inherited speech and language disorder. *Proceedings of the National Academy of Sciences*, *95*, 12695–12700.

Varin, V. P., & Petrov, A. G. (2009). A hydrodynamic model of human cochlea. *Computational Mathematics and Mathematical Physics*, *49*, 1632–1647.

Vater, M., Meng, J., & Fox, R. C. (2004). Hearing organ evolution and specialization: Early and later mammals. In G. A. Manley, A. N. Popper, & R. R. Fay (Eds.), *Evolution of the vertebrate auditory system* (pp. 256–288). New York, NY: Springer-Verlag.

Vázques-Salazar, A., & Lazcano, A. (2018). Early life: Embracing the RNA world. *Current Biology*, *28*, R220–R222.

Verdon, V., Schwartz, S., Lovblad, K-O., Hauert, C-A., & Vuilleumier, P. (2010). Neuroanatomy of hemispatial neglect and its functional components: A study using voxel-based lesion-symptom mapping. *Brain*, *133*, 880–894.

Vereecke, E., D'Août, K., De Clercq, D., Van Elsacker, L., & Aerts, P. (2003). Dynamic plantar pressure distribution during terrestrial locomotion of bonobos (*Pan paniscus*). *American Journal of Physical Anthropology*, *120*, 373–383.

Verfaellie, M., Koseff, P., & Alexander, M. P. (2000). Acquisition of novel semantic information in amnesia: Effects of lesion location. *Neuropsychologia*, *38*, 484–492.

Verhoef, B-E., Vogels, R., & Janssen, P. (2016). Binocular depth processing in the ventral visual pathway. *Philosophical Transactions of the Royal Society, B*, *371*, article 2050259.

Vesia, M., Barnett-Cowan, M., Elahi, B., Jegatheeswaran, G., Isayama, R., Neva, J. L., . . . Chen, R. (2017). Human dorsomedial parieto-motor circuit specifies grasp during the planning of goal-directed hand actions. *Cortex*, *92*, 175–186.

Viala, D. (2006). Evolution and behavioral adaptation of locomotor pattern generators in vertebrates. *Comptes Rendus Palevol*, *5*, 667–674.

Videan, E. N., & McGrew, W. C. (2002). Bipedality in chimpanzee (*Pan troglodytes*) and bonobo (*Pan paniscus*): Testing hypotheses on the evolution of bipedalism. *American Journal of Physical Anthropology*, *118*, 184–190.

Vingerhoets, G. (2014). Contribution of the posterior parietal cortex in reaching, grasping, and using objects and tools. *Frontiers in Psychology*, *5*, article 151.

Völter, C. J., & Call, J. (2014). Younger apes and human children plan their moves in a maze task. *Cognition*, *130*, 186–203.

von der Heydt, R., Peterhans, E., & Dürsteler, M. R. (1992). Periodic-pattern-selective cells in monkey visual cortex. *The Journal of Neuroscience*, *12*, 1416–1434.

Vulliemoz, S., Raineteau, O., & Jabaudon, D. (2005). Reaching beyond the midline: Why are human brains cross wired? *Lancet Neurology*, *4*, 87–99.

Wacewicz, S., Zywickzynski, P., & Orzechowski, S. (2016). Visible movements of the orofacial area: Evidence for gestural or multimodal theories of language evolution? *Gesture*, *15*, 250–282.

Waguespack, N. M., & Surovell, T. A. (2003). Clovis hunting strategies, or how to make out on plentiful resources. *American Antiquity*, *68*, 333–352.

Walker, A., & Shipman, P. (2005). *The ape in the tree: An intellectual & natural history of proconsul*. Cambridge, MA: Belknap Press of Harvard University Press.

Wall, J. D., & Brandt, D. Y. C. (2016). Archaic admixture in human history. *Current Opinion in Genetics & Development*, *41*, 93–97.

Wang, D. Y-C., Kumar, S., & Hedges, S. B. (1999). Divergence time estimates for the early history of animal phyla and the origin of plants, animals and fungi. *Proceedings of the Royal Society of London, B, 266*, 163–171.

Wang, W-J., & Crompton, R. H. (2004). The role of load-carrying in the evolution of modern proportions. *Journal of Anatomy, 204*, 417–430.

Ward, C. V. (2002). Interpreting the posture and locomotion of *Australopithecus afarensis*: Where do we stand? *Yearbook of Physical Anthropology, 45*, 185–215.

Ward, C. V., Kimbel, W. H., & Johanson, D. C. (2011). Complete fourth metatarsal and arches in the foot of *Australopithecus afarensis*. *Science, 331*, 750–753.

Warren, D. M., Stern, M., Duggirala, R., Dyer, T. D., & Almasy, L. (2006). Heritability and linkage analysis of hand, foot, and eye preference in Mexican Americans. *Laterality, 11*, 508–524.

Watts, D. P., & Mitani, J. C. (2002). Hunting behavior of chimpanzees at Ngogo, Kibale National Park, Uganda. *International Journal of Primatology, 23*, 1–28.

Wei, C., & Pohorille, A. (2015). M2 proton channel: Toward a model of a primitive proton pump. *Origins of Life and Evolution of Biospheres, 45*, 241–248.

Welch, J. J., & Bromham, L. (2005). Molecular dating when rates vary. *Trends in Ecology and Evolution, 20*, 320–327.

Wende, K. C., Nagels, A., Blos, J., Stratmann, M., Chatterjee, A., Kircher, T., & Straube, B. (2013). Differences and commonalities in the judgment of causality in physical and social contexts: An fMRI study. *Neuropsychologia, 51*, 2572–2580.

Wesley, M. J., Fernandez-Carriba, S., Hostetter, A., Pilcher, D., Poss, S., & Hopkins, W. D. (2002). Factor analysis of multiple measures of hand use in captive chimpanzees: An alternative approach to the assessment of handedness in nonhuman primates. *International Journal of Primatology, 23*, 1155–1168.

Westall, F. (2004). Early life on earth: The ancient fossil record. In P. Ehrenfreund et al. (Eds.), *Astrobiology: Future perspectives* (pp. 287–316). Dordrecht: Kluwer Academic Publishers.

Whelan, N. V., Kocot, K. M., Moroz, L. L., & Halanych, K. M. (2015). Error, signal, and the placement of Ctenophora sister to all other animals. *Proceedings of the National Academy of Sciences, 112*, 5773–5778.

White, J. (1974). The yogi in the lab. In J. White (Ed.), *Frontiers of consciousness* (pp. 77–93). New York, NY: The Julian Press, Inc.

White, S., Gowlett, J. A. J., & Grove, M. (2014). The place of Neanderthals in hominin phylogeny. *Journal of Anthropological Archaeology, 35*, 32–50.

White, T. D. (2002). Earliest hominids. In W. C. Hartwig (Ed.), *The primate fossil record* (pp. 407–417). Cambridge, UK: Cambridge University Press.

White, T. D., Asfaw, B., Beyene, Y., Haile-Selassie, Y., Lovejoy, C. O., Suwa, G., & WoldeGabriel, G. (2009). *Ardipithecus ramidus* and the paleobiology of early hominids. *Science, 326*, 64.

Whiten, A., Goodall, J., McGrew, W. C., Nishida, T., Reynolds, V., Sugiyama, Y., . . . Boesch, C. (2001). Charting cultural variation in chimpanzees. *Behaviour, 138*, 1481–1516.

Wickelgren, W. A. (1973). The long and the short of memory. *Psychological Bulletin, 80*, 425–438.

Wilde, S., Timpson, A., Kirsanow, K., Kaiser, E., Kayser, M., Unterländer, M., . . . Burger, J. (2014). Direct evidence for positive selection of skin, hair, and eye pigmentation in Europeans during the last 5,000 y. *Proceedings of the National Academy of Sciences, 111*, 4832–4837.

Wildman, D. E., Uddin, M., Liu, G., Grossman, L. I., & Goodman, M. (2003). Implications of natural selection in shaping 99.4% nonsynonymous DNA identity between humans and chimpanzees: Enlarging genus *Homo*. *Proceedings of the National Academy of Sciences, 100*, 7181–7188.

Wilke, C., Kavanagh, E., Donnellan, E., Waller, B. M., Machanda, Z. P., & Slocombe, K. E. (2017). Production of and responses to unimodal and multimodal signals in wild chimpanzees, *Pan troglodytes schweinfurthii*. *Animal Behaviour, 123*, 305–316.

Wilkins, J., Schoville, B. J., Brown, K. S., & Chazan, M. (2012). Evidence for early hafted hunting technology. *Science, 338*, 942–946.

Wilkinson, R. D., Steiper, M. E., Soligo, C., Martin, R. D., Yang, Z., & Tavaré, S. (2011). Dating primate divergences through an integrated analysis of palaeontological and molecular data. *Systematic Biology, 60*, 16–31.

Williams, M. F. (2002). Primate encephalization and intelligence. *Medical Hypotheses, 58,* 284–290.

Williams, S. A., & Russo, G. A. (2015). Evolution of the hominoid vertebral column: The long and the short of it. *Evolutionary Anthropology, 24,* 15–32.

Williams, T. A., Heaps, S. E., Cherlin, S., Nye, T. M. W., Boys, R. J., & Embley, T. M. (2015). New substitution models for rooting phylogenetic trees. *Philosophical Transactions of the Royal Society, 370,* article 20140336.

Wilson, A. C. V. (2007). *The divergence of genes in ciliary photoreceptors* (Master's thesis), University of California at Santa Barbara.

Wilson, D. A., & Tomonaga, M. (2018). Visual discrimination of primate species based on faces in chimpanzees. *Primates, 59,* 243–251.

Wixted, J. T., & Ebbesen, E. B. (1997). Genuine power curves in forgetting: A quantitative analysis of individual subject forgetting functions. *Memory & Cognition, 25,* 731–739.

Wokke, M. E., Vandenbroucke, A. R. E., Scholte, H. S., & Lamme, V. A. F. (2013). Confuse your illusion: Feedback to early visual cortex contributes to perceptual completion. *Psychological Science, 24,* 63–71.

Wolbers, T., Weiller, C., & Büchel, C. (2004). Neural foundations of emerging route knowledge in complex spatial environments. *Cognitive Brain Research, 21,* 401–411.

Wolpert, L. (2007). *Six impossible things before breakfast: The evolutionary origins of belief.* New York, NY: W. W. Norton & Company.

Wolpoff, M. H., Hawks, J., Senut, B., Pickford, M., & Ahern, J. (2006). An ape or *the* ape: Is the Toumaï cranium TM 266 a hominid? *PaleoAnthropology, 2006,* 36–50.

Wong, P., & Kaas, J. H. (2010). Architectonic subdivisions of neocortex in the galago (*Otolemur garnetti*). *The Anatomical Record, 293,* 1033–1069.

Wood, B. A. (1992). Evolution of australopithecines. In S. Jones, R. Martin, & D. Pilbeam (Eds.), *The Cambridge encyclopedia of human evolution* (pp. 231–240). Cambridge, UK: Cambridge University Press.

Wood, B. A., & Aiello, L. C. (1998). Taxonomic and functional implications of mandibular scaling in early hominins. *American Journal of Physical Anthropology, 105,* 523–538.

Wood, B. A., & Boyle, E. K. (2016). Hominin taxic diversity: Fact or fantasy? *Yearbook of Physical Anthropology, 159,* S37-S78.

Wood, B. A., & Constantino, P. (2004). Human origins: Life at the top of the tree. In J. Cracraft & M. J. Donohue (Eds.), *Assembling the tree of life* (pp. 517–535). Oxford: Oxford University Press.

Woollett, K., & Maguire, E. A. (2011). Acquiring "the Knowledge" of London's layout drives structural brain changes. *Current Biology, 21,* 2109–2114.

Wright, A. A., Santiago, H. C., Sands, S. F., Kendrick, D. F., & Cook, R. G. (1985). Memory processing of serial lists by pigeons, monkeys, and people. *Science, 229,* 287–289.

Xing, J., Wang, H., Han, K., Ray, D. A., Huang, C. H., Chemnick, L. G., . . . Batzer, M. A. (2005). A mobile element based phylogeny of old world monkeys. *Molecular Phylogenetics and Evolution, 37,* 872–880.

Yamada, E. S., Silveira, L. C. L., Perry, V. H., & Franco, E. C. S. (2001). M and P retinal ganglion cells of the owl monkey: Morphology, size and photoreceptor convergence. *Vision Research, 41,* 119–131.

Yartsev, M. M., Witter, M. P., & Ulanovsky, N. (2011). Grid cells without theta oscillations in the entorhinal cortex of bats. *Nature, 479,* 103–107.

Young, N. M., Capellini, T. D., Roach, N. T., & Alemseged, Z. (2015). Fossil hominin shoulders support an African ape-like last common ancestor of humans and chimpanzees. *Proceedings of the National Academy of Sciences, 112,* 11829–11834.

Yuehai, K., Bing, S., Xiufeng, S., Daru, L., Lifeng, C., Hongyu, L., . . . Jianzhong, J. (2001). African origins of modern humans in East Asia: A tale of 12,000 Y chromosomes. *Science, 292,* 1151–1153.

Zachar, I., Szilágyi, A., Számado, S., & Szathmáry, E. (2018). Farming the mitochondrial ancestor as a model of endosymbiotic establishment by natural selection. *Proceedings of the National Academy of Sciences, 115,* E1504–E1510.

Zaehle, T., Jordan, K., Wüstenberg, T., Baudewig, J., Dechent, P., & Mast, F. W. (2007). The neural basis of the egocentric and allocentric spatial frame of reference. *Brain Research, 1137*, 92–103.

Zahnle, K. J., Catling, D. C., & Claire, M. W. (2013). The rise of oxygen and the hydrogen hourglass. *Chemical Geology, 362*, 26–34.

Zahnle, K. J., Schaefer, L., & Fegley, B. (2010). Earth's earliest atmospheres. *Cold Spring Harbor Perspectives in Biology, 2*, article a004895.

Zaksas, D., & Pasternak, T. (2005). Area MT neurons respond to visual motion distant from their receptive fields. *Journal of Neurophysiology, 94*, 4156–4167.

Zeitoun, V., Barriel, V., & Widianto, H. (2016). Phylogenetic analysis of the calvaria of *Homo floresiensis*. *Comptes Rendus Palevol, 15*, 555–568.

Zentall, T. R. (2006). Imitation: Definitions, evidence, and mechanisms. *Animal Cognition, 9*, 335–353.

Zhang, L., Peritz, A., & Meggers, E. (2005). A simple glycol nucleic acid. *Journal of the American Chemical Society, 127*, 4174–4175.

Zhang, Z., & Cui, L. (2016). Oxygen requirements for the Cambrian explosion. *Journal of Earth Science, 27*, 187–195.

Zhou, C., Wu, S., Martin, T., & Luo, Z-X. (2013). A Jurassic mammaliaform and the earliest mammalian evolutionary adaptations. *Nature, 500*, 163–167.

Zilhão, J. (2007). The emergence of ornaments and art: An archaeological perspective on the origins of "behavioral modernity". *Journal of Archaeological Research, 15*, 1–54.

Zollikofer, C. P. E., de León, M. S. P., Lieberman, D. E., Guy, F., Pilbeam, D., Likius, A., . . . Brunet, M. (2005). Virtual cranial reconstruction of *Sahelanthropus tchadensis*. *Nature, 434*, 755–759.

Additional picture credits

Most picture sources can be located by web searching the title-license-author triads shown under pictures in the text. The following are additional credits for pictures requiring them.

Figures 6.2, 6.3, 6.4, 6.5, and 6.6. Blausen.com staff (2014). "Medical gallery of Blausen Medical 2014". *WikiJournal of Medicine 1 (2)*. DOI:10.15347/wjm/2014.010. ISSN 2002-4436.

Figure 8.1. John R. Hutchinson, Karl T. Bates, Julia Molnar, Vivian Allen, and Peter J. Makovicky, (2011). A computational analysis of limb and body dimensions in Tyrannosaurus rex with implications for locomotion, ontogeny, and growth. *PLoS ONE 6(10)*: e26037.

Figure 9.6. Fidelis T. Masao, Elgidius B. Ichumbaki, Marco Cherin, Angelo Barili, Giovanni Boschian, Dawid A. Iurino, et al. (2016). New footprints from Laetoli (Tanzania) provide evidence for marked body size variation in early hominins. *eLife, 2016(5)*: e19568

Figure 12.7. "Response times from Study 1". Reprinted by permission from Springer Nature Customer Service Centre GmbH: Springer Nature, *Perception & Psychophysics*, What dot clusters and bar graphs reveal: Subitizing is fast counting and subtraction, David B. Boles, Jeffrey B. Phillips, Somer M. Givens, 2007.

Figure 14.2. "Architectonic map of the human and macaque monkey prefrontal cortex", appearing in Petrides M, Pandya DN (2009). Distinct parietal and temporal pathways to the homologues of Broca's area in the monkey. *PLoS Biol 7(8)*: e1000170. doi:10.1371/journal.pbio.1000170, p. 2.

Figure 15.3. Adaptation of "The RM was used in order to construct the connectivity matrices of the two species", appearing in Goulas A, Bastiani M, Bezgin G, Uylings HBM, Roebroeck A, et al. (2014). Comparative analysis of the macroscale structural connectivity in the macaque and human brain. *PLoS Comput Biol 10(3)*: e1003529. doi:10.1371/journal.pcbi.1003529, p. 3.

Index

Note: Page numbers in **bold** indicate a table and page numbers in *italics* indicate a figure on the corresponding page.